CONTEMPORARY MATHEMATICS

722

Arithmetic Geometry: Computation and Applications

16th International Conference
Arithmetic, Geometry, Cryptography, and Coding Theory
June 19–23, 2017
Centre International de Rencontres Mathématiques,
Marseille, France

Yves Aubry
Everett W. Howe
Christophe Ritzenthaler
Editors

AMS AMERICAN MATHEMATICAL SOCIETY

2010 *Mathematics Subject Classification.* Primary 11G20, 11G30, 11G32, 11G40, 11T71, 14G10, 14H40, 14Q05, 20C20, 20G41.

Library of Congress Cataloging-in-Publication Data

Names: International Conference on Arithmetic, Geometry, Cryptography and Coding Theory (16th : 2017 : Marseille, France) | Aubry, Yves, 1965- editor. | Howe, Everett W., editor. | Ritzenthaler, Christophe, 1976- editor.

Title: Arithmetic geometry : computation and applications : 16th International Conference on Arithmetic, Geometry, Cryptography, and Coding Theory, June 19-23, 2017, Centre International de Rencontres Mathematiques, Marseille, France / Yves Aubry, Everett W. Howe, Christophe Ritzenthaler, editors.

Description: Providence, Rhode Island : American Mathematical Society, [2019] | Series: Contemporary mathematics ; volume 722 | Includes bibliographical references.

Identifiers: LCCN 2018037194 | ISBN 9781470442125 (alk. paper)

Subjects: LCSH: Coding theory–Congresses. | Geometry, Algebraic–Congresses. | Cryptography–Congresses. | Number theory–Congresses. | AMS: Number theory – Arithmetic algebraic geometry (Diophantine geometry) – Curves over finite and local fields. msc | Number theory – Arithmetic algebraic geometry (Diophantine geometry) – Curves of arbitrary genus or genus $\neq 1$ over global fields. msc | Number theory – Arithmetic algebraic geometry (Diophantine geometry) – Dessins d'enfants, Belyi theory. msc | Number theory – Arithmetic algebraic geometry (Diophantine geometry) – L-functions of varieties over global fields; Birch-Swinnerton-Dyer conjecture. msc | Number theory – Finite fields and commutative rings (number-theoretic aspects) – Algebraic coding theory; cryptography. msc | Algebraic geometry – Arithmetic problems. Diophantine geometry – Zeta-functions and related questions. msc | Algebraic geometry – Computational aspects in algebraic geometry – Curves. msc | Group theory and generalizations – Representation theory of groups – Modular representations and characters. msc | Group theory and generalizations – Linear algebraic groups and related topics – Exceptional groups. msc

Classification: LCC QA268 .I57 2017 | DDC 512.7/4–dc23

LC record available at https://lccn.loc.gov/2018037194

Contemporary Mathematics ISSN: 0271-4132 (print); ISSN: 1098-3627 (online)

DOI: https://doi.org/10.1090/conm/722

∞ The paper used in this book is acid-free and falls within the guidelines
established to ensure permanence and durability.
Visit the AMS home page at `https://www.ams.org/`

10 9 8 7 6 5 4 3 2 1   24 23 22 21 20 19

Contents

Preface

The 16th edition of the $\mathsf{AGC^2T}$ conference (Arithmetic, Geometry, Cryptography, and Coding Theory) took place at CIRM (Centre International de Rencontres Mathématiques) in Marseille, France, on June 19–23, 2017. This international conference has been a major event in the area of arithmetic geometry and its applications since 1987, and more than 94 participants joined us to celebrate its 30th anniversary. We thank all of them for creating a stimulating research environment during our week together.

The topics of the talks extended from algebraic number theory to diophantine geometry, and from curves and abelian varieties over finite fields to applications in codes and cryptography. We especially thank the speakers — Jeff Achter, Elise Barelli, Irene Bouw, Nils Bruin, Wouter Castryck, Alina Cojocaru, Mrinmoy Datta, Lucile Devin, Iwan Duursma, Elena Egorova, Sudhir Ghorpade, Marc Hindry, Nathan Kaplan, Valentijn Karemaker, Daniel Katz, Kiran Kedlaya, Pınar Kılıçer, Gilles Lachaud, Philippe Lebacque, Elisa Lorenzo García, Stefano Marseglia, Ivan Pogildiakov, Bjorn Poonen, Rachel Pries, Matthieu Rambaud, Alice Silverberg, Prasant Singh, Andrew Sutherland, Medhi Tibouchi, Andrey Trepalin, Michael Tsfasman, John Voight, Felipe Voloch, Marius Vuille, and Chia-Fu Yu — for their lectures.

As with any anniversary, it was a time for joy and exuberance, and some of the participants' memories of earlier editions of $\mathsf{AGC^2T}$ were recorded with the kind assistance of Stéphanie Vareilles and Guillaume Hennenfent.[1] It was also a moment for recollection and reflection, as the "$\mathsf{AGC^2T}$ family" had lost one of its youngest and most brilliant members in the person of Alexey Zykin, who passed away with his wife Tatyana Makarova in a tragic accident a few months before the conference. With Alexey's friends and colleagues, we celebrated one more time his constant enthusiasm in all aspects of his research and life.

Gilles Lachaud, one of the founding fathers of the conference and one of its mainstays, passed away in February 2018. The next edition of the conference will have time set aside for memories of Gilles and for celebrations of his legacy. For now, we are honored to present one of his papers in this volume.

The editors would like to thank the staff of CIRM (Olivia Barbarroux, Muriel Milton, and Laure Stefanini) and of the Institut de Mathématiques de Marseille (Jessica Bouanane and Corinne Roux) for their remarkable professionalism and their friendly and constant help. We would also like to thank Christine Thivierge at the American Mathematical Society for guiding us through the *Contemporary Mathematics* production process. And of course, we give our deepest thanks to the

[1] The video is available at `https://youtu.be/KWklMv4Yya8`.

authors of the papers in this volume for their mathematical creativity and for their patience with the editors.

Toulon, France — Yves Aubry
San Diego, California, U.S.A. — Everett Howe
Rennes, France — Christophe Ritzenthaler
June 2018

Contemporary Mathematics
Volume **722**, 2019
https://doi.org/10.1090/conm/722/14534

Hasse–Witt and Cartier–Manin matrices: A warning and a request

Jeffrey D. Achter and Everett W. Howe

ABSTRACT. Let X be a curve in positive characteristic. A *Hasse–Witt matrix* for X is a matrix that represents the action of the Frobenius operator on the cohomology group $H^1(X, \mathcal{O}_X)$ with respect to some basis. A *Cartier–Manin matrix* for X is a matrix that represents the action of the Cartier operator on the space of holomorphic differentials of X with respect to some basis. The operators that these matrices represent are adjoint to one another, so Hasse–Witt matrices and Cartier–Manin matrices are related to one another, but there seems to be a fair amount of confusion in the literature about the exact nature of this relationship. This confusion arises from differences in terminology, from differing conventions about whether matrices act on the left or on the right, and from misunderstandings about the proper formulæ for iterating semilinear operators. Unfortunately, this confusion has led to the publication of incorrect results. In this paper we present the issues involved as clearly as we can, and we look through the literature to see where there may be problems. We encourage future authors to clearly distinguish between Hasse–Witt and Cartier–Manin matrices, in the hope that further errors can be avoided.

Prologue

An example. Consider the genus-2 hyperelliptic curve X over $\mathbb{F}_{125}$ with affine model

$$y^2 = f(x) = x^5 + x^4 + \alpha^{92}x^3 + \alpha^{18}x^2 + \alpha^{56}x\,, \tag{0.1}$$

where $\alpha \in \mathbb{F}_{125}$ satisfies $\alpha^3+3\alpha+3 = 0$. Let us compute the 5-rank of (the Jacobian of) X.

On one hand, we can follow Yui [**14**] and compute the effect of the Cartier operator on the space of regular one-forms. Let c_m be the coefficient of x^m in $f(x)^{(5-1)/2}$. Yui [**14**, p. 381] constructs a matrix (denoted A in her paper, but

2010 *Mathematics Subject Classification.* Primary 11G20, 14Q05; Secondary 14G10, 14G15, 14G17.

Key words and phrases. Cartier–Manin, Hasse–Witt, p-rank, zeta-function, semi-linear operator.

JDA's work partially supported by NSA grant H98230-16-1-0046.

denoted here by Y to prevent a conflict of notation later on) given by

$$\begin{aligned} Y &= \begin{pmatrix} c_{5\cdot 1-1} & c_{5\cdot 1-2} \\ c_{5\cdot 2-1} & c_{5\cdot 2-2} \end{pmatrix} \\ &= \begin{pmatrix} \alpha^{41} & \alpha^{105} \\ 2 & \alpha^{95} \end{pmatrix}. \end{aligned}$$

We compute as well that the image of Y under the Frobenius automorphism σ of $\mathbb{F}_{125}$ is given by

$$\begin{aligned} Y^\sigma &= \begin{pmatrix} c^\sigma_{5\cdot 1-1} & c^\sigma_{5\cdot 1-2} \\ c^\sigma_{5\cdot 2-1} & c^\sigma_{5\cdot 2-2} \end{pmatrix} \\ &= \begin{pmatrix} \alpha^{81} & \alpha^{29} \\ 2 & \alpha^{103} \end{pmatrix}, \end{aligned}$$

and the product $Y \cdot Y^\sigma$ is

$$Y \cdot Y^\sigma = \begin{pmatrix} \alpha^{32} & \alpha^{104} \\ \alpha^{22} & \alpha^{94} \end{pmatrix}.$$

Since this last matrix has rank one, according to Yui's Lemma E [**14**, p. 387] we should be able to conclude that X has 5-rank one.

On the other hand, X is actually supersingular; indeed, its L-polynomial is $(1 + 125T^2)^2$, and thus the only slope of its normalized 5-adic Newton polygon is 1/2. In particular, X has 5-rank zero.

Our aim in this note is to tease out the source of this dissonance.

Genesis of this project. We noticed this discrepancy while attempting to obtain numerical data in support of some earlier work [**1**]. Moreover, we found that one of us invoked an erroneous formula in a separate project [**63**] (see Section 5.2).

Works such as Yui's 1978 paper [**14**], as well as its antecedents (including works by Manin [**7, 8**]) and consequents, rely on the construction and analysis of certain semilinear operators. Since the work of Hasse and Witt [**4**], it has been understood that there is such an operator, acting on some subquotient of the de Rham cohomology of a given curve X in characteristic p, that encodes information about the p-torsion group scheme of the Jacobian of X. The ideas of Hasse and Witt are beautifully clear, but one must navigate around several potential sources of error in order to arrive at a correct formula. Indeed, one must decide whether to work with the summand $H^0(X, \Omega^1_X)$ or the quotient $H^1(X, \mathcal{O}_X)$ of $H^1_{\mathrm{dR}}(X)$; this choice, in turn, determines whether the operator in question is σ-linear or σ^{-1}-linear, where σ is the p-th powering map on the base field. One is given a further opportunity to make a "sign error" when one chooses bases for these vector spaces and then decides whether the semilinear operator acts on the right or on the left.[1] Given these multiple opportunities for mistake, it is hardly surprising that there are occasional misstatements in the literature.

Conversations with others suggest to us that the community has an interest in (re)documenting these semilinear methods, especially in view of the continuing

[1]Of course, there is no literal "sign" to get wrong in any of the formulæ we discuss, but the terminology is suggestive of the fact that two such errors will typically cancel one another out. We will continue to use the the term "sign error" in this sense throughout the paper.

expansion of the role of computing in arithmetic geometry. With this backdrop, we offer the following survey of Cartier–Manin and Hasse–Witt matrices.

In Section 1 we review basic facts about the representation of semilinear operators by matrices. In Section 2 we define the Cartier operator on the space of holomorphic differentials of a curve X and the Frobenius operator on the cohomology group $H^1(X, \mathcal{O}_X)$, in its guise as a quotient group of the space of répartitions of X. The Cartier–Manin and Hasse–Witt matrices represent these two operators. In Section 3 we follow the work of Manin **[9, 10]** and Yui **[14]** to explicitly calculate the Cartier–Manin matrix of a hyperelliptic curve, and we resolve the problem posed by the example in our Prologue. In Section 4 we review the papers of Manin and Yui, keeping a watchful eye out for sign errors. We close in Section 5 with a review of the literature that cites Manin and Yui, to see whether any sign errors have propagated. Fortunately, there are only a few papers that contain results or examples that are in error.

Of course, it is unpleasant to find any errors at all in published papers. We have a suggestion for the community, which we hope will help prevent this type of sign error in the future: Please be careful with terminology. If you are working with the Cartier operator on differentials, refer to the matrix representation as the *Cartier–Manin* matrix; if you are working with the Frobenius operator on $H^1(X, \mathcal{O}_X)$, refer to the matrix representation as the *Hasse–Witt* matrix. These matrices are *related* to one another, but they are not *equal* to one another, and they represent semilinear operators with different properties.

Acknowledgments. We thank Yuri Manin, Noriko Yui, Arsen Elkin, Pierrick Gaudry, Takehiro Hasegawa, Rachel Pries, Andrew Sutherland, Saeed Tafazolian, Doug Ulmer, Felipe Voloch, and Yuri Zarhin, as well as the referees, for their comments on draft versions of this paper.

1. Matrices and semilinear algebra

We start with some notation concerning the use of matrices to represent semilinear algebra.

Let K be a field; we work exclusively with finite-dimensional K-vector spaces.

1.1. Bases, matrices, and linear operators. Let W be a vector space with basis $\mathcal{C} = \{w_1, \cdots, w_n\}$. Any $w \in W$ is expressible as $w = \sum c_i w_i$; let $[w]_\mathcal{C}$ denote the corresponding column vector

$$[w]_\mathcal{C} = \begin{pmatrix} c_1 \\ c_2 \\ \vdots \\ c_n \end{pmatrix}.$$

Now let V be an m-dimensional vector space with chosen basis $\mathcal{B}$, and let $f \colon W \to V$ be a linear transformation. Define numbers a_{ij} by

$$f(w_j) = \sum_{i=1}^{m} a_{ij} v_i \,.$$

The matrix of f relative to the chosen bases $\mathcal{C}$ and $\mathcal{B}$ is

$$[f]_{\mathcal{B} \leftarrow \mathcal{C}} = (a_{ij}) \in \mathrm{Mat}_{m \times n}(K) \,.$$

Matrix multiplication is defined so that, with this notation,

$$[f(w)]_{\mathcal{B}} = [f]_{\mathcal{B}\leftarrow\mathcal{C}} \cdot [w]_{\mathcal{C}} \,.$$

Change of basis works as follows. Let $f\colon V \to V$ be an endomorphism, and let $\mathcal{B}$ and $\mathcal{D}$ be two different bases for V. Then

$$[f]_{\mathcal{D}\leftarrow\mathcal{D}} = [\mathrm{id}]_{\mathcal{D}\leftarrow\mathcal{B}}\, [f]_{\mathcal{B}\leftarrow\mathcal{B}}\, [\mathrm{id}]_{\mathcal{B}\leftarrow\mathcal{D}} \,;$$

if $S = [\mathrm{id}]_{\mathcal{B}\leftarrow\mathcal{D}}$, then

$$[f]_{\mathcal{D}\leftarrow\mathcal{D}} = S^{-1}\, [f]_{\mathcal{B}\leftarrow\mathcal{B}}\, S\,.$$

(Of course, if one prefers that matrices act on the right, then one consistently writes elements of the vector space as row vectors, and the matrix that represents the action of a linear operator is the *transpose* of the matrix described above.)

1.2. Semilinear algebra. Let ϵ be an automorphism of K. Now suppose that $f\colon V \to V$ is ϵ-linear, in the sense that for $a \in K$ and $v \in V$,

$$f(av) = a^{\epsilon} f(v)\,.$$

Naturally, f is determined by its effect on a basis, but the use of the matrices changes a little bit. Let $\mathcal{B} = \{v_1, \cdots, v_n\}$ be a basis, and again define numbers a_{ij} by

$$f(v_j) = \sum_i a_{ij} v_i\,.$$

If $v = \sum_j c_j v_j$ then

$$f(v) = \sum_j f(c_j v_j) = \sum_j c_j^{\epsilon} f(v_j) = \sum_j \Big(\sum_i a_{ij} v_i\Big) c_j^{\epsilon}$$

and so

$$[f(v)]_{\mathcal{B}} = [f]_{\mathcal{B}\leftarrow\mathcal{B}} \cdot [v]_{\mathcal{B}}^{\epsilon}\,,$$

where B^{ϵ} is the matrix obtained by applying ϵ to each entry of B.

Similarly, change of basis is accomplished with ϵ-twisted conjugacy:

$$\begin{aligned} [f]_{\mathcal{D}\leftarrow\mathcal{D}} &= [\mathrm{id}]_{\mathcal{D}\leftarrow\mathcal{B}}\, [f]_{\mathcal{B}\leftarrow\mathcal{B}}\, [\mathrm{id}]_{\mathcal{B}\leftarrow\mathcal{D}}^{\epsilon} \\ &= S^{-1}\, [f]_{\mathcal{B}\leftarrow\mathcal{B}}\, S^{\epsilon}. \end{aligned}$$

If we suppress our subscripts for a moment, then the iterates of f are represented by

$$\begin{aligned} [f\circ f] &= [f]\,[f]^{\epsilon} \\ [f^{\circ r}] &= [f]\,[f]^{\epsilon}\,[f]^{\epsilon^2} \cdots [f]^{\epsilon^{r-1}}\,. \end{aligned}$$

(Again, if one wants matrices to act on the right, then the highest iterate of ϵ is applied to the *leftmost* factor in the r-fold product.)

1.3. Adjointness. Let V^* be the dual vector space of V and let $(\cdot,\cdot)\colon V \times V^* \to K$ be the natural pairing. Continue to let $f\colon V \to V$ be ϵ-linear, and let $\delta = \epsilon^{-1}$. The adjoint f^* of f with respect to the pairing $(\cdot,\cdot)$ is δ-linear and is characterized by the relation

$$(v, f^*w^*) = (fv, w^*)^\delta$$

for all $v \in V$ and $w^* \in V^*$. Let $\mathcal{B}^* = \{v_1^*, \cdots, v_n^*\}$ be the basis dual to $\mathcal{B}$. Since for $1 \le j, \ell \le n$ we have

$$(v_j, f^*v_\ell^*) = (fv_j, v_\ell^*)^\delta = \Big(\sum_i a_{ij}v_i, v_\ell^*\Big)^\delta = a_{\ell j}^\delta\,,$$

we find that

$$f^*v_\ell^* = \sum_j a_{\ell j}^\delta v_j^*$$

and therefore

$$[f^*]_{\mathcal{B}^*\leftarrow\mathcal{B}^*} = \big([f]_{\mathcal{B}\leftarrow\mathcal{B}}^\delta\big)^\mathsf{T} \tag{1.1}$$

where $^\mathsf{T}$ indicates the transpose of a matrix.

2. Hasse–Witt and Cartier–Manin matrices

We record here some properties of the Frobenius and Cartier operators and their representations by Hasse–Witt and Cartier–Manin matrices, deferring a complete exposition to, for example, Serre **[13]**. Let k be a perfect field of characteristic $p > 0$. Let $\sigma\colon k \to k$ be the Frobenius automorphism, and let τ be its inverse. Finally, let X/k be a smooth, projective curve of genus $g > 0$.

2.1. Cohomology groups. The Hodge to de Rham spectral sequence gives a canonical exact sequence

$$0 \longrightarrow H^0(X, \Omega_X^1) \longrightarrow H^1_{\mathrm{dR}}(X) \longrightarrow H^1(X, \mathcal{O}_X) \longrightarrow 0\,.$$

There is a canonical duality between the g-dimensional k-vector spaces $H^0(X, \Omega_X^1)$ and $H^1(X, \mathcal{O}_X)$. This duality is realized by the cup product and the trace map:

$$H^0(X, \Omega_X^1) \times H^1(X, \mathcal{O}_X) \longrightarrow H^1(X, \Omega_X^1) \xrightarrow{\ \sim\ } k\,.$$

If k is algebraically closed, Serre **[13**, § 8**]** gives the following explicit description of this pairing. Let $\mathcal{R} = \mathcal{R}(X)$ be the ring of répartitions on X — that is, the subring of $\prod_{P \in X(k)} k(X)$ consisting of those elements $\{r_P\}$ for which, for all but finitely many P, the function r_P is regular at P. Let $\mathcal{R}(0)$ be the subring consisting of those répartitions such that each r_P is regular at P. Then there is an isomorphism

$$H^1(X, \mathcal{O}_X) \cong \frac{\mathcal{R}}{\mathcal{R}(0) + k(X)}\,,$$

where we view $k(x)$ as a subring of $\mathcal{R}$ via the diagonal embedding. The duality between this space and $H^0(X, \Omega_X^1)$ then admits the description

$$\begin{aligned} H^0(X, \Omega_X^1) \times H^1(X, \mathcal{O}_X) &\longrightarrow k \\ (\omega, r) &\longmapsto \textstyle\sum_{P \in X(k)} \mathrm{res}_P(r_P\omega)\,, \end{aligned} \tag{2.1}$$

where res_P denotes the residue at the point P.

2.2. The Cartier operator and the Cartier–Manin matrix. Cartier [**2**] (see also Katz [**5**, §7]) defines an operator on the de Rham complex of a smooth proper variety of arbitrary dimension. In the special case of a curve X, this gives rise to a map from $H^0(X, \Omega^1_X)$ to itself. We follow here the explicit description given by Serre [**13**, §10].

Let P be a closed point on X and let t be a uniformizing parameter at P. Then the functions $1, t, \cdots, t^{p-1}$ form a p-basis for the local ring $\mathcal{O}_{X,P}$, that is, a basis for $\mathcal{O}_{X,P}$ as a module over $\mathcal{O}^p_{X,P}$. Any 1-form holomorphic at P admits an expression

$$\omega = \left(\sum_{j=0}^{p-1} f_j^p t^j\right) dt$$

for certain $f_j \in \mathcal{O}_{X,P}$, and one declares that

$$\mathcal{C}(\omega) = f_{p-1}\, dt\,.$$

The value of $\mathcal{C}(\omega)$ does not depend on the choice of uniformizer t, and the map $\mathcal{C}$ can be extended to give a map $\Omega^1_{k(X)/k} \to \Omega^1_{k(X)/k}$.

It is not hard to see that, for ω, ω_1, and ω_2 in $\Omega^1_{k(X)/k}$ and for $f \in k(X)$, one has

$$\begin{aligned} \mathcal{C}(\omega_1 + \omega_2) &= \mathcal{C}(\omega_1) + \mathcal{C}(\omega_2) \\ \mathcal{C}(f^p\omega) &= f\mathcal{C}(\omega)\,. \end{aligned}$$

In particular, the Cartier operator restricts to give a τ-linear operator

$$H^0(X, \Omega^1_X) \xrightarrow{\ \mathcal{C}\ } H^0(X, \Omega^1_X)\,.$$

(Yui [**14**] refers to this as the *modified* Cartier operator.) A matrix associated to $\mathcal{C}$ and a choice of basis for $H^0(X, \Omega^1_X)$ is called a *Cartier*, or *Cartier–Manin*, matrix for X.

2.3. The Frobenius operator and the Hasse–Witt matrix. There is also a Frobenius operator

$$H^1(X, \mathcal{O}_X) \xrightarrow{\ \mathcal{F}\ } H^1(X, \mathcal{O}_X)$$

which, under the isomorphism $H^1(X, \mathcal{O}_X) \cong \mathcal{R}/(\mathcal{R}(0) + k(X))$, takes the class of a répartition $r = \{r_P\}$ to the class of $\{r_P^p\}$. In particular, $\mathcal{F}$ is a σ-linear operator. Following Serre, we call any matrix associated to $\mathcal{F}$ and a choice of basis a *Hasse–Witt* matrix for X.

Like the Cartier operator, the Frobenius operator admits a generalization to varieties of arbitrary dimension. For a smooth variety for which the Hodge to de Rham spectral sequence degenerates at E_1, Katz defines [**66**, (2.3.4.1.3), p. 27] a σ-linear operator on each cohomology group of the structure sheaf. In the special case of a smooth projective hypersurface Y/k of dimension n, Katz gives an explicit formula for the action of this operator on $H^n(Y, \mathcal{O}_Y)$ in terms of a defining polynomial for Y [**66**, Algorithm (2.3.7.14), p. 35].

2.4. Adjointness. Serre goes on to show [**13**, Proposition 9, p. 40] that $\mathcal{F}$ and $\mathcal{C}$ are adjoint with respect to the pairing $(\cdot,\cdot)$ of (2.1), in the sense (see Section 1.3) that

$$(\omega, \mathcal{F}r) = (\mathcal{C}\omega, r)^\sigma. \tag{2.2}$$

By (1.1), if B is a Cartier–Manin matrix for X, then $(B^\sigma)^\intercal$ is a Hasse–Witt matrix for X. Conversely, if A is a Hasse–Witt matrix for X, then $(A^\tau)^\intercal$ is a Cartier–Manin matrix for X.

2.5. Zeta functions. Now suppose X is a curve over $\mathbb{F}_q$, the field with $q = p^e$ elements. The zeta function of X has the form

$$Z_{X/\mathbb{F}_q}(T) = \frac{L(T)}{(1-T)(1-qT)},$$

where $L(T) \in \mathbb{Z}[T]$. The e-fold iterate of $\mathcal{F}$ is $\mathbb{F}_q$-linear, and its characteristic polynomial satisfies the congruence

$$\operatorname{charpoly}_{\mathcal{F}^e}(T) \equiv L(T) \bmod p$$

([**7**, Theorem 1], [**8**, Theorem 1], [**6**, Théorème 3.1]). Consequently, if A is any Hasse–Witt matrix for X, then

$$\det(\mathrm{id} - AA^\sigma \cdots A^{\sigma^{e-1}} T) \equiv L(T) \bmod p.$$

Using (1.1), we find that $\mathcal{C}^e$ and $\mathcal{F}^e$ are adjoint $\mathbb{F}_q$-linear operators. In particular, if B is any Cartier–Manin matrix for X, then

$$\det(\mathrm{id} - BB^\tau \cdots B^{\tau^{e-1}} T) \equiv L(T) \bmod p.$$

Similarly, the characteristic polynomial $\chi_{X/\mathbb{F}_q}(T)$ of the relative Frobenius endomorphism of $\operatorname{Jac} X$ satisfies

$$\chi_{X/\mathbb{F}_q}(T) \equiv (-1)^g T^g \det([\mathcal{F}^e] - T\cdot \mathrm{id}) \equiv (-1)^g T^g \det([\mathcal{C}^e] - T\cdot \mathrm{id}) \bmod p.$$

3. Cartier–Manin matrices for hyperelliptic curves

We use the methods of Manin [**9, 10**] and Yui [**14**] to give a formula for a Cartier–Manin matrix of a hyperelliptic curve. We then use this formula to compute such a matrix for the curve (0.1), and independently compute a Hasse–Witt matrix to verify our work.

3.1. An explicit formula. Let k be a perfect field of odd characteristic p, and let X/k be a hyperelliptic curve of genus g with affine equation $y^2 = f(x)$, where $f(x) \in k[x]$ is square-free of degree $2g+1$ or $2g+2$.

As a basis for $H^0(X, \Omega^1_X)$ we choose

$$\mathcal{B} = \left\{ \omega_i = x^{i-1}\frac{dx}{y} : 1 \le i \le g \right\}. \tag{3.1}$$

If we write $f(x)^{\frac{p-1}{2}} = \sum c_m x^m$, we obtain the following equalities of differentials on X:

$$\begin{aligned}\frac{dx}{y} &= \frac{(y^2)^{\frac{p-1}{2}}}{y^p} dx \\ &= \frac{f(x)^{\frac{p-1}{2}}}{y^p} dx \\ &= y^{-p}\Big(\sum_{m\geq 0} c_m x^m\Big) dx\,.\end{aligned}$$

We find that

$$\omega_j = x^{j-1}\frac{dx}{y} = y^{-p}\Big(\sum_{m\geq 0} c_m x^{m+j-1}\Big) dx\,.$$

If we apply the Cartier operator to ω_j, the only terms that will make a contribution are the terms where $m+j-1 \equiv p-1 \bmod p$. In particular, we only need consider m of the form $ip-j$, for $i = 1, \ldots, g$. We find that

$$\begin{aligned}\mathcal{C}(\omega_j) &= \mathcal{C}\Big(y^{-p}\Big(\sum_{i=1}^{g} c_{ip-j}x^{ip-p}\Big)x^{p-1}\,dx\Big) \\ &= \sum_{i=1}^{g}\mathcal{C}\Big(\big(c_{ip-j}^{\tau}x^{i-1}/y\big)^p\, x^{p-1}\,dx\Big) \\ &= \sum_{i=1}^{g} c_{ip-j}^{\tau}x^{i-1}/y\,dx \\ &= \sum_{i\geq 1} c_{ip-j}^{\tau}\omega_i\,.\end{aligned}$$

If we let $B\in \mathrm{Mat}_g(k)$ be the matrix with entries $B_{ij} = c_{ip-j}^{\tau}$, then left-multiplication by B calculates the effect of $\mathcal{C}$ in the basis $\mathcal{B}$.

3.2. The example, revisited. We reconsider the curve (0.1) and the associated matrix Y. Then

$$B = Y^{\tau} = \begin{pmatrix}\alpha^{33} & \alpha^{21}\\ 2 & \alpha^{19}\end{pmatrix}.$$

We compute the effect of the second iterate of the Cartier operator as

$$[\mathcal{C}^{\circ 2}]_{\mathcal{B}\leftarrow\mathcal{B}} = [\mathcal{C}][\mathcal{C}]^{\tau} = BB^{\tau} = \begin{pmatrix}0 & 0\\ 0 & 0\end{pmatrix};$$

this reflects the supersingularity of our original curve.

For thoroughness, we will use direct computation to find the Hasse–Witt matrix for this example as well. Let k be an algebraic closure of $\mathbb{F}_{125}$. By the strong approximation theorem, the vector space $H^1(X, \mathcal{O}_X) \cong \mathcal{R}/(\mathcal{R}(0) + k(X))$ can be represented by the classes of répartitions supported only at the point at infinity ∞ on the curve X. In fact, the répartitions $r = \{r_P\}_{P\in X(k)}$ and $s = \{s_P\}_{P\in X(k)}$ defined by

$$r_P = \begin{cases} 2y/x & \text{if } P = \infty;\\ 0 & \text{otherwise}\end{cases} \qquad\text{and}\qquad s_P = \begin{cases} 2y/x^2 & \text{if } P = \infty;\\ 0 & \text{otherwise}\end{cases}$$

give a basis for $\mathcal{R}/(\mathcal{R}(0)+k(X))$ that is dual to the basis $\{\omega_1,\omega_2\}$ of $H^0(X,\Omega^1_X)$ given in (3.1) under the pairing (2.1); we see this as follows. Let $z = x^2/y$, so that z is a uniformizing parameter for X at ∞. We compute that

$$\begin{aligned} \omega_1 &= dx/y = \big(3z^2 + O(z^4)\big)\,dz \\ \omega_2 &= x\,dx/y = \big(3 + 3z^2 + O(z^4)\big)\,dz \\ r_\infty &= 2y/x = 2z^{-3} + 3z^{-1} + O(z) \\ s_\infty &= 2y/x^2 = 2z^{-1}. \end{aligned}$$

It follows easily that $(\omega_1, r) = (\omega_2, s) = 1$ and $(\omega_1, s) = (\omega_2, r) = 0$.

We compute also that

$$\begin{aligned} r_\infty^5 &= (2x^5 + 4x^4 + \alpha^2x^3 + \alpha^{69}x^2 + \alpha^{77}x + \alpha^{94})y + \alpha^{41}r_\infty + \alpha^{105}s_\infty + O(z) \\ s_\infty^5 &= 2y + 2r_\infty + \alpha^{95}s_\infty + O(z), \end{aligned}$$

and it follows that the Hasse–Witt matrix for our curve X is given by

$$A = \begin{pmatrix} \alpha^{41} & 2 \\ \alpha^{105} & \alpha^{95} \end{pmatrix}.$$

As expected, we see that A is the transpose of Yui's matrix Y, that $B = (A^\tau)^\intercal$, and that $AA^\sigma = \left(\begin{smallmatrix} 0 & 0 \\ 0 & 0 \end{smallmatrix}\right)$.

3.3. A generalization. Garcia and Tafazolian generalize Manin and Yui's computation, and calculate a matrix [**3**, p. 212] such that left-multiplication by this matrix gives the effect of the n-th iterate of the Cartier operator in terms of the basis $\mathcal{B}$; the (i,j) entry of their matrix is the p^n-th root of the coefficient of x^{ip^n-j} in the polynomial $f(x)^{(p^n-1)/2}$. The penultimate displayed equation on page 212 of their paper shows this matrix acting on the right, but the formulæ presented elsewhere in in their paper make it clear that it acts on the left.

4. Hasse–Witt matrices through the ages

As noted in the introduction, Hasse and Witt [**4**] showed that various properties of a curve X can be read off from the action of Frobenius on $H^1(X,\mathcal{O}_X)$, the equivalence classes of répartitions of the curve, and they associated a matrix to this semilinear operator. In the paper in which he defined his operator on differential forms, Cartier [**2**] already noted a connection to the Hasse–Witt matrix of the curve; Serre [**13**, § 10] explains this well. Over the years, different authors have made this connection more and more computationally explicit. In this section, we focus on the work of Manin and of Yui, because their papers are the ones referred to most often when present-day authors write about computational aspects of the Cartier operator.

4.1. The work of Manin. Manin published three works relevant to our discussion here. We treat them each in turn.

In the first of these works [**7**] (available also in an English translation [**8**]), Manin develops explicit formulæ for computing the action of $\mathcal{F}$ on $H^1(X,\mathcal{O}_X)$. On one hand, the definition of the matrix A in the second displayed equation on page 153 of [**7**] assumes a *right* action.[2] This is further emphasized in the first

[2] The second displayed equation on page 245 of the English translation.

displayed equation on page 154.[3] On the other hand, the change of basis formula in the last displayed formula on page 153, and the formula for the g-fold iterate of $\mathcal{F}$ on page 154, are valid provided matrices act on the *left.*[4]

The main result of this work ([**7**, Theorem 1, p. 155], [**8**, Theorem 1, p. 247]) considers a curve X over a field with $q = p^e$ elements, and relates the characteristic polynomial of the Frobenius endomorphism of $\operatorname{Jac} X$ to the characteristic polynomial of a matrix representing the linear, e-fold iterate $\mathcal{F}^e$. The theorem as stated is correct, but only if we take A to be the matrix representing the Frobenius endomorphism of $H^1(X, \mathcal{O}_X)$ *acting on the left.* However, since the matrix A as defined in the text before the theorem is taken to act on the *right*, the theorem is incorrect if it is taken in the larger context of the paper.

In the second paper we would like to discuss, Manin [**9, 10**] reconsiders some of these operators. He works with the Cartier operator $\mathcal{C}$, observes that it is τ-linear, and that it acts on the space $H^0(X, \Omega^1_X)$. He explicitly calculates a basis for the space of differentials on a particular hyperelliptic curve and computes the action of the Cartier operator in terms of this basis, using the same techniques that we reproduce here in Section 3.1. No matrices are written down, so there are no obvious sign errors in this paper. Note, however, that in this paper Manin considers the Cartier operator on $H^0(X, \Omega^1_X)$, while in the preceding paper he considered the Frobenius operator on $H^1(X, \mathcal{O}_X)$.

In Section IV.5.2 of his paper on formal groups [**11, 12**], Manin computes an operator that he *calls* the Hasse–Witt matrix — and thus, in theory, should represent the action of $\mathcal{F}$ on $H^1(X, \mathcal{O}_X)$ — but which actually *represents* the action of $\mathcal{C}$ on $H^0(X, \Omega^1_X)$, as in the paper discussed in the preceding paragraph. The formula Manin uses for iterates of this operator implicitly (and incorrectly) assumes that it is σ-linear. This leads to errors in Section IV.5.2; there are several problems with the displayed group of equations that deduce conditions on the formal group of a curve's Jacobian from conditions on the equation of the curve ([**11**, p. 86], [**12**, p. 79]). It seems to us that this paper may be the original source of a recurrent conflation in the literature of "Hasse–Witt" and "Cartier–Manin" matrices.

4.2. The work of Yui. Yui [**14**] analyzes hyperelliptic curves with affine model $y^2 = f(x)$, and computes the Cartier operator $\mathcal{C}$ on $H^0(X, \Omega^1_X)$. (We remind the reader that Yui refers to the object we call the *Cartier operator* as the *modified Cartier operator*, and that she denotes it by $\mathcal{C}'$.) In Theorem 2.1 [**14**, p. 382] and Theorem 2.2 [**14**, p. 384], the formula for iterates is appropriate for a σ-linear operator, but $\mathcal{C}$ is τ-linear. Moreover, Lemma D [**14**, p. 386] exploits the semilinear adjointness (1.1) between $\mathcal{C}$ and $\mathcal{F}$, but overlooks the transpose necessary for such matrix calculations. Because of sign errors like these, Theorem 2.2 [**14**, p. 384] and Lemma E [**14**, p. 387] are incorrect; the curve we discussed in the Prologue gives a counterexample to both.

Although several explicit examples are worked out in Yui's paper, none of them can detect these inconsistencies. Indeed, in Example 3.3 [**14**, p. 391] both $\mathcal{C}$ and $\mathcal{F}$ are diagonalized by the natural basis, which hides ambiguity between left- and right-multiplication. Moreover, both this example and Example 5.4 [**14**, p. 400] are

[3]The final displayed equation on page 245 of the English translation.

[4]The third displayed formula on page 245, and the g-fold iterate formula on the top of page 246, of the English translation.

worked out for curves over $\mathbb{F}_p$, in which case σ- and τ-linear operators are simply linear.

Yui writes at the end of the paper's introduction that the article stemmed from her working through Manin's papers [**9**–**12**], so some of the sign errors in Yui's paper are reflections of Manin's earlier ambiguities between left actions and right actions and between σ-linear and τ-linear operators. This paper also encourages the unfortunate conflation of the concepts of the Hasse–Witt matrix and the Cartier–Manin matrix that began with Manin; we have already noted Lemma D [**14**, p. 386], which says that the two matrices are "identified" with one another.

5. Subsequent developments

Explicit computational methods are becoming increasingly useful in arithmetic geometry, and this utility is reflected in the large number of citations of the articles of Manin and Yui that we discussed in the preceding section. Indeed, by consulting MathSciNet and the Web of Science, we found 92 works that refer to Yui's paper [**14**] or Manin's paper on Hasse–Witt matrices [**7**, **8**], and by personal knowledge we found one more. These works are listed below in a separate section of our bibliography.

It is somewhat worrisome to see so many citations, because — as we have noted above — these papers of Manin and Yui contain sign errors that invalidate some of their results. To determine whether these sign errors have propagated to other papers, we went through the 93 articles we found to see how they applied the results of Manin and Yui. Of course, we could not go through all of these articles with great care; for the most part, we limited ourselves to looking at how they made use of the work of Manin and Yui described above, and it is possible we missed some subtleties.

In the vast majority of these works, we did not find any obvious errors stemming from the citation of the papers of Manin and Yui. For example:

- Sometimes the papers of Manin and Yui were given as general references (for the computation of Hasse–Witt matrices or for something else), and no particular results from the papers were used.
- In some cases, specific results from Manin or Yui *were* quoted, but either they were not applied, or they did not contain any sign errors, or the sign errors were silently corrected.
- In some cases, statements containing sign errors (quoted from Manin or Yui or elsewhere, or derived independently) *were* applied to specific examples, but in these examples the sign errors in the general formulæ did not lead to errors in the specific cases. Incorrect formulæ might not lead to errors, for example,
  - if the genus of the curve is 1;
  - or, more generally, if the Hasse–Witt matrix is diagonal, so that A commutes with all of its Galois conjugates;
  - or if the base field is $\mathbb{F}_p$, so that no iteration is necessary;
  - or if the base field is $\mathbb{F}_{p^2}$, so that $A \cdot A^\sigma = A \cdot A^\tau$;
  - or in a number of other situations.

But in eight of these papers, incorrect results *were* used in ways that we felt required further investigation. We look at these papers here.

5.1. Combining a theorem of Manin with a formula of Yui. The paper of Gaudry and Harley [**51**], as well as the papers of Bostan, Gaudry, and Schost [**27**, **28**], all quote a result of Manin ([**7**, Theorem 1, p. 155], [**8**, Theorem 1, p. 247]; see also §2.5) that relates the mod-p reduction of the Weil polynomial of a curve over $\mathbb{F}_{p^e}$ to the characteristic polynomial of a matrix

$$H_\pi = HH^{(p)} \cdots H^{(p^{e-1})},$$

where H is the Hasse–Witt matrix for the curve. As we noted earlier, Manin's theorem is only correct as written if we take our matrices to act on the left. However, the papers of Bostan, Gaudry, Harley, and Schost under discussion take H to be the matrix computed by Yui [**14**, p. 381]. Yui does intend for this matrix to act on the left, but it represents the Cartier operator on differentials, not the Frobenius operator on répartitions, so Yui's matrix must be transposed to give the Hasse–Witt matrix. In other words, the naïve combination of Yui's matrix with Manin's theorem gives incorrect results.

This can be seen very concretely. Consider the genus-2 curve X over $\mathbb{F}_{27}$ defined by $y^2 = x^5 + a^2x^2 + ax$, where $a^3 - a + 1 = 0$. On one hand, the matrices H and H_π from the cited papers are

$$H = \begin{bmatrix} a^2 & a \\ 1 & 0 \end{bmatrix} \quad \text{and} \quad H_\pi = HH^{(3)}H^{(9)} = \begin{bmatrix} a^{12} & a^{14} \\ a^{15} & a^{15} \end{bmatrix},$$

and the characteristic polynomial $\kappa(t)$ of H_π is $t^2 + t + 1$. On the other hand, the characteristic polynomial of Frobenius for X is $\chi(t) = t^4 + 6t^3 + 52t^2 + 162t + 729$, and it is visibly *not* the case that $\chi(t) \equiv (-1)^2 t^2 \kappa(t) \bmod 3$, as the cited theorems claim.

However, we suspect that Bostan, Gaudry, and Schost must have implemented the computation of H_π with the matrices in the opposite order (or they transposed H, or something similar), because the example they present [**27**, §5] satisfies the basic sanity check that several randomly-chosen points on the Jacobian are annihilated by the integer they give as the order of the Jacobian.

Likewise, Gaudry and Harley present an example [**51**, §7.2] of a computation over $\mathbb{F}_{p^4}$ in which they explicitly mention the order of the Jacobian modulo p computed by Manin's result, and the numerical value they get shows that their computation must have involved either transposing H or computing H_π with the factors reversed.

5.2. Supersingular genus-2 curves. We found three papers that use Yui's computation of the iterated Cartier operator to determine when a genus-2 curve is supersingular.

Elkin [**42**, §9] gives a characterization of supersingular genus-2 curves that includes a sign error. This incorrect characterization does not affect the main part of his work (for example, Theorems 1.1, 1.6, and 1.7 [**42**, pp. 54–56]), but we have not checked to see whether it affects the validity of his examples [**42**, §9].

Howe [**63**] uses Yui's Lemma E [**14**, p. 387] in the proof of his Theorem 2.1 [**63**, p. 51], which claims that all supersingular genus-2 curves over a field of characteristic 3 can be put into a certain standard form. The proof as written is invalid, because the criterion for supersingularity has a sign error; however, the proof can easily be repaired by using the correct criterion, and one can check that the theorem as stated is true.

Zarhin [**108**] also studies supersingular genus-2 curves in characteristic 3. In the proof of his Lemma 6.1 [**108**, p. 629] he correctly characterizes when a genus-2 curve is supersingular in terms of a matrix that specifies the action of the Cartier operator. Unfortunately, in a later paper [**15**, §5, p. 213] he provides a "correction" to this proof that replaces the correct characterization with an incorrect one. Fortunately, this did not require changing the statement of the result he was proving; the statement of his Lemma 6.1 [**108**, p. 629] is correct.

5.3. Genus-3 curves of p-rank 0. We found one paper, by Elkin and Pries [**44**], that uses Yui's results to compute the moduli space of hyperelliptic genus-3 curves of p-rank 0 in characteristic 3 and characteristic 5. The notation in their Lemma 2.2 [**44**, p. 246] is ambiguous, but when they apply this lemma in the proofs of Lemmas 3.3 and 3.6 [**44**, pp. 248 and 250] they multiply the matrices in the wrong order. This invalidates their calculations of the defining equations of the moduli spaces. Pries reports that Theorem 4.2 [**44**, p. 251] still holds.

5.4. Supersingularity versus superspeciality. Yui's 1986 paper [**105**] cites her 1978 paper [**14**], as well as a paper of Nygaard [**80**], in the course of the proof of Theorem 2.5 [**105**, p. 113]. In particular, Yui cites these papers to show that a curve over $\overline{\mathbb{F}}_p$ has supersingular Jacobian (that is, its Jacobian is isogenous to a power of a supersingular elliptic curve) if and only if the Cartier operator on its differentials is zero. In fact, Nygaard shows that the vanishing of the Cartier operator is equivalent to the Jacobian being *superspecial* (that is, *isomorphic* to a power of a supersingular elliptic curve) [**80**, Theorem 4.1, p. 388]. Furthermore, Yui herself gives examples showing that while the vanishing of the Cartier operator implies that the curve is supersingular, the converse is not true [**14**, Example 5.4, p. 400]. Thus, Theorem 2.5 [**105**, p. 113] is incorrect.

6. Conclusion

As we noted, most of the 93 papers that cite Manin [**7**,**8**] or Yui [**14**] do not seem to have inherited any errors in their main results. However, it might be prudent for authors who have used results from these 93 papers to double check that the results they quoted are indeed free of sign errors.

We conclude by repeating our supplication from the introduction: Please be careful with terminology, and make a clear distinction between the Cartier operator on differentials (represented by the Cartier–Manin matrix) and the Frobenius operator on $H^1(X, \mathcal{O}_X)$ (represented by the Hasse–Witt matrix). We hope that if such care is taken, there will be no need in the future for another paper like this one.

References

[1] J. D. Achter and E. W. Howe, *Split abelian surfaces over finite fields and reductions of genus-2 curves*, Algebra Number Theory **11** (2017), no. 1, 39–76, DOI 10.2140/ant.2017.11.39. MR3602766

[2] P. Cartier, *Une nouvelle opération sur les formes différentielles* (French), C. R. Acad. Sci. Paris **244** (1957), 426–428. MR0084497

[3] A. Garcia and S. Tafazolian, *Certain maximal curves and Cartier operators*, Acta Arith. **135** (2008), no. 3, 199–218, DOI 10.4064/aa135-3-1. MR2457195

[4] H. Hasse and E. Witt, *Zyklische unverzweigte Erweiterungskörper vom Primzahlgrade p über einem algebraischen Funktionenkörper der Charakteristik p* (German), Monatsh. Math. Phys. **43** (1936), no. 1, 477–492, DOI 10.1007/BF01707628. MR1550551

[5] N. M. Katz, *Nilpotent connections and the monodromy theorem: Applications of a result of Turrittin*, Inst. Hautes Études Sci. Publ. Math. **39** (1970), 175–232. MR0291177

[6] N. M. Katz, Une formule de congruence pour la fonction ζ. In *Groupes de monodromie en géométrie algébrique. II* (French), Lecture Notes in Mathematics, Vol. 340, Springer-Verlag, Berlin-New York, 1973, pp. 401–438. Séminaire de Géométrie Algébrique du Bois-Marie 1967–1969 (SGA 7 II); Dirigé par P. Deligne et N. Katz. MR0354657

[7] J. I. Manin, *The Hasse-Witt matrix of an algebraic curve* (Russian), Izv. Akad. Nauk SSSR Ser. Mat. **25** (1961), 153–172. MR0124324

[8] J. I. Manin, *The Hasse–Witt matrix of an algebraic curve*, Amer. Math. Soc. Transl. (2), **45** (1965), 245–264. Translated by J. W. S. Cassels.

[9] J. I. Manin, *On the theory of Abelian varieties over a field of finite characteristic* (Russian), Izv. Akad. Nauk SSSR Ser. Mat. **26** (1962), 281–292. MR0139611

[10] J. I. Manin, *On the theory of Abelian varieties over a field of finite characteristic*, Amer. Math. Soc. Transl. (2), **50** (1966), 127–140. Translated by G. Wagner.

[11] J. I. Manin, *Theory of commutative formal groups over fields of finite characteristic* (Russian), Uspehi Mat. Nauk **18** (1963), no. 6 (114), 3–90. MR0157972

[12] J. I. Manin, *Theory of commutative formal groups over fields of finite characteristic*, Russian Math. Surveys, **18** (1963), no. 6, 1–83.

[13] J.-P. Serre, *Sur la topologie des variétés algébriques en caractéristique p* (French), Symposium internacional de topología algebraica International symposium on algebraic topology, Universidad Nacional Autónoma de México and UNESCO, Mexico City, 1958, pp. 24–53. MR0098097

[14] N. Yui, *On the Jacobian varieties of hyperelliptic curves over fields of characteristic $p > 2$*, J. Algebra **52** (1978), no. 2, 378–410, DOI 10.1016/0021-8693(78)90247-8. MR0491717

[15] Y. G. Zarhin, *Homomorphisms of abelian varieties* (English, with English and French summaries), Arithmetic, geometry and coding theory (AGCT 2003), Sémin. Congr., vol. 11, Soc. Math. France, Paris, 2005, pp. 189–215. MR2182844

For the reader's convenience, we gather together here a list of all of the papers that we are aware of that cite Manin's 1961 paper [**7**, **8**] or Yui's 1978 paper [**14**]. We omit Yui's paper [**14**] itself, even though it cites Manin [**8**].

Works that cite Manin (1961) or Yui (1978)

[16] A. Adolphson, *The U_p-operator of Atkin on modular functions of level three*, Illinois J. Math. **24** (1980), no. 1, 49–60. MR550651

[17] A. Álvarez, *The p-rank of the reduction* mod p *of Jacobians and Jacobi sums*, Int. J. Number Theory **10** (2014), no. 8, 2097–2114, DOI 10.1142/S1793042114500705. MR3273477

[18] N. Anbar and P. Beelen, *A note on a tower by Bassa, Garcia and Stichtenoth*, Funct. Approx. Comment. Math. **57** (2017), no. 1, 47–60, DOI 10.7169/facm/1615. MR3704225

[19] M. Asada, *On the action of the Frobenius automorphism on the pro-l fundamental group*, Math. Z. **199** (1988), no. 1, 15–28, DOI 10.1007/BF01160206. MR954748

[20] M. H. Baker, *Cartier points on curves*, Internat. Math. Res. Notices **2000**, no. 7, 353–370, DOI 10.1155/S1073792800000209. MR1749740

[21] S. Ballet, C. Ritzenthaler, and R. Rolland, *On the existence of dimension zero divisors in algebraic function fields defined over $\mathbb{F}_q$*, Acta Arith. **143** (2010), no. 4, 377–392, DOI 10.4064/aa143-4-4. MR2652586

[22] E. Ballico, *On the automorphisms of surfaces of general type in positive characteristic. II* (English, with English and Italian summaries), Atti Accad. Naz. Lincei Cl. Sci. Fis. Mat. Natur. Rend. Lincei (9) Mat. Appl. **5** (1994), no. 1, 63–68. MR1273894

[23] A. Bassa and P. Beelen, *The Hasse-Witt invariant in some towers of function fields over finite fields*, Bull. Braz. Math. Soc. (N.S.) **41** (2010), no. 4, 567–582, DOI 10.1007/s00574-010-0026-8. MR2737317

[24] M. Bauer, M. J. Jacobson Jr., Y. Lee, and R. Scheidler, *Construction of hyperelliptic function fields of high three-rank*, Math. Comp. **77** (2008), no. 261, 503–530, DOI 10.1090/S0025-5718-07-02001-7. MR2353964

[25] M. Bauer, E. Teske, and A. Weng, *Point counting on Picard curves in large characteristic*, Math. Comp. **74** (2005), no. 252, 1983–2005, DOI 10.1090/S0025-5718-05-01758-8. MR2164107

[26] J.-B. Bost, *Algebraization, transcendence, and D-group schemes*, Notre Dame J. Form. Log. **54** (2013), no. 3–4, 377–434, DOI 10.1215/00294527-2143961. MR3091663

[27] A. Bostan, P. Gaudry, and É. Schost, *Linear recurrences with polynomial coefficients and computation of the Cartier-Manin operator on hyperelliptic curves*, Finite fields and applications, Lecture Notes in Comput. Sci., vol. 2948, Springer, Berlin, 2004, pp. 40–58, DOI 10.1007/978-3-540-24633-6_4. MR2092621

[28] A. Bostan, P. Gaudry, and É. Schost, *Linear recurrences with polynomial coefficients and application to integer factorization and Cartier-Manin operator*, SIAM J. Comput. **36** (2007), no. 6, 1777–1806, DOI 10.1137/S0097539704443793. MR2299425

[29] I. I. Bouw, C. Diem, and J. Scholten, *Ordinary elliptic curves of high rank over $\overline{\mathbb{F}}_p(x)$ with constant j-invariant*, Manuscripta Math. **114** (2004), no. 4, 487–501, DOI 10.1007/s00229-004-0476-7. MR2081948

[30] A. Buium and J. F. Voloch, *Reduction of the Manin map modulo p*, J. Reine Angew. Math. **460** (1995), 117–126. MR1316574

[31] B. Cais, J. S. Ellenberg, and D. Zureick-Brown, *Random Dieudonné modules, random p-divisible groups, and random curves over finite fields*, J. Inst. Math. Jussieu **12** (2013), no. 3, 651–676, DOI 10.1017/S1474748012000862. MR3062874

[32] G. Cardona and E. Nart, *Zeta function and cryptographic exponent of supersingular curves of genus 2*, Pairing-based cryptography—Pairing 2007, Lecture Notes in Comput. Sci., vol. 4575, Springer, Berlin, 2007, pp. 132–151, DOI 10.1007/978-3-540-73489-5_8. MR2423637

[33] J. W. S. Cassels, *Diophantine equations with special reference to elliptic curves*, J. London Math. Soc. **41** (1966), 193–291, DOI 10.1112/jlms/s1-41.1.193. MR0199150

[34] P. Cassou-Noguès, T. Chinburg, B. Erez, and M. J. Taylor, *Derived category invariants and L-series*, J. Lond. Math. Soc. (2) **92** (2015), no. 2, 265–283, DOI 10.1112/jlms/jdv027. MR3404024

[35] W. Castryck, M. Streng, and D. Testa, *Curves in characteristic 2 with non-trivial 2-torsion*, Adv. Math. Commun. **8** (2014), no. 4, 479–495, DOI 10.3934/amc.2014.8.479. MR3290950

[36] J.-P. Cherdieu, *Remarks on the zeta function of some diagonal hyperelliptic curves*, J. Pure Appl. Algebra **190** (2004), no. 1–3, 31–43, DOI 10.1016/j.jpaa.2003.10.017. MR2043320

[37] G. Cornelissen, F. Oort, T. Chinburg, C. Gasbarri, D. Glass, C. Lehr, M. Matignon, R. Pries, and S. Wewers, *Problems from the Workshop on Automorphisms of Curves*, Rend. Sem. Mat. Univ. Padova **113** (2005), 129–177. MR2168985

[38] B. Ditters and S. Hoving, *Sur la composante connexe du module de Tate covariant de la famille des courbes, donnée par l'équation $y^2 = 1 + \mu x^N$* (French, with English summary), C. R. Acad. Sci. Paris Sér. I Math. **306** (1988), no. 14, 621–624. MR941639

[39] B. Ditters and S. J. Hoving, *On the connected part of the covariant Tate p-divisible group and the ζ-function of the family of hyperelliptic curves $y^2 = 1 + \mu x^N$ modulo various primes*, Math. Z. **200** (1989), no. 2, 245–264, DOI 10.1007/BF01230285. MR978298

[40] E. J. Ditters, *On the classification of commutative formal group laws over p-Hilbert domains and a finiteness theorem for higher Hasse-Witt matrices*, Math. Z. **202** (1989), no. 1, 83–109, DOI 10.1007/BF01180685. MR1007741

[41] I. Dolgachev and D. Lehavi, *On isogenous principally polarized abelian surfaces*, Curves and abelian varieties, Contemp. Math., vol. 465, Amer. Math. Soc., Providence, RI, 2008, pp. 51–69, DOI 10.1090/conm/465/09100. MR2457735

[42] A. Elkin, *Hyperelliptic Jacobians with real multiplication*, J. Number Theory **117** (2006), no. 1, 53–86, DOI 10.1016/j.jnt.2005.05.002. MR2204735

[43] A. Elkin, *The rank of the Cartier operator on cyclic covers of the projective line*, J. Algebra **327** (2011), 1–12, DOI 10.1016/j.jalgebra.2010.09.046. MR2746026

[44] A. Elkin and R. Pries, *Hyperelliptic curves with a-number 1 in small characteristic*, Albanian J. Math. **1** (2007), no. 4, 245–252. MR2367217

[45] A. Elkin and R. Pries, *Ekedahl-Oort strata of hyperelliptic curves in characteristic 2*, Algebra Number Theory **7** (2013), no. 3, 507–532, DOI 10.2140/ant.2013.7.507. MR3095219

[46] J. Estrada Sarlabous, *On the Jacobian varieties of Picard curves defined over fields of characteristic $p > 0$*, Math. Nachr. **152** (1991), 329–340, DOI 10.1002/mana.19911520125. MR1121242

[47] S. Farnell and R. Pries, *Families of Artin-Schreier curves with Cartier-Manin matrix of constant rank*, Linear Algebra Appl. **439** (2013), no. 7, 2158–2166, DOI 10.1016/j.laa.2013.06.012. MR3090462

[48] F. Fité and A. V. Sutherland, *Sato-Tate groups of $y^2 = x^8 + c$ and $y^2 = x^7 - cx$*, Frobenius distributions: Lang-Trotter and Sato-Tate conjectures, Contemp. Math., vol. 663, Amer. Math. Soc., Providence, RI, 2016, pp. 103–126, DOI 10.1090/conm/663/13351. MR3502940

[49] E. Furukawa, M. Kawazoe, and T. Takahashi, *Counting points for hyperelliptic curves of type $y^2 = x^5 + ax$ over finite prime fields*, Selected areas in cryptography, Lecture Notes in Comput. Sci., vol. 3006, Springer, Berlin, 2004, pp. 26–41, DOI 10.1007/978-3-540-24654-1_3. MR2094719

[50] S. D. Galbraith, *Mathematics of public key cryptography*, Cambridge University Press, Cambridge, 2012. MR2931758

[51] P. Gaudry and R. Harley, *Counting points on hyperelliptic curves over finite fields*, Algorithmic number theory (Leiden, 2000), Lecture Notes in Comput. Sci., vol. 1838, Springer, Berlin, 2000, pp. 313–332, DOI 10.1007/10722028_18. MR1850614

[52] A. Ghosh and K. Ward, *The number of roots of polynomials of large degree in a prime field*, Int. Math. Res. Not. IMRN **2015**, no. 4, 898–926, DOI 10.1093/imrn/rnt227. MR3340341

[53] D. Glass and R. Pries, *Hyperelliptic curves with prescribed p-torsion*, Manuscripta Math. **117** (2005), no. 3, 299–317, DOI 10.1007/s00229-005-0559-0. MR2154252

[54] H. Goodson, *A complete hypergeometric point count formula for Dwork hypersurfaces*, J. Number Theory **179** (2017), 142–171, DOI 10.1016/j.jnt.2017.03.018. MR3657161

[55] H. Goodson, *Hypergeometric functions and relations to Dwork hypersurfaces*, Int. J. Number Theory **13** (2017), no. 2, 439–485, DOI 10.1142/S1793042117500269. MR3606631

[56] R. Granger, F. Hess, R. Oyono, N. Thériault, and F. Vercauteren, *Ate pairing on hyperelliptic curves*, Advances in cryptology—EUROCRYPT 2007, Lecture Notes in Comput. Sci., vol. 4515, Springer, Berlin, 2007, pp. 430–447, DOI 10.1007/978-3-540-72540-4_25. MR2449224

[57] P. Guerzhoy, *The Ramanujan differential operator, a certain CM elliptic curve and Kummer congruences*, Compos. Math. **141** (2005), no. 3, 583–590, DOI 10.1112/S0010437X04000995. MR2135278

[58] D. Harvey and A. V. Sutherland, *Computing Hasse-Witt matrices of hyperelliptic curves in average polynomial time*, LMS J. Comput. Math. **17** (2014), suppl. A, 257–273, DOI 10.1112/S1461157014000187. MR3240808

[59] D. Harvey and A. V. Sutherland, *Computing Hasse-Witt matrices of hyperelliptic curves in average polynomial time, II*, Frobenius distributions: Lang-Trotter and Sato-Tate conjectures, Contemp. Math., vol. 663, Amer. Math. Soc., Providence, RI, 2016, pp. 127–147, DOI 10.1090/conm/663/13352. MR3502941

[60] T. Hasegawa, *Some remarks on superspecial and ordinary curves of low genus*, Math. Nachr. **286** (2013), no. 1, 17–33, DOI 10.1002/mana.201010024. MR3019501

[61] K.-i. Hashimoto and N. Murabayashi, *Shimura curves as intersections of Humbert surfaces and defining equations of QM-curves of genus two*, Tohoku Math. J. (2) **47** (1995), no. 2, 271–296, DOI 10.2748/tmj/1178225596. MR1329525

[62] W. A. Hawkins Jr., *The étale cohomology of p-torsion sheaves. I*, Trans. Amer. Math. Soc. **301** (1987), no. 1, 163–188, DOI 10.2307/2000333. MR879568

[63] E. W. Howe, *Supersingular genus-2 curves over fields of characteristic 3*, Computational arithmetic geometry, Contemp. Math., vol. 463, Amer. Math. Soc., Providence, RI, 2008, pp. 49–69, DOI 10.1090/conm/463/09046. MR2459989

[64] T. Ibukiyama, T. Katsura, and F. Oort, *Supersingular curves of genus two and class numbers*, Compositio Math. **57** (1986), no. 2, 127–152. MR827350

[65] F. A. Izadi and V. K. Murty, *Counting points on an abelian variety over a finite field*, Progress in cryptology—INDOCRYPT 2003, Lecture Notes in Comput. Sci., vol. 2904, Springer, Berlin, 2003, pp. 323–333, DOI 10.1007/978-3-540-24582-7_24. MR2092391

[66] N. M. Katz, *Algebraic solutions of differential equations (p-curvature and the Hodge filtration)*, Invent. Math. **18** (1972), 1–118, DOI 10.1007/BF01389714. MR0337959

[67] H. A. W. M. Kneppers, *The Hasse-Witt matrix of a formal group*, Math. Z. **189** (1985), no. 2, 151–165, DOI 10.1007/BF01175041. MR779214

[68] T. Kodama and T. Washio, *On class numbers of hyperelliptic function fields with Hasse-Witt-invariant zero*, Arch. Math. (Basel) **49** (1987), no. 3, 208–213, DOI 10.1007/BF01271660. MR906735

[69] T. Kodama and T. Washio, *Hasse-Witt matrices of Fermat curves*, Manuscripta Math. **60** (1988), no. 2, 185–195, DOI 10.1007/BF01161928. MR924086

[70] T. Kodama and T. Washio, *A family of hyperelliptic function fields with Hasse-Witt-invariant zero*, J. Number Theory **36** (1990), no. 2, 187–200, DOI 10.1016/0022-314X(90)90072-Y. MR1072464

[71] M. Kudo and S. Harashita, *Superspecial curves of genus 4 in small characteristic*, Finite Fields Appl. **45** (2017), 131–169, DOI 10.1016/j.ffa.2016.12.001. MR3631358

[72] C. Lennon, *Trace formulas for Hecke operators, Gaussian hypergeometric functions, and the modularity of a threefold*, J. Number Theory **131** (2011), no. 12, 2320–2351, DOI 10.1016/j.jnt.2011.05.005. MR2832827

[73] D. J. Madden, *Arithmetic in generalized Artin-Schreier extensions of $k(x)$*, J. Number Theory **10** (1978), no. 3, 303–323, DOI 10.1016/0022-314X(78)90027-6. MR506641

[74] K. Matsuo, J. Chao, and S. Tsujii, *An improved baby step giant step algorithm for point counting of hyperelliptic curves over finite fields*, Algorithmic number theory (Sydney, 2002), Lecture Notes in Comput. Sci., vol. 2369, Springer, Berlin, 2002, pp. 461–474, DOI 10.1007/3-540-45455-1_36. MR2041104

[75] K. Matsuo, J. Chao, and S. Tsujii, *Baby step giant step algorithms in point counting of hyperelliptic curves*, IEICE Trans. Fundamentals **E86-A** (2003), no. 5, 1127–1134.

[76] B. Mazur, *Frobenius and the Hodge filtration*, Bull. Amer. Math. Soc. **78** (1972), 653–667, DOI 10.1090/S0002-9904-1972-12976-8. MR0330169

[77] L. Miller, *Curves with invertible Hasse-Witt-matrix*, Math. Ann. **197** (1972), 123–127, DOI 10.1007/BF01419588. MR0314849

[78] L. Miller, *Über gewöhnliche Hyperflächen. I*, J. Reine Angew. Math. **282** (1976), 96–113, DOI 10.1515/crll.1976.282.96. MR0407024

[79] L. Miller, *Über gewöhnliche Hyperflächen. II*, J. Reine Angew. Math. **283/284** (1976), 402–420, DOI 10.1515/crll.1976.283-284.402. MR0407025

[80] N. O. Nygaard, *Slopes of powers of Frobenius on crystalline cohomology*, Ann. Sci. École Norm. Sup. (4) **14** (1981), no. 4, 369–401 (1982). MR654203

[81] N. O. Nygaard, *On supersingular abelian varieties*, Algebraic geometry (Ann Arbor, Mich., 1981), Lecture Notes in Math., vol. 1008, Springer, Berlin, 1983, pp. 83–101, DOI 10.1007/BFb0065702. MR723711

[82] L. D. Olson, *Hasse invariants and anomalous primes for elliptic curves with complex multiplication*, J. Number Theory **8** (1976), no. 4, 397–414, DOI 10.1016/0022-314X(76)90087-1. MR0450191

[83] Ë. Onishi, *Generalized Bernoulli-Hurwitz numbers and universal Bernoulli numbers* (Russian, with Russian summary), Uspekhi Mat. Nauk **66** (2011), no. 5(401), 47–108. MR2919272

[84] Ë. Onishi, *Generalized Bernoulli-Hurwitz numbers and universal Bernoulli numbers*, Russian Math. Surveys **66** (2011), no. 5, 871–932, DOI 10.1070/RM2011v066n05ABEH004763. MR2919272

[85] A. Pacheco, *A note on relations between the zeta-functions of Galois coverings of curves over finite fields*, Canad. Math. Bull. **33** (1990), no. 3, 282–285, DOI 10.4153/CMB-1990-046-x. MR1077097

[86] R. J. Pries, *Jacobians of quotients of Artin-Schreier curves*, Recent progress in arithmetic and algebraic geometry, Contemp. Math., vol. 386, Amer. Math. Soc., Providence, RI, 2005, pp. 145–156, DOI 10.1090/conm/386/07221. MR2182774

[87] R. Pries, *The p-torsion of curves with large p-rank*, Int. J. Number Theory **5** (2009), no. 6, 1103–1116, DOI 10.1142/S1793042109002560. MR2569747

[88] R. Pries and K. Stevenson, *A survey of Galois theory of curves in characteristic p*, WIN—women in numbers, Fields Inst. Commun., vol. 60, Amer. Math. Soc., Providence, RI, 2011, pp. 169–191. MR2777804

[89] H.-G. Rück, *Class groups and L-series of function fields*, J. Number Theory **22** (1986), no. 2, 177–189, DOI 10.1016/0022-314X(86)90068-5. MR826950

[90] P. Sarkar and S. Singh, *A simple method for obtaining relations among factor basis elements for special hyperelliptic curves*, Appl. Algebra Engrg. Comm. Comput. **28** (2017), no. 2, 109–130, DOI 10.1007/s00200-016-0299-2. MR3614746

[91] J. H. Silverman, *The arithmetic of elliptic curves*, 2nd ed., Graduate Texts in Mathematics, vol. 106, Springer, Dordrecht, 2009. MR2514094

[92] G. Sohn, *Computing the number of points on genus 3 hyperelliptic curves of type $Y^2 = X^7 + aX$ over finite prime fields*, J. Appl. Math. Inform. **32** (2014), no. 1–2, 17–26, DOI 10.14317/jami.2014.017. MR3156260

[93] G. Sohn and H. Kim, *Explicit bounds of polynomial coefficients and counting points on Picard curves over finite fields*, Math. Comput. Modelling **49** (2009), no. 1–2, 80–87, DOI 10.1016/j.mcm.2008.03.012. MR2480034

[94] K.-O. Stöhr and J. F. Voloch, *A formula for the Cartier operator on plane algebraic curves*, J. Reine Angew. Math. **377** (1987), 49–64, DOI 10.1515/crll.1987.377.49. MR887399

[95] F. J. Sullivan, *p-torsion in the class group of curves with too many automorphisms*, Arch. Math. (Basel) **26** (1975), 253–261, DOI 10.1007/BF01229737. MR0393035

[96] Y. Sung, *Rational points over finite fields on a family of higher genus curves and hypergeometric functions*, Taiwanese J. Math. **21** (2017), no. 1, 55–79, DOI 10.11650/tjm.21.2017.7724. MR3613974

[97] S. Tafazolian, *A family of maximal hyperelliptic curves*, J. Pure Appl. Algebra **216** (2012), no. 7, 1528–1532, DOI 10.1016/j.jpaa.2012.01.019. MR2899820

[98] Y. Takeda, *Groups of Russell type and Tango structures*, Affine algebraic geometry, CRM Proc. Lecture Notes, vol. 54, Amer. Math. Soc., Providence, RI, 2011, pp. 327–334. MR2768648

[99] Y. Takeda and K. Yokogawa, *Pre-Tango structures on curves*, Tohoku Math. J. (2) **54** (2002), no. 2, 227–237. MR1904950

[100] Y. Takizawa, *Some remarks on the Picard curves over a finite field*, Math. Nachr. **280** (2007), no. 7, 802–811, DOI 10.1002/mana.200410515. MR2321141

[101] D. L. Ulmer, *On universal elliptic curves over Igusa curves*, Invent. Math. **99** (1990), no. 2, 377–391, DOI 10.1007/BF01234424. MR1031906

[102] R. C. Valentini, *Hyperelliptic curves with zero Hasse-Witt matrix*, Manuscripta Math. **86** (1995), no. 2, 185–194, DOI 10.1007/BF02567987. MR1317743

[103] T. Washio, *On class numbers of algebraic function fields defined by $y^2 = x^5 + ax$ over* $\mathrm{GF}(p)$, Arch. Math. (Basel) **41** (1983), no. 6, 509–516, DOI 10.1007/BF01198580. MR731634

[104] N. Yui, *On the Jacobian variety of the Fermat curve*, J. Algebra **65** (1980), no. 1, 1–35, DOI 10.1016/0021-8693(80)90236-7. MR578793

[105] N. Yui, *The arithmetic of the product of two algebraic curves over a finite field*, J. Algebra **98** (1986), no. 1, 102–142, DOI 10.1016/0021-8693(86)90018-9. MR825138

[106] N. Yui, *Jacobi quartics, Legendre polynomials and formal groups*, Elliptic curves and modular forms in algebraic topology (Princeton, NJ, 1986), Lecture Notes in Math., vol. 1326, Springer, Berlin, 1988, pp. 182–215, DOI 10.1007/BFb0078046. MR970289

[107] L. Zapponi, *On the 1-pointed curves arising as étale covers of the affine line in positive characteristic*, Math. Z. **258** (2008), no. 4, 711–727, DOI 10.1007/s00209-007-0192-6. MR2369052

[108] Y. G. Zarhin, *Non-supersingular hyperelliptic Jacobians* (English, with English and French summaries), Bull. Soc. Math. France **132** (2004), no. 4, 617–634. MR2131907

Department of Mathematics, Colorado State University, Fort Collins, Colorado 80523

Email address: `j.achter@colostate.edu`

URL: `http://www.math.colostate.edu/~achter`

Center for Communications Research, 4320 Westerra Court, San Diego, California 92121-1967

Email address: `however@alumni.caltech.edu`

URL: `http://ewhowe.com/`

Contemporary Mathematics
Volume **722**, 2019
https://doi.org/10.1090/conm/722/14531

Analogues of Brauer-Siegel theorem in arithmetic geometry

Marc Hindry

This paper is dedicated to our late friend Alexey Ivanovich Zykin.

Abstract. We explain an analogy between the classical Brauer-Siegel theorem, a statement relating asymptotically the class number, regulator of units and discriminant of a number field, but whose proof involves the Dedekind zeta function, and similar statement involving arithmetic invariants of algebraic varieties over a finite or global field. We develop precisely the analogy for surfaces over a finite field and for abelian varieties over a global field, surveying some recent results. We also formulate a quite general question along these lines, and develop the case of projective hypersurfaces.

Introduction

Many problems in quantitative arithmetic geometry lead to estimates for the volume of lattices arising from geometry or arithmetic. We will focus on three examples: *a*) the lattice of units of a number field K, via the logarithmic embedding; *b*) the Mordell-Weil lattice of an abelian variety A, over a global field K, equipped with the Néron-Tate height pairing; *c*) The Néron-Severi lattice of a surface S defined over a finite field, equipped with the intersection pairing. The corresponding volume (or square of the volume) is then defined respectively as the regulator of units R_K, the Néron-Tate regulator $\mathrm{Reg}(A/K)$ and the Néron-Severi regulator $\mathrm{Reg}(S/\mathbb{F}_q)$. The classical Brauer-Siegel theorem taught us that instead of looking directly for estimates of R_K, we should consider the product $h_K R_K$, where $h_k = \sharp\mathrm{Pic}\,(\mathcal{O}_K)$ is the class number. We thus replace the quantities R_K, $\mathrm{Reg}(A/K)$ and $\mathrm{Reg}(S/\mathbb{F}_q)$ by respectively the products $h_K R_K$, $\sharp\text{Ш}(A/K)\mathrm{Reg}(A/K)$ and $\sharp\mathrm{Br}(S/\mathbb{F}_q)\mathrm{Reg}(S/\mathbb{F}_q)$, where $\text{Ш}(A/K)$ is the Shafarevich-Tate group and $\mathrm{Br}(S/\mathbb{F}_q)$ is the Brauer group, both being conjecturally finite, although this is known only in special cases. The classical Brauer-Siegel theorem can be stated for a family of number field of fixed degree (with self-explanatory notation detailed in the first section) as:

$$\lim_{\Delta_F\to\infty} \frac{\log\left(\sharp\mathrm{Pic}\,(\mathcal{O}_F)R_F\right)}{\log\sqrt{\Delta_F}} = 1.$$

2010 *Mathematics Subject Classification.* Primary 11-XX, 14E20, 11Mxx, 11Rxx, 11R42, 14-XX, 14Gxx, 14Kxx.

Key words and phrases. Diophantine geometry, algebraic geometry, analytic number theory, global fields, algebraic curves and surfaces, abelian varieties, zeta and L-function, Brauer-Siegel theorem, Birch & Swinnerton-Dyer conjecture, Bloch-Kato conjecture.

The author is supported by ANR-17-CE40-0012 Flair.

We (try to) generalise this and reach statements with the following shape, for a family of algebraic varieties X:

$$\lim_{H(X/K)\to\infty} \frac{\log\left(\sharp\text{Ш}(X/K)\text{Reg}(X/K)\right)}{\log H(X/K)} = ??.$$

The first *motto* is to estimate the volume of a finitely generated group, a *regulator* often denoted $\text{Reg}(X/K)$, *and* the size of an associated finite group describing local-to-global obstructions, often denoted $\text{Ш}(X/K)$, and compare their product with a measure of the arithmetic complexity denoted $H(X/K)$ (discriminant of a number field, Faltings height of an abelian variety, geometric genus of a surface).

The second *motto* is the use of zeta and L-functions and their special values, as a sophisticated local-to-global tool, to carry out this programme.

The organisation of the paper is as follows. The first section reviews the classical Brauer-Siegel theorem for number fields and the subsequent analogues for a function field and generalisations due to Tsfasman, Vlăduţ and Zykin. The second section gives a quick introduction to L-functions associated to (smooth projective) algebraic varieties over global fields, their special values and analytic estimates. The third section displays the case of surfaces over finite fields and abelian varieties over global fields. The fourth section is highly speculative and contains a tentative guess of what would be the *Brauer-Siegel ratio* of a motive and its asymptotic behaviour. The last section contains a summary and a discussion of upper bounds and lower bounds (the latter being much more mysterious) for the Brauer-Siegel ratio in several instances.

Thanks. The author is grateful to Richard Griffon and Michael Tsfasman for many helpful comments and conversations.

Dedication. Alexey Zykin left us much too early in April 2017, depriving us from a brillant mathematician and a wonderful friend. The title of his PhD was the *asymptotic properties of global fields* (Marseille, 2009 and Moscow, 2010) and that theme was one of his favourite among many interests in mathematics. In particular he published a paper "*On the generalizations of the Brauer-Siegel theorem*" in the proceedings of the 2009 edition of the *Arithmetic, geometry, cryptography and coding theory* conference, which is closely related to the topic of the present article. It is therefore very fitting to dedicate this work to the memory of Alexey, I will add a thought for his wife Tanya, who departed the same fateful day. We will miss them forever, but they will remain present in our hearts.

1. Classical Brauer-Siegel theorems

1.1. Number fields. The original Brauer-Siegel theorem (see for example [**La70**]) is a result about asymptotics of number fields: one considers a family $\mathcal{F} = \{K_i\}_{i\in I}$ of number fields, where the (absolute value of the) discriminant $\Delta_i = \Delta_{K_i}$ goes to infinity. In addition to the discriminant, the quantities describing the arithmetic of a number field K are:

- the number of real embeddings $r_1 = r_1(K)$, the number of pairs of conjugate complex embeddings $r_2 = r_2(K)$ and the degree $n = n(K) = r_1(K) + 2r_2(K)$.
- the class number $h = h_K = \sharp\text{Pic}\,(\mathcal{O}_K)$.

- the regulator of units R_K which represents the volume of the lattice $\mathcal{O}_K^\times$ modulo torsion (see [**La70**] for a precise definition); the rank of the unit lattice is $r = r(K) = r_1(K) + r_2(K) - 1$.
- the number of roots of unity $w_K = \sharp\mathbb{G}_m(\mathcal{O}_K)_{\rm tor}$.

They all intervene in the celebrated *class number formula* describing the residue of the Dedekind zeta function at $s = 1$:

$$\zeta_K^*(1) := \lim_{s\to 1}(s-1)\zeta_K(s) = \frac{h_K R_K}{\sqrt{\Delta_K}} \times \frac{2^{r_1}(2\pi)^{r_2}}{w_K}. \tag{1.1}$$

Invoking the functional equation relating $\zeta_K(s)$ and $\zeta_K(1-s)$, we may restate this in a more suggestive way:

(1) The order of $\zeta_K(s)$ at $s = 0$ is equal to the rank of $\mathcal{O}_K^\times$.
(2) The group $\mathrm{Pic}\,(\mathcal{O}_K)$ is finite.
(3) The leading term at $s = 0$ is given by:

$$\zeta_K^*(0) := \lim_{s\to 1} s^{-r}\zeta_K(s) = -\frac{h_K R_K}{w_K}. \tag{1.2}$$

In order to formulate the Brauer-Siegel theorem and compare it with other statements we introduce the *Brauer-Siegel ratio* as follows.

DEFINITION 1.1. The *Brauer-Siegel ratio* of a number field K is the quantity:

$$\mathfrak{Bs}(K) := \frac{\log\,(h_K R_K)}{\log\sqrt{\Delta_K}}.$$

The *Brauer-Siegel ratio* of a family of number fields $\mathcal{F} = \{K_i\}_{i\in I}$ is the limit (if it exists):

$$\mathfrak{Bs}(\mathcal{F}) := \lim_{\mathcal{F}} \mathfrak{Bs}(K_i)$$

When the limit does not exist, we replace it by lim inf and lim sup; alternatively limits will always exist when we restrict to an appropriate subfamily.

A simple form of the Brauer-Siegel theorem is obtained by considering the family $\mathcal{F}_n$ of all number fields of degree n.

THEOREM 1.2. *[Brauer-Siegel Theorem] Let $\mathcal{F}_n$ be the family of all number fields of degree n, then*

$$\mathfrak{Bs}(\mathcal{F}_n) := \lim_{K\in\mathcal{F}_n} \mathfrak{Bs}(K) = 1. \tag{1.3}$$

One of the most interesting corollary is already obtained when considering real and complex imaginary quadratic fields.

COROLLARY 1.3. Let $K_d = \mathbb{Q}(\sqrt{d})$ where d runs over negative (resp. positive) square-free integers, then

$$\log h_{K_d} \sim \log\sqrt{d} \qquad (\text{resp. } \log\,(h_{K_d}\log\epsilon_d) \sim \log\sqrt{d}) \tag{1.4}$$

where ϵ_d denotes the fundamental unit in K_d.

The proof of Theorem 1.2 combines the class number formula (1.1) with analytic estimates involving an upper bound estimate for the residue of $\zeta_K(s)$:

$$\zeta_K^*(1) \le c(n_K)\,(\log\Delta_K)^{n_K-1} \tag{1.5}$$

and a (much harder) lower bound estimate for the residue at $s = 1$ of $\zeta_K(s)$:

$$\zeta_K^*(1) \ge c(n_K,\epsilon)\Delta_K^{-\epsilon}. \tag{1.6}$$

The lower estimate is not effective but, under Generalised Riemann Hypothesis (GRH) or even under the more innocent looking hypothesis that $\zeta_K(s)$ does not vanish on the segment $[1-1/\log\Delta_K, 1[$, it can be sharpened to give (with an effective constant) :

$$\zeta_K^*(1) \geq \frac{c(n_K)}{\log \Delta_K}. \quad \textit{(GRH)} \tag{1.7}$$

More explicitly, it is known that the zeta function $\zeta_K(s)$ has at most one zero in a disk of center 1 and radius $c/\log\Delta_K$; further such a zero – if it exists – must be real and thus must belong to the segment $(1-c/\log\Delta_K, 1)$. Stark [**St74**] has shown that such a so-called Siegel zero may exist only when K contains a quadratic field.

The asymptotic behaviour

$$\lim_{K\in\mathcal{F}} \mathfrak{Bs}(K) = \lim_{K\in\mathcal{F}} \frac{\log(h_K R_K)}{\log\sqrt{\Delta_K}} = 1 \tag{1.8}$$

remains true under the following weaker conditions

- $\lim \frac{n_K}{\log \Delta_K} = 0$,
- and $K/\mathbb{Q}$ is Galois or GRH.

Tsfasman-Vlăduţ found a far reaching generalisation of this when hypotheses are weakened. In order to state it neatly, let us introduce the following definition concerning a family $\mathcal{F} = \{K_i\}$ of number fields.

DEFINITION 1.4. A family $\mathcal{F} = \{K_i\}$ is called *asymptotically exact* if the following limits all exist:

$$\text{(for } q \text{ power of a prime)} \quad \phi_q := \lim_{K\in\mathcal{F}} \frac{\sharp\{\mathfrak{p}\in\mathcal{P}_K \mid N\mathfrak{p} = q\}}{\log\sqrt{\Delta_K}}. \tag{1.9}$$

$$\phi_{\mathbb{R}} := \lim_{K\in\mathcal{F}} \frac{r_1(K)}{\log\sqrt{\Delta_K}} \quad \text{and} \quad \phi_{\mathbb{C}} := \lim_{K\in\mathcal{F}} \frac{r_2(K)}{\log\sqrt{\Delta_K}}. \tag{1.10}$$

Note that, from any family, we can extract a sub-family which will be asymptotically exact, and observe that the previous condition $\lim \frac{n_K}{\log\Delta_K} = 0$ is equivalent to all ϕ_v's vanish.

THEOREM 1.5. *(Tsfasman-Vlăduţ, Zykin* [**TsVl97, Zy09, Zy09c**]*) Let $\mathcal{F} = \{K_i\}$ be an asymptotically exact sequence of number fields. Then*

$$\lim_{K\in\mathcal{F}} \frac{\log(h_K R_K)}{\log\sqrt{\Delta_K}} = 1 + \sum \phi_q \log\left(\frac{q}{q-1}\right) - \phi_{\mathbb{R}}\log 2 - \phi_{\mathbb{C}}\log(2\pi) \tag{1.11}$$

when the sequence is normal or almost normal or satisfies GRH.
Further, under GRH, the following "basic inequalities" hold:

$$\sum \phi_q \left(\frac{\log q}{\sqrt{q}-1}\right) + \phi_{\mathbb{R}}\left(\log(2\sqrt{2\pi}) + \frac{\pi}{4} + \frac{\gamma}{2}\right) + \phi_{\mathbb{C}}(\log(8\pi)+\gamma) \leq 1 \tag{1.12}$$

where γ is Euler's constant.

1.2. Function fields. We have formulated things over number fields, but everything goes over function fields over a finite field, in fact most of the time proofs are easier, e.g. GRH is known. Let us look first at the case of a smooth geometrically irreducible projective curve C over a finite field $\mathbb{F}_q$. The analogue of the class number formula for the zeta function of the curve is

$$\zeta_C^*(1) = \lim_{s\to 1}(s-1)\zeta_C(s) = \frac{h_C}{q^{g(C)-1}} \cdot \frac{(\log q)^{-1}}{q-1} \tag{1.13}$$

where $g(C)$ is the genus of the curve, $h_C = \sharp\mathrm{Pic}^0(C)$ is the number of $\mathbb{F}_q$-points of the Jacobian of C. Notice $q-1 = \sharp\mathbb{F}_q^\times = \sharp\mathbb{G}_m(\mathbb{F}_q)_{\mathrm{tor}}$. Further note that, when comparing Hurwitz formula and the discriminant formula, we see that $g-1$ (or $(g-1)\log q$ is the analogue of $\log\sqrt{\Delta_K}$, but when considering asymptotics with g going to infinity, we may replace $g-1$ by g. Weil's theorem implies that

$$\left(1-q^{-1/2}\right)^{2g} \leq \frac{h_C}{q^g} \leq \left(1+q^{-1/2}\right)^{2g} \tag{1.14}$$

but to prove the full analogue of Brauer-Siegel requires some extra hypothesis. Recall that the *gonality* of a curve C, denoted $n(C)$, is defined as the minimum of the degrees of finite morphisms $\phi : C \to \mathbb{P}^1$; in our context $n(C) = [\mathbb{F}_q(C) : \mathbb{F}_q(t)]$ with $\mathbb{F}_q(t) = \phi^*(\mathbb{F}_q(\mathbb{P}^1))$.

THEOREM 1.6. *(Inaba* [**In50**]*) Let C_i be a family of curves over $\mathbb{F}_q$ with $g_i = g(C_i)$ going to infinity and bounded gonality, then*

$$\lim \mathfrak{Bs}(C_i/\mathbb{F}_q) := \lim \frac{\log h_{C_i}}{g(C_i)\log q} = 1.$$

The following generalisation is formally closer to the original Brauer-Siegel and is worked out in [**GoLu78**]. Let S be a finite set of geometric points (or places) on a projective curve $C/\mathbb{F}_q$, denote $\mathcal{O}_{C,S}$ the ring of functions with poles contained in S. It is known that the group of units $\mathcal{O}_{C,S}^\times$ has rank $r = |S|-1$. If $\epsilon_1, \dots, \epsilon_r$ denote generators of the group of units modulo torsion and ord_i the valuations associated to the places in S and d_i the degree of the places, then the *S-Regulator* is defined as the determinant:

$$R_{C,S} := \left|\det\left(d_i \mathrm{ord}_i \epsilon_j\right)_{1\leq i,j\leq r}\right| \tag{1.15}$$

(notice that the last column "$i = r+1$" is omitted but the product formula $\sum_i d_i \mathrm{ord}_i \epsilon_j = 0$ guarantees that the resulting determinant is independent of this choice).

THEOREM 1.7. *(Gogia, Luthar* [**GoLu78**]*) Let $C_i/\mathbb{F}_q$ be a family of curves with gonality n_i and genus g_i, let S_i be the finite set of places of C_i above a fixed set of places of $C_0 = \mathbb{P}^1$. Assume that the ratio n_i/g_i tends to zero, then*

$$\lim \mathfrak{Bs}(\mathcal{O}_{C_i,S_i}) := \lim \frac{\log(h_{C_i,S}R_{C_i,S_i})}{g\log q} = 1.$$

The analogues over function field of Theorem 1.5 and inequality (1.12) are quite interesting. For an asymptotically exact family of function fields $K = \mathbb{F}_q(C_i)$ (i.e. such that all limits $\phi_{q^m} := \lim \sharp C_i(\mathbb{F}_{q^m})/g(C_i)$ exist), the analogue states that (see [**TsVl97**]):

$$\lim_i \mathfrak{Bs}(K_i) = 1 + \frac{1}{\log q}\sum_{m=1}^{\infty} \phi_{q^m} \log\left(\frac{q^m}{q^m-1}\right) \tag{1.16}$$

The inequality corresponding to (1.12) is the following:

$$\sum_{m=1}^{\infty} \frac{m\phi_{q^m}}{q^{\frac{m}{2}}-1} \leq 1. \tag{1.17}$$

It refines the Drinfeld-Vlăduţ bound [**DrVl83**] which states that $\phi_q \leq \sqrt{q}-1$ or:

$$\sharp C_i(\mathbb{F}_q) \leq (\sqrt{q}-1)g_i + o(g_i).$$

For a nice survey and more on the Brauer-Siegel theorems for number fields and function fields see [**LeZy11**].

2. Zeta and L-functions

We will also consider (smooth projective) varieties over number fields K; their reduction modulo good prime ideals can be viewed as a family of varieties over varying $\mathbb{F}_q$; there is a close connection with varieties over $K=\mathbb{F}_q(B)$ (where B will be a smooth irreducible projective curve) obtained by considering $\mathcal{X}/\mathbb{F}_q$ a smooth projective variety of dimension $n+1$, equipped with a fibration $\pi:\mathcal{X}\to B$, such that the generic fiber X is a smooth projective variety over $K=\mathbb{F}_q(B)$. Conversely if we are given X/K, we will call a *model* of X such a variety $\mathcal{X}$.

It is customary to call K a *global field* if it is either a number field – a finite extension of $\mathbb{Q}$, or a function field $\mathbb{F}_q(B)$ – a finite extension of $\mathbb{F}_q(t)$.

2.1. Zeta and L-functions associated to algebraic varieties.

Consider a smooth projective algebraic variety X over a global field K. When K is a number field we refer to it as the number field case and write $B := \mathrm{spec}\,(\mathcal{O}_K)$; when K is a function field over a finite field $\mathbb{F}_q$, we refer to it as the function field case and denote $K=\mathbb{F}_q(B)$, where B is a smooth projective curve. The separable closure of K is denoted $\bar{K}$ and the absolute Galois group $G_K=\mathrm{Gal}(\bar{K}/K)$. We write $\mathcal{P}_K$ for the set of places of K and, for a finite place $v\in\mathcal{P}_K$, we denote p_v the residual characteristic and q_v the cardinality of the residual field. The following definition and further foundation material can be found, for example, in Serre [**Se65, Se69**].

DEFINITION 2.1. Let $\mathcal{X}$ be a scheme of finite type and presentation over $\mathbb{Z}$, the zeta function of $\mathcal{X}$ is the function

$$\zeta(\mathcal{X},s) := \prod_{x\in|\mathcal{X}|}\left(1-N(x)^{-s}\right)^{-1} \tag{2.1}$$

where $|\mathcal{X}|$ denotes the set of closed points and $N(x)$ the norm of a closed point (i.e. the cardinality of the finite residual field).

If $\mathcal{X}$ is actually a variety X of dimension n over $\mathbb{F}_q$, then $\zeta(\mathcal{X},s)=Z(X/\mathbb{F}_q,q^{-s})$ where $Z(X/\mathbb{F}_q,T) = \prod_{i=0}^{2n} P_i(X,T)^{(-1)^{i+1}}$ is Weil's Zeta function and P_i is the characteristic polynomial of the q-Frobenius acting on $H^i(X)$. When $\mathcal{X}$ is flat over $\mathbb{Z}$ we get $\zeta(\mathcal{X},s)=\prod_p\zeta(\mathcal{X}_p,s)=\prod_p Z(\mathcal{X}_p/\mathbb{F}_p,p^{-s})$, where $\mathcal{X}_p$ is the fibre above p.

We abbreviate $H^j(X)$ or $H^j(X,\mathbb{Q}_\ell) := H^j_{\acute{e}t}(X\times_K\bar{K},\mathbb{Q}_\ell)$ and denote the associated Galois representation by $\rho=\rho_{X,j,\ell}: G_K\to \mathrm{GL}(H^j(X,\mathbb{Q}_\ell))$. We will also use the classical notation $H^j(X,\mathbb{Q}_\ell)(n)=H^j(X,\mathbb{Q}_\ell(n))$ to designate Tate twists. For all places $v\in\mathcal{P}_K$, the *Frobenius element* Frob_v is actually a conjugacy class modulo the inertia, but the next definitions are independent of the choice of this element.

$$L_v(H^j(X),s) := \det\left(1-\rho(\mathrm{Frob}_v)q_v^{-s}\,|H^j(X,\mathbb{Q}_\ell)^{I_v}\right)^{-1} \tag{2.2}$$

$$L(H^j(X),s) = \prod_{v\in\mathcal{P}_K} L_v(H^j(X),s) = \prod_{v\in\mathcal{P}_K}\prod_{i=1}^{b_j(X)}(1-\alpha_{v,i}q_v^{-s})^{-1} \tag{2.3}$$

where $b_j(X) = \dim H^j(X)$ and for almost all v we have $|\alpha_{v,j}| = q_v^{j/2}$. We say that the L-function has *weight* $w = j$. In the finite field case there is an integer $b = b_j(X)$ (the Betti number) such that if we set $\Lambda(s) = q^{bs/2}L(s)$ then we have

$$\Lambda(s) = \pm\Lambda(w-s). \tag{2.4}$$

When X is a variety over a function field $K = \mathbb{F}_q(B)$, one can show in many cases that $L(s)$ is actually a polynomial $\prod_{i=1}^{b(X)}(1-\beta_i q^{-s})$ with now $|\beta_i| = q^{(w+1)/2}$, hence the same is true with $b(X)$ essentially equal to the *conductor* and with a shift:

$$\Lambda(s) = \pm\Lambda(w+1-s). \tag{2.5}$$

In the number field case one must add archimedean factors called Γ-factors and denoted $L_\infty(H^j(X),s)$; these are defined in terms of the Hodge numbers of X (see [**Se69**] for a precise description). Then there is an integer $F = F_j(X)$, called the *analytic conductor*, such if we set $\Lambda(s) = F^{s/2}L_\infty(H^j(X),s)\cdot L(H^j(X),s)$ then the *expected* functional equation is as follows.

CONJECTURE 2.2. Define $\Lambda(H^j(X),s) = F^{s/2}L_\infty(H^j(X),s)L(H^j(X),s)$, then $\Lambda(H^j(X),s)$ has analytical continuation to the whole complex plane, with at most a pole at $s = 1+j/2$ and $s = j/2$, if j is even, and satisfies the following functional equation:

$$\Lambda(H^j(X),s) = \pm\Lambda(H^j(X),j+1-s).$$

The Euler product defining $L(H^j(X),s)$ converges absolutely for $\Re s > 1+j/2$ and does not vanish on this half-plane; by the (conjectural) functional equation it does not vanish on the half-plane $\Re s < j/2$ except at the "trivial zeroes" provoked by the Γ-factor $L_\infty(H^j(X),s)$. The behaviour in the critical strip $j/2 \leq \Re s \leq 1+j/2$ is more mysterious. The celebrated generalised Riemann hypothesis (GRH) states the following.

CONJECTURE 2.3. The function $L(H^j(X),s)$ does not vanish for $\Re s > (1+j)/2$. In other words, apart from the "trivial zeroes", the function $L(H^j(X),s)$ has all its zeroes on the line $\Re s = (j+1)/2$.

This is known in the function field case thanks to Grothendieck and Deligne but, in the number field case, it is not known for a single example.

2.2. Special values of L-functions. For a zeta function or L-function, it is customary to define the *special value* at a point (which will always be an integer) as the leading coefficient in the Taylor expansion at this point. That is, when $L(s)$ is meromorphic around $s = n$ and has order r at $s = n$, we set:

$$L^*(n) := \lim_{s\to n}(s-n)^{-r}L(s).$$

The order at an integer $s = n$ of the L-function of weight w is well known for $n > 1+w/2$: the Euler product is convergent and the function has neither zero nor pole. When $n < w/2$ the functional equation (when conjectured or when proven) provides the order of vanishing in terms of the Hodge numbers. The orders of vanishing at $1+w/2$ and $w/2$ (when $j = w$ is even) are related via the functional equation and predicted by the Tate conjecture.

Conjecture 2.4. (Tate [**Ta65**]) Let X be a smooth projective algebraic variety over a global field K. Let $C^k(X/K)$ denote the image of algebraic cycles of codimension k, under the cycle map, in the cohomology group $H^{2k}(X)(k)$, and G_K the absolute Galois group, then:

(1) The following induced map (known to be injective) is an isomorphism:

$$C^k(X/K) \otimes \mathbb{Q}_\ell \to H^{2k}(X, \mathbb{Q}_\ell(k))^{G_K}.$$

(2) The function $L(H^{2k}(X), s)$ has a pole of order $\operatorname{rk} C^k(X/K)$ at $s = 1 + k$.

Analogously, when X is defined over a finite field, then $\zeta(X, s)$ has a pole of order $\operatorname{rk} C^k(X)$ at $s = k$.

For example, for a surface S, the conjecture predicts that $L(H^2(S), s)$ has a pole at $s = 1$ of order the rank of the Néron-Severi group. For another analytic point of view on Tate's conjecture see [**HiPaWa05**].

The order of $L(s)$ at the center of symmetry $\frac{w+1}{2}$ (when w is odd) is more mysterious and predicted by the Birch & Swinnerton-Dyer conjecture (reviewed in the next section) for $w = 1$ and the Bloch-Kato conjecture for odd $w > 1$.

We have already used the notation $\zeta_K^*(1)$ to denote the residue of $\zeta_K(s)$ at $s = 1$. The class number formula has been vastly generalized, at least conjecturally. Deligne [**De77**] has proposed a conjecture for the value up to a rational number of $L(s)$ at a *critical integer* (the value is a suitable period), Beilinson has extended the conjectured formulae to all integers, still up to a rational number (the value is the product of a suitable period by a regulator linked with algebraic cycles) and finally Bloch-Kato have made these conjectures more precise by specifying the exact (conjectural) value; a prototype for these formula is the conjecture of Birch & Swinnerton-Dyer predicting the special value of the L-function of an abelian variety over a global field [**Bei85, BlKa90, De77, Ta66**]. The Deligne conjecture applies at $s = m$ where $L_\infty(s)$ neither has poles at $s = m$ nor at $s = j + 1 - m$; the Beilinson conjecture applies at all integer m except at the central value $m = \frac{j+1}{2}$.

There are also conjectures by Lichtenbaum [**Li84**] for the special value at integers of zeta functions of varieties over finite fields.

2.3. Classical estimates. If we assume analytic continuation and functional equation, we can write very useful estimates for the growth of $L(s)$; if we assume further Generalized Riemann Hypothesis, these estimates can be sharpened.

The Phragmén-Lindelöf principle gives a convexity property: over number field it says that if $\psi(\sigma)$ is the infimum of the numbers $\rho > 0$ such that there exists a constant $c = c(\rho)$ with $|L(\sigma + it)| \leq c \max(1, |t|)^\rho$ for all $t \in \mathbb{R}$, then $\psi(\sigma)$ is a convex function. It is clear that $\psi(\sigma) = 0$ for σ greater than σ_a, the abscissa of absolute convergence; the functional equation gives the value of $\psi(\sigma)$ for $\sigma < \sigma_a - 1$ and applying convexity gives a bound inside the critical strip.

Granting GRH we can consider $\log L(s)$ in half of the critical strip and apply Pragmén-Lindelöf principle to it. Over finite fields or function fields, this is usually sufficient, over number fields one can use the Borel-Caratheodory theorem and Hadamard three circle lemma to prove generalisations of the classical statement "Riemann hypothesis implies Lindelöf hypothesis".

A very useful tool is also provided by the so called explicit formulae, a nice survey over the number field $\mathbb{Q}$ is given in [**Me86**], which shows (conditionally) that the rank of an abelian variety with conductor F is $O(\log F / \log\log F)$, whereas an efficient formulation in positive characteristic is provided in [**Br92**], where it is applied for example to show (unconditionally) that the rank of an abelian variety over $\mathbb{F}_q(B)$ with conductor of degree d is $O(d/\log d)$. This latter upper bound is optimal as shown by examples of families of elliptic curves constructed by Ulmer [**Ul02**].

3. Abelian varieties and surfaces

3.1. Surfaces and Artin-Tate conjecture. Les S be a smooth projective surface defined over a finite field $\mathbb{F}_q$. We denote its *Brauer group* by $\mathrm{Br}(S) = H^2_{\acute{e}t}(S, \mathbb{G}_m)$, its Néron-Severi group by $\mathrm{NS}(S)$. The Néron-Severi group is defined as the quotient of the divisor group – formal linear combinations of irreducible divisors, with integral coefficients – by the algebraic equivalence relation. The torsion subgroup, denoted $\mathrm{NS}(S)_{\mathrm{tor}}$ is finite and the quotient $\mathrm{NS}(S)/\mathrm{NS}(S)_{\mathrm{tor}}$ is isomophic to the group of divisors modulo numerical equivalence. The Néron-Severi group is equipped with the intersection pairing: if C, C' are two (classes of) curves, we denote $(C \cdot C')$ their intersection number and define the Néron-Severi regulator as

$$\mathrm{Reg}(S/\mathbb{F}_q) := \left|\det\left((C_i \cdot C_j)\log q\right)_{1\le i,j\le\rho}\right| \in \mathbb{Z}(\log q)^\rho \tag{3.1}$$

where $C_1, \dots, C_\rho$ form a $\mathbb{Z}$-basis of $\mathrm{NS}(S)$ modulo torsion. The *geometric genus* of S is defined as $p_g(S) = \dim H^0(S, \Omega^2_S) = \dim H^2(S, \mathcal{O}_S)$. For example the geometric genus of a smooth surface of degree d in $\mathbb{P}^3$ is $p_g = (d-1)(d-2)(d-3)/6$.

In characteristic zero the tangent space of the Picard variety of a variety X is the cohomology vector space $H^1(X, \mathcal{O}_X)$, in particular $\dim \mathrm{Pic}\,(X) = \dim H^1(X, \mathcal{O}_X)$. However in positive characteristic the Picard scheme is not necessarily reduced. To mesure the "defect of smoothness" the following invariant is introduced.

$$\delta(S) := \dim H^1(S, \mathcal{O}_S) - \dim \mathrm{Pic}\,(S). \tag{3.2}$$

Finally we consider the L-function (the part associated with $H^2(S)$) defined as:

$$L_2(S/\mathbb{F}_q, s) = \det\left(1 - \mathrm{Frob}_S \cdot q^{-s} \mid H^2_{\acute{e}t}(S \times \bar{\mathbb{F}}_q, \mathbb{Q}_\ell)\right). \tag{3.3}$$

The analogue of the class number formula is:

CONJECTURE 3.1. (Artin-Tate conjecture [**Ta66**]) Let S be a surface over $\mathbb{F}_q$

(1) The order of the zero of $L_2(S/\mathbb{F}_q, s)$ at $s = 1$ is the rank of the Néron-Severi group $\mathrm{NS}(S)$.
(2) The Brauer group $\mathrm{Br}(S/\mathbb{F}_q)$ is finite.
(3) The special value at $s = 1$ of the (main part of) zeta function is given by:

$$L_2^*(S/\mathbb{F}_q, 1) = \frac{\sharp\mathrm{Br}(S/\mathbb{F}_q)\mathrm{Reg}(S/\mathbb{F}_q)}{q^{p_g(S)}} \cdot \frac{q^{\delta(S)}}{\sharp\mathrm{NS}(S)^2_{\mathrm{tor}}} \tag{3.4}$$

Note that the conjecture is usually presented using $\chi(S, \mathcal{O}_S)$ instead of the geometric genus. One can easily establish a translation using the relation $\chi(S, \mathcal{O}_S) - 1 + \dim \mathrm{Pic}\,(S) = p_g(S) - \delta(S)$; we have chosen $p_g(S)$ as the main parameter because it is simpler to define an analogue for hypersurfaces in $\mathbb{P}^n$, which we will be considering further down.

Remark that it is possible to modify the formulae into a formula involving only rational numbers, i.e. give an equality in $\mathbb{Q}^\times$. Indeed by construction $\mathrm{Reg}(S/\mathbb{F}_q) = a(\log q)^\rho$ with $a \in \mathbb{Q}^\times$ and $\rho = \mathrm{rk}\,\mathrm{NS}(S/\mathbb{F}_q)$; on the other hand, putting $T = q^{-s}$, we have $L_2(S,s) = (1-qT)^\rho M(S,T)$ so $L_2(S,s) = (1-q^{1-s})^\rho M(S,q^{-s})$ and $L_2^*(S,1) = (\log q)^\rho M(S,q^{-1})$ and, by construction $M(S,q^{-1}) \in \mathbb{Q}^\times$.

It is known that $\mathrm{rk}\,\mathrm{NS}(S) \leq \mathrm{ord}_{s=1} L_2(S,s)$ and further that equality implies the full Artin-Tate conjecture [**Mi75**]. In particular the Tate conjecture 2.4 for the surface S implies the Artin-Tate conjecture for the surface.

It is natural to expect that, under appropriate conditions, for a family of surfaces with growing genus

$$\delta(S) = o(p_g(S)) \quad \text{and} \quad \frac{\log \sharp \mathrm{NS}(S)_{\mathrm{tor}}}{p_g(S)\log q} = o(1)? \tag{3.5}$$

This is indeed true for example for surfaces in $\mathbb{P}^3$, because in this case the torsion part of $\mathrm{NS}(S)$ is trivial and the Picard variety and $h^1(S,\mathcal{O}_S)$ are trivial, hence $\delta(S) = 0$. Application of standard complex analysis yields (see section 5.2 where a more general result is proved) the following upper bound for the special value.

LEMMA 3.2. *Let S vary in a family of smooth surfaces in $\mathbb{P}^3$, then*

$$\log |L_2^*(S/\mathbb{F}_q,1)| \leq o(p_g(S)\log q). \tag{3.6}$$

Thus, under mild conditions, we expect that for a family of surfaces with growing genus

$$\mathfrak{Bs}(S_i) := \frac{\log\left(\sharp \mathrm{Br}(S_i/\mathbb{F}_q)\mathrm{Reg}(S_i/\mathbb{F}_q)\right)}{p_g(S_i)\log q} \leq 1 + o(1)?$$

For the time being we refrain from discussing lower bounds in general, but see the last section (section 5.3). Instead we quote a nice result of Griffon [**Gri18**] which exhibits a family of surfaces for which the full analogue of Brauer-Siegel is proven unconditionally. In particular in the following example, the veracity of Tate conjecture for the Fermat surfaces, shown by Shioda, implies that the Brauer group of these surfaces is finite.

THEOREM 3.3. *(Griffon [**Gri18**]) Let S_d be the Fermat surface defined in $\mathbb{P}^3/\mathbb{F}_q$ by the equation $x_0^d + x_1^d + x_2^d + x_3^d = 0$, where d is coprime with q.*

$$\lim_d \mathfrak{Bs}(S_d) = \lim_d \frac{\log\left(\sharp \mathrm{Br}(S_d/\mathbb{F}_q)\mathrm{Reg}(S_d/\mathbb{F}_q)\right)}{\log q^{p_g(S_d)}} = 1 \tag{3.7}$$

Notice that, since $\log q^{p_g(S_d)} \sim \log q \cdot \frac{d^3}{6}$, we have an asymptotic growth of type $\log\left(\sharp \mathrm{Br}(S_d/\mathbb{F}_q)\mathrm{Reg}(S_d/\mathbb{F}_q)\right) \sim \log q \cdot \frac{d^3}{6}$, which says, for example, that either the Brauer group or the intersection regulator is exponentially large with respect to the input p_g or d. It also implies the only known upper bound for the regulator:

$$\mathrm{Reg}(S_d/\mathbb{F}_q) \leq C_\epsilon q^{d^3\left(\frac{1}{6}+\epsilon\right)}.$$

Combining this with lemma 3.2 we obtain.

COROLLARY 3.4. Let S vary in the family of surfaces in $\mathbb{P}^3/\mathbb{F}_q$, assume $\mathrm{Br}(S/\mathbb{F}_q)$ is finite, then

$$\limsup \mathfrak{Bs}(S) = \limsup \frac{\log\left(\sharp \mathrm{Br}(S/\mathbb{F}_q)\mathrm{Reg}(S/\mathbb{F}_q)\right)}{\log q^{p_g(S)}} = 1. \tag{3.8}$$

The value of the liminf is discussed in the last section (section 5.3), where it is shown to be ≥ 0, under the same hypotheses.

3.2. Abelian varieties and Birch & Swinnerton-Dyer conjecture. Items of this section are discussed in [**Hi07, HiPa16, Gri16, KuTs08, Zy09a, Zy09b**]. We now consider an abelian variety A defined over a global field K and treat parallelly the number field case and function field case. The Mordell-Weil group $A(K)$ is finitely generated and can thus be written $\mathbb{Z}^r \times A(K)_{\mathrm{tor}}$. Denote $\check{A}$ the dual abelian variety, there is a canonical pairing $< .,. >: A(K) \times \check{A}(K) \to \mathbb{R}$ which is non degenerate modulo torsion. This pairing leads to the definition of the *Néron-Tate regulator*, by picking $P_1, \dots P_r$ (resp. $\check{P}_1, \dots \check{P}_r$) a basis of $A(K)$ (resp. a basis of $\check{A}(K)$) modulo torsion and setting

$$\mathrm{Reg}(A/K) = \left|\det(< P_i, \check{P}_j >)_{1\leq i,j\leq r}\right| . \tag{3.9}$$

The *Shafarevich-Tate group* is defined as

$$Ш(A/K) := \ker\left\{H^1(G_K, A) \to \prod_v H^1(G_{K_v}, A_{K_v})\right\}, \tag{3.10}$$

it sits in the descent exact sequence

$$0 \to A(K)/nA(K) \to \mathrm{Sel}^n(A, K) \to Ш(A/K)[n] \to 0,$$

where the n-th *Selmer group* is defined as

$$\mathrm{Sel}^n(A/K) := \ker\left\{H^1(G_K, A[n]) \to \prod_v H^1(G_{K_v}, A_{K_v}[n])\right\}.$$

The *height* of an abelian variety is defined as follows, the Néron model $\pi : \mathcal{A} \to B$ over $B = \mathrm{spec}\,\mathcal{O}_K$ in the number field case and over B the curve such that $K = \mathbb{F}_q(B)$ in the function field case, is equipped with the neutral section $e : B \to \mathcal{A}$ and yields a line bundle $\omega_A := e^*\Omega^{\dim A}_{\mathcal{A}/B}$ on B. One then defines $\deg(\omega_A)$ as the degree of a line bundle on the curve B or the Arakelov degree in the number field case, where the metric are given by $|\eta|^2_\sigma = \int_{A_\sigma(\mathbb{C})|} |\eta \wedge \bar{\eta}|$. The Faltings height is then $h_{\mathrm{Falt}}(A/K) = [K : \mathbb{Q}]^{-1} \deg_{\mathrm{Arak}} \omega_A$ and finally:

DEFINITION 3.5. The exponential height of an abelian variety A/K is:

- (number field case) $H(A/K) = \exp(h_{\mathrm{Falt}}(A/K))$.
- (function field case) $H(A/K) = q^{\deg \omega_A}$.

In the number field case there is also a real period $\Omega_{A/K}$ which, in case $K = \mathbb{Q}$, is simply $\int_{A(\mathbb{R})} \eta$ for a Néron differential (a generator of the sections of $\Omega_{\mathcal{A}/\mathbb{Z}}$). For coherence of notations in what follows, we introduce the notation, where $d = \dim A$ and g is the genus of B:

$$\tilde{H}(A/K) = \begin{cases} \Omega^{-1}_{A/K} & \text{(number field case)} \\ H(A/K)q^{d(g-1)} & \text{(function field case)} \end{cases}$$

The notation $\tilde{H}$ is motivated by the fact that $\tilde{H}(A/K)$ is exactly the term appearing in the Birch & Swinnerton-Dyer conjectural formula. For function fields it is, by definition and up to a constant, the height $H(A/K)$, while for number fields we have the following comparison (see [**Hi07**], Lemma 3.7) $H(A/K) \ll \Omega_A^{-1} \ll H(A)(\log H(A/K))^{\dim A/2}$. In particular asymptotically $\log H(A/K) \sim \log \tilde{H}(A/K)$.

Finally we introduce the *local Tamagawa numbers* (that were originally called *fudge factors* by Birch and Swinnerton-Dyer) $c_v(A/K) = \sharp\Phi_v(\mathbb{F}_v)$, where Φ_v is the component group of the Néron model at v. With all these notations we may now state:

CONJECTURE 3.6. (Birch & Swinnerton-Dyer conjecture [**Ta66**]) Let A/K be an abelian variety over a global field K.

(1) The order of the zero of $L(A/K,s)$ at $s=1$ is the rank of the Mordell-Weil group $A(K)$.
(2) The Shafarevich-Tate group $\text{Ш}(A/K)$ is finite.
(3) The special value of the L-function is given by:

$$L^*(A/K,1) := \lim_{s\to 1}\frac{L(A/K,s)}{(s-1)^r} = \frac{\sharp\text{Ш}(A/K)\text{Reg}(A/K)}{\tilde{H}(A/K)} \times \frac{\prod_v c_v(A/K)}{\sharp(\check{A}\times A)(K)_{\text{tor}}}. \tag{3.11}$$

This conjecture is actually linked with Artin-Tate conjecture (see [**Ta66, Gor79**]; it is far from settled, but in the function field case the situation is better; the following results, starting with Tate and Milne and culminating with Kato-Trihan [**Ta66, Mi75, Sc82, Ba92, KaTr03**], have been established:

THEOREM 3.7. [**Ta66, Mi75, Sc82, Ba92, KaTr03**] *Let A be an abelian variety over a function field K, then*

(1) $\text{rk}\, A(K) \le \text{ord}_{s=1}L(A/K,s)$.
(2) *Equality in the previous item is equivalent to* $\text{Ш}(A/K)$ *finite and even equivalent to the finiteness of a primary component* $\text{Ш}(A/K)[\ell^\infty]$ *for one* ℓ *(which may be equal to* p*). Further, if these statements are true then the full conjecture* 3.6 *is true.*

The analogy between the class number formula and the expected BSD formula is apparent, it can be strengthened by noting that if we define $\text{Ш}(\mathcal{O}_K^\times)$ by replacing the abelian variety A in the definition of $\text{Ш}(A/K)$ in (3.10) by the multiplicative group of units:

$$\text{Ш}(\mathcal{O}_K^\times) := \ker\left\{H^1(G_K,\mathcal{O}_{\bar{K}}^\times) \to \prod_v H^1(G_{K_v},\mathcal{O}_{\bar{K}_v}^\times)\right\},$$

then it can be shown (known to the experts) that $\text{Ш}(\mathcal{O}_K^\times) \cong \text{Pic}\,(\mathcal{O}_K)$.

The natural analogue of the Brauer-Siegel ratio is therefore:

$$\mathfrak{Bs}(A/K) = \frac{\log\left(\sharp\text{Ш}(A/K)\text{Reg}(A/K)\right)}{\log H(A/K)}. \tag{3.12}$$

We have shown in [**HiPa16**] that both $\mathcal{T}(A/K) := \prod_v c_v(A/K)$ and the cardinality of $(\check{A}\times A)(K)_{\text{tor}}$ are $O_\epsilon(H(A/K)^\epsilon)$. This is one of the ingredients in the proof of the next theorem.

THEOREM 3.8. [**HiPa16**] *Let A_i be a family of abelian varieties over K, of fixed dimension. Assume the Birch & Swinnerton-Dyer conjecture and GRH are true for A_i then we have*

$$\mathfrak{Bs}(A_i/K) = \frac{\log\left(\sharp\text{Ш}(A_i/K)\text{Reg}(A_i/K)\right)}{\log H(A_i/K)} \le 1 + o(1) \tag{3.13}$$

As mentioned, in the function field case, it is enough to assume finiteness of the Shafarevich-Tate group. The full analogue of Brauer-Siegel theorem has been shown for several families of elliptic curves. Let $\mathcal{F}$ be one of the following families (more examples are given in [**Gri16**]) where for simplicity it is assumed that $\mathrm{car}(\mathbb{F}_q) \neq 2, 3$ and that d remains coprime with q (notice though that the latter condition can be dropped [**Gri17c**]):

- E_d: $y^2 + xy = x^3 - t^d$ (see [**HiPa16**])
- E_d: $y^2 + xy - t^d y = x^3$ (see [**Gri17b**])
- E_d: $y^2 = x(x+1)(x+t^d)$ (see [**Gri17c**])

In a recent preprint [**Ul18**] Ulmer has given a new and more algebraic proof covering these and more examples.

THEOREM 3.9. *Let $\mathcal{F}$ be among the previous families of elliptic curves over $\mathbb{F}_q(t)$, then, for any member $E \in \mathcal{F}$, the group $\text{Ш}(E/K)$ is finite and:*

$$\lim_{E \in \mathcal{F}} \mathfrak{Bs}(E/K) = 1.$$

COROLLARY 3.10. Consider the family of abelian varieties over the function field K, of fixed dimension, for which $\text{Ш}(A/K)$ is finite, then

$$\limsup_{H(A/K) \to \infty} \mathfrak{Bs}(A/K) = 1. \tag{3.14}$$

The same conclusion holds over K a number field, albeit assuming BSD and GRH conjectures. Combining this with the result from [**HiPa16**] which states that $\mathrm{Reg}(A/K) \geq c_\epsilon H(A)^{-\epsilon}$ we get

$$0 \leq \liminf \mathfrak{Bs}(A/K) \leq \limsup \mathfrak{Bs}(A/K) = 1. \tag{3.15}$$

We will briefly discuss at the end of the paper what should be the liminf.

Instead of varying the abelian variety, a very interesting and quite distinct problem is obtained by fixing the abelian variety and making the field K grow. This problem has been attacked by Kunyavski and Tsfasman [**KuTs08**], who arrive at the following statement

CONJECTURE 3.11. [**KuTs08**] Let E be an elliptic curve over $\mathbb{F}_q$ and let $K_i = \mathbb{F}_q(X_i)$ be a sequence of function fields of genus g_i, such that the limits

$$\beta_m = \lim \frac{\sharp X_i(\mathbb{F}_{q^m})}{g_i}$$

all exist. Consider the family of elliptic curves $E_i = E \times_{\mathbb{F}_q} K_i$, then $\text{Ш}(E/K_i)$ is finite and

$$\lim \mathfrak{Bs}(E_i/K_i) := \lim \frac{\log(\text{Ш}(E_i/K_i)\mathrm{Reg}(E_i/K_i))}{g_i \log q} = 1 - \frac{1}{\log q} \sum_{m=1}^{\infty} \beta_m \log \frac{\sharp E(\mathbb{F}_{q^m})}{q^m} \tag{3.16}$$

Finiteness of Ш in this context is known [**Mi68**]. This conjecture is stated as Theorem 1.2 in [**KuTs08**] but the authors and Zykin realised there is a gap in the proof (see errata of [**KuTs08**]).

4. Generalisations

This speculative section is essentially descriptive and destined to put the previous examples into a broader, albeit exceedingly conjectural, context.

4.1. The motives $H^j(X)(i)$. Instead of describing the theory of general pure motives (see [**An04, JKS94**] for general reference) we will focus on the interesting case of $M = H^j(X)(i)$ for a smooth projective variety $X/\mathbb{Q}$. Such a "motive" has several incarnations and structures associated: ℓ-adic cohomology groups $H^j_{\acute{e}t}(X_{\bar{\mathbb{Q}}}, \mathbb{Q}_\ell(i))$ with its Galois action, the de Rham cohomology $H^j_{dR}(X/\mathbb{Q})$ and its filtration, the singular or Betti cohomology $H^j(X(\mathbb{C}), \mathbb{Q}(2\pi\sqrt{-1})^i)$ and the associated Hodge decomposition.

In general a motive has a dual motive $\check{M}$ and an L-function $L(M,s)$ which is defined by an Euler product:

$$L(M,s) = \prod_p L_p(M,s) = \prod_p P_p(M,p^{-s})^{-1} = \prod_p \prod_{i=1}^{d} (1 - \alpha_{p,i}p^{-s})^{-1}$$

where $P_p(M,T) \in \mathbb{Z}[T]$ and or almost all p we have $|\alpha_{p,j}| = p^{w/2}$; the integer w is called the *weight*, thus for example $H^j(X)(i)$ has weight $j - 2i$. The product converges absolutely for $\Re(s) > 1 + w/2$ and is expected to have analytic continuation to the whole complex plane, with at most a pole at $s = 1 + w/2$ and $w/2$ when w is even, and satisfy a functional equation relating $L(M,s)$ and $L(\check{M}, 1-s) = L(\check{M}(1), -s)$. Therefore the following definition provides examples where we recover (conjecturally) functional equations of the previous type.

DEFINITION 4.1. A pure motive of weight w is polarised if $\check{M} \cong M(w)$.

EXAMPLE 4.2. Let X be a smooth projective variety, the motive $M := H^j(X)(i)$ is pure of weight $w = j - 2i$. It is a polarised motive, in fact Poincaré duality is tantamount to the relation $\check{M} \cong M(j-2i) = H^j(X)(j-i)$.

The L–function $L(H^j(X), s)$ is thus expected to have a functional equation relating it to $L(H^j(X), j+1-s)$. For even j, we'll consider $M = H^j(X)(\frac{j}{2}+1)$ which has weight -2 and for which $s = 0$ is the integer near the central value of $L(M,s)$. For odd j, we'll consider $M = H^j(X)(\frac{j+1}{2})$ which has weight -1 and for which $s = 0$ is the central value of $L(M,s)$.

4.2. Bloch-Kato conjecture. We assume K is a number field in this section. The Bloch-Kato conjecture predicts the special value of the L-function of a pure motive M, denoted $L(M,s)$, at the integer $s = 0$. Since $L(M(n), s) = L(M, s+n)$, we see that it actually predicts the behaviour of $L(M,s)$ at all integers. One of the motivation for shifting weights is that, for example, motives and representations of weight zero are stable under operations like $\otimes$, Symm^r, Λ^r, etc. Just like BSD conjecture, which it generalises, the first part of the Bloch-Kato conjecture predicts the order of vanishing of $L(M,s)$ at $s = 0$. Notice the place of the zero is shifted but this fits with BSD conjecture because, in the natural example $M = V_p(A)(1)$, we have $L(M,s) = L(A, s+1)$ and hence $L^*(M,0) = L^*(A,1)$.

Let V be a p-adic incarnation, then the *Bloch-Kato Selmer group* is defined as

$$H^1_f(G_{\mathbb{Q}}, V) := \ker\left\{ H^1(G_{\mathbb{Q}}, V) \to \prod_v H^1(G_{\mathbb{Q}_v}, V)/H^1_f(G_{\mathbb{Q}_v}, V) \right\}$$

where $H^1_f(G_{\mathbb{Q}_v}, V)$ is $\ker(H^1(G_{\mathbb{Q}_v}, V) \to H^1(G_{\mathbb{Q}_v^{\mathrm{unr}}}, V)$ for $v \neq p$ and $H^1_f(G_{\mathbb{Q}_p}, V)$ is $\ker(H^1(G_{\mathbb{Q}_v}, V) \to H^1(G_{\mathbb{Q}_v}, B_{\mathrm{crys}} \otimes V)$ for $v = p$ (here $\mathbb{Q}_v^{\mathrm{unr}}$ is the maximal unramified extension of $\mathbb{Q}_v$ and B_{crys} is Fontaine's ring). When $V = V_p(M)$ is the realisation of M, we will also write $H^1_f(G_K, M)$ for $H^1_f(G_K, V)$.

CONJECTURE 4.3 (Bloch-Kato conjecture, first part).

$$\mathrm{ord}_{s=0}L(M,s) = \dim H^1_f(G_K, \check{M}(1)) - \dim H^0(G_K, \check{M}(1)) \tag{4.1}$$

To see the connexion with previous examples, note that $\dim H^0(G_K, M) = 0$ unless the representation $V = V_p(M)$ contains $\mathbb{Q}_p(1)$, and that we have the following isomorphisms [**Bel09**]:

(1) For a number field K, the Kummer map induces an isomorphism $\mathcal{O}_K^* \otimes_{\mathbb{Z}} \mathbb{Q}_p \cong H^1_f(G_K, \mathbb{Q}_p(1))$, thus we know that $\dim H^1_f(G_K, \mathbb{Q}_p(1)) = \mathrm{rk}\,\mathcal{O}_K^*$.

(2) For E/K an elliptic curve, we have an exact sequence

$$0 \to E(K) \otimes_{\mathbb{Z}_p} \mathbb{Q}_p \to H^1_f(G_K, V_p(E)) \to Ш(E/K) \otimes \mathbb{Q}_p \to 0$$

thus whenever $Ш(E/K)[p^\infty]$ is finite, the prediction coincides with Birch & Swinnerton-Dyer prediction.

(3) For $M = H^2(S)(1)$ the motive of a surface S, we have $H^1_f(G_K, H^2(S)(1)) = 0$ because $H^2(S)(1)$ has weight zero and thus the Bloch-Kato conjecture predicts a pole of order $\dim H^0(G_K, H^2(S)(1)) \cong H^2(S)(1)^{G_K}$ which, according to Tate's conjecture (conjecture 2.4) should be the image of $\mathrm{NS}(S) \otimes_{\mathbb{Z}} \mathbb{Q}_p$ in $H^2(S)(1)$; more precisely the cycle map produces an inclusion $\mathrm{NS}(S) \otimes_{\mathbb{Z}} \mathbb{Q}_p \hookrightarrow H^2(S)(1)^{G_K}$ and Tate conjectures that the inclusion is an isomorphism. Thus the Bloch-Kato conjecture is compatible with the second conjecture of Tate which predicts a pole of order $\rho(S) = \mathrm{rk}\,\mathrm{NS}(S)$.

Bloch and Kato then define local and global groups of *rational points* $M(\mathbb{Q}_v)$ and $M(\mathbb{Q})$, Haar measures μ_v on $M(\mathbb{Q}_v)$, a Shafarevic-Tate group $Ш(M)$ and conjecture that, if S is the set of places of bad reduction and $L_S(M,s)$ is the L-function with Euler factors at $v \in S$ removed, we have equation (5.15.1) in [**BlKa90**]:

$$L_S^*(V,0) = \sharp Ш(M)\mu_\infty(M(\mathbb{R})/M(\mathbb{Q}))\frac{\prod_{p\in S}\mu_p(M(\mathbb{Q}_p))}{\sharp H^0(G_{\mathbb{Q}}, \check{M} \otimes \mathbb{Q}/\mathbb{Z}(1))}.$$

Choosing a decomposition $M(\mathbb{Q}) = M'(\mathbb{Q}) \oplus M(\mathbb{Q})_{\mathrm{tor}}$, adding the missing factors and denoting the *fudge factors* $c_p = \mu_p(M(\mathbb{Q}_p))L_p(M,0)$ we get the formula

$$L^*(V,0) = \sharp Ш(M)\mu_\infty(M(\mathbb{R})/M'(\mathbb{Q}))\frac{\prod_{p\in S} c_p}{\sharp(\check{M}(1) \times M)(\mathbb{Q})_{\mathrm{tor}}}.$$

It is possible along the lines of [**Bl84**] or [**Bei85**] to interpret $\mu_\infty(M(\mathbb{R})/M'(\mathbb{Q}))$ as the product of a period, which we write Ω_M and a regulator related to a (conjectural) cycle pairing, which we write $\mathrm{Reg}(M)$. The full Bloch-Kato conjecture can then be given the following form, for a *motive* M:

$$L^*(M,0) = \mathrm{Reg}(M)\sharp Ш(M)\Omega_M \times \frac{\prod_p c_p}{\sharp\{M(\mathbb{Q})_{\mathrm{tor}} \times M^*(1)(\mathbb{Q})_{\mathrm{tor}}\}}. \tag{4.2}$$

By analogy with the case of abelian varieties, we define the *archimedean height* of M by:

$$\tilde{H}(M) := \frac{1}{\Omega_M} \tag{4.3}$$

and we recover a more familiar looking (conjectural) formula which we formally state as:

CONJECTURE 4.4 (Bloch-Kato conjecture, final part).

$$L^*(M,0) = \frac{\mathrm{Reg}(M)\sharp\text{Ш}(M)}{\tilde{H}(M)} \cdot \frac{\prod_p c_p}{\sharp\{M(\mathbb{Q})_{\mathrm{tor}} \times M^*(1)(\mathbb{Q})_{\mathrm{tor}}\}}. \tag{4.4}$$

Of course, in order to make sense, the conjecture assumes analytic continuation of the L-function and finiteness of $\text{Ш}(M)$.

4.3. Brauer-Siegel type conjecture for motives. The conjectural formula (4.4) suggests looking at the following problems. We need a notion of limited family of motives; we define this naively as a subfamily of motives of fixed weight and Hodge numbers. The correct parameter for a family of motives has to be its height and a convenient definition has been proposed by Kato [**Ka14**], we denote it $h(M) = h_{\mathrm{Kato}}(M)$. We will also use $H(M) := \exp(h(M))$. One should check that

$$-\log \Omega_M \sim h(M)?$$

Estimates for the special value (see next section) will show (see section 5.2) that, assuming GRH, we have $L^*(M,0) \leq c_\epsilon F_M^\epsilon$ for all $\epsilon > 0$, where F_M is the analytic conductor. Note that without GRH, but with analytic continuation and functional equation, one can prove weaker inequalities of the type $L^*(M,0) \leq c_1 F_M^{c_2}$. It is natural to expect an inequality of the shape

$$\log F_M \leq c \cdot h(M)?$$

and therefore

$$L^*(M,0) \leq c_\epsilon H(M)^\epsilon?$$

Similarily one would expect

$$\prod_p c_p \leq c_\epsilon H(M)^\epsilon? \qquad \text{and} \qquad \sharp\{M(\mathbb{Q})_{\mathrm{tor}} \times M^*(1)(\mathbb{Q})_{\mathrm{tor}}\} \leq c_\epsilon H(M)^\epsilon?$$

One could even ask for a uniform bound for the torsion, but this is not even known for abelian varieties, only for elliptic curves.

All this leads to the interesting conjecture

CONJECTURE 4.5. (Generalised Brauer-Siegel upper estimate) Let M vary in a limited family of motives; assume that $\text{Ш}(M)$ is finite, then:

$$\mathrm{Reg}(M)\sharp\text{Ш}(M) \leq c_\epsilon H(M)^{1+\epsilon}? \tag{4.5}$$

This would naturally imply $\mathrm{Reg}(M) \leq c_\epsilon H(M)^{1+\epsilon}$ and, provided we verify that $\mathrm{Reg}(M) \geq c_\epsilon H(M)^{-\epsilon}$, would also imply $\sharp\text{Ш}(M) \leq c_\epsilon H(M)^{1+\epsilon}$. By analogy we introduce the *Brauer-Siegel ratio* of a motive M as

$$\mathfrak{Bs}(M) := \frac{\log\left(\sharp\text{Ш}(M)\mathrm{Reg}(M)\right)}{\log H(M)} \tag{4.6}$$

We also postulate that the following asymptotics should hold:

$$0 \leq \liminf_{H(M)\to\infty} \mathfrak{Bs}(M) \leq \limsup_{H(M)\to\infty} \mathfrak{Bs}(M) = 1? \tag{4.7}$$

5. Theorems and conjectures of Brauer-Siegel type

5.1. Brauer-Siegel ratio. We recapitulate in the form of Table 1 the analogies we underlined. The last row displays the various Brauer-Siegel ratios we defined. We emphasize again that the formulae of Birch & Swinnerton-Dyer, Artin-Tate, and Dirichlet justify the respective choices of ζ and L-functions.

TABLE 1. Analogies: number fields, curves, surfaces and motives

num. field F	curve $C/\mathbb{F}_q$	ell. curve $E/\mathbb{Q}$	surface $S/\mathbb{F}_q$	motive M
$\zeta_F(s)$	$\zeta_C(s)$	$L(E,s)$	$L(H^2(S),s)$	$L(M,s)$
$\zeta_F^*(1)$	$\zeta_C^*(1)$	$L^*(E,1)$	$L^*(S,1)$	$L^*(M,0)$
R_F	$\mathrm{Reg}=1$	$\mathrm{Reg}(E/\mathbb{Q})$	$\mathrm{Reg}(S/\mathbb{F}_q)$	$\mathrm{Reg}(M)$
$\mathrm{Pic}\,(\mathcal{O}_F)$	$\mathrm{Pic}^0(C)$	$Ш(E/\mathbb{Q})$	$\mathrm{Br}(S/\mathbb{F}_q)$	$Ш(M)$
$\sqrt{\Delta_F}$	$q^{g(C)-1}$	$H(E/\mathbb{Q})$	$q^{p_g(S)}$	$H(M)$
$2^{r_1}(2\pi)^{r_2}$	$(\log q)^{-1}$	$\prod_p c_p(E)$	$q^{\delta(S)}$	$\prod_p c_p(M)$
$\mathbb{G}_m(\mathcal{O}_F)_{\mathrm{tor}}$	$\mathbb{G}_m(\mathbb{F}_q)_{\mathrm{tor}}$	$E(\mathbb{Q})^2_{\mathrm{tor}}$	$\mathrm{NS}(S/\mathbb{F}_q)^2_{\mathrm{tor}}$	$(M\times \check{M}(1))(\mathbb{Q})_{\mathrm{tor}}$
$\frac{\log(h_F R_F)}{\log\sqrt{\Delta_F}}$	$\frac{\log \sharp\mathrm{Pic}^0(C)}{\log q^{g(C)}}$	$\frac{\log\left(\sharp Ш(E/\mathbb{Q})\mathrm{Reg}(E/\mathbb{Q})\right)}{\log H(E/\mathbb{Q})}$	$\frac{\log(\sharp\mathrm{Br}(S/\mathbb{F}_q)\mathrm{Reg}(S/\mathbb{F}_q))}{\log q^{p_g(S)}}$	$\frac{\log\left(\sharp Ш(M)\mathrm{Reg}(M)\right)}{\log H(M)}$

5.2. Upper bounds. To illustrate the available analytic techniques we prove or sketch the proof of two results, one for varieties over finite field, the other over global fields. The next proposition contains as a special case the upper bound stated for surfaces in section 3.1 (lemma 3.2). We consider smooth hypersurfaces X over $\mathbb{F}_q$ in $\mathbb{P}^{n+1}$, it is well known that there is a polynomial $Q(X,T)=\prod_{j=1}^{b'}(1-\beta_j T)$ with $|\beta_j|=q^{n/2}$ such that:

$$Z(X,T)=\frac{Q(X,T)^{(-1)^{n+1}}}{(1-T)(1-qT)\dots(1-q^nT)}$$

and $L(H^n(X),s)=P(X,q^{-s})$ where

$$P(X,T)=\begin{cases} Q(X,T) & \text{if } n \text{ is odd}\\ (1-q^{n/2}T)Q(X,T) & \text{if } n \text{ is even}\end{cases}$$

We set $b=\deg P(X,T)$, thus $b=\deg Q$ (resp. $b=\deg Q+1$) when n is odd (resp. when n is even).

PROPOSITION 5.1. There is a constant $c=c(n)$ such that for all smooth hypersurface $X/\mathbb{F}_q$ in $\mathbb{P}^{n+1}$, of degree $d=d(X)$, we have

$$\left|L^*\left(H^n(X),\frac{n}{2}\right)\right|\le \exp\left(cd^{n+1}\frac{\log\log d}{\log d}\right) \tag{5.1}$$

In particular, as d goes to infinity we have $\log\left|L^*\left(H^n(X),\frac{n}{2}\right)\right|=o(p_g(X))$.

PROOF. We first use the elementary inequality $\sharp X(\mathbb{F}_{q^m})\le d\sharp\mathbb{P}^{\dim X}(\mathbb{F}_{q^m})$ to obtain, for $s=\sigma+i\tau$ with $\sigma>n$:

$$|\log Z(X,s)|=\left|\sum_{m=1}^{\infty}\sharp X(\mathbb{F}_{q^m})\frac{q^{-ms}}{m}\right|\le \log Z(X,\sigma)\le d\log Z(\mathbb{P}^n,\sigma).$$

Thus, in the same range

$$|\log L(H^n(X),s)| \ll d\log(1-q^{n-\sigma})$$

which, when $\sigma = n+\delta$ with small δ gives

$$|\log L(H^n(X), n+\delta+it)| \ll d|\log\delta|. \tag{5.2}$$

On the other hand, the expression as a polynomial in q^{-s} gives, when $\sigma > n/2$

$$|\log L(H^n(X),s)| \leq b\log\left(\frac{1}{1-q^{\frac{n}{2}-\sigma}}\right)$$

which for $\sigma = \frac{n}{2}+\delta$ with small δ gives

$$|\log L(H^n(X), \frac{n}{2}+\delta+it)| \ll b|\log\delta| \tag{5.3}$$

Applying Phragmén-Lindelöf to $L(s) = L(H^n(X),s)$ and the strip $\frac{n}{2}+\delta \leq \sigma \leq n+\delta$ provides:

$$|\log L(H^n(X),\sigma+it)| \ll d^{\frac{2\sigma-n-2\delta}{n}} b^{\frac{2n-2\sigma+2\delta}{n}}|\log\delta| \tag{5.4}$$

Recalling that $b = d^{-1}((d-1)^{n+2}+(-1)^{n+1})$, we see that is is asymptotically d^{n+1} and applying (5.4) with $\delta = \rho/2$, we get

$$\log L(H^n(X), \frac{n}{2}+\rho+it)| \ll d^{n+1-\rho}|\log\rho| \tag{5.5}$$

Using now the functional equation $L(s) = \pm q^{b(\frac{n}{2}-s)}L(n-s)$ we get

$$|L(H^n(X), \frac{n}{2}-\rho+it)| \leq \exp(cd^{n+1-\rho}|\log\rho| + \rho b\log q)$$

We now choose $\rho = \log\log d/\log d$ and, using again Phragmén-Lindelöf, this time for $L(H^n(X),s)$ in the strip $\frac{n}{2}-\rho \leq \sigma \leq \frac{n}{2}+\rho$, we get

$$|L(H^n(X),\sigma+it)| \leq \exp\left(cd^{n+1}\frac{\log\log d}{\log d}\right)$$

Let r denote the order of vanishing of the L-function at $s = \frac{n}{2}$. To finish the proof, we invoke the explicit formula as in [**Br92**] (the proof given there applies with appropriate modification) to claim that $r = O(d^{n+1}/\log d)$. Applying Cauchy's inequality to the circle of center $\frac{n}{2}$ and radius ρ, we obtain

$$|L^*(H^n(X), \frac{n}{2})| \leq \exp\left(cd^{n+1}\left(\frac{\log\log d}{\log d} + r\frac{\log\log d}{d^{n+1}}\right)\right),$$

which implies the desired estimate.

□

Next we work with a global field K and assume that we have a "limited" family of L-functions, say $L(M,s)$, which satisfy the following

(1) The L-function has analytical continuation and expected functional equation.
(2) The L-function satisfies the generalised Riemann hypothesis.

PROPOSITION 5.2. When conditions (1), and (2) above are satisfied, we have the estimate:

$$|L^*(M,0)| \leq c_\epsilon F_M^\epsilon \tag{5.6}$$

PROOF. The proof starts with the estimate, valid for $\sigma > 1 + w/2$, giving: $|\log L(M, \sigma + it)| \leq d \log \zeta(\sigma - w/2)$, and is similar to the classical "Riemann hypothesis implies Lindelöf hypothesis" and is given for abelian varieties in [**Hi07**] over number field and [**HiPa16**] over function field. □

5.3. Lower bounds. In positive characteristic we discuss two lower bounds for the special value $L(H^j(X), s)$ at the central point. One is a simple corollary of Liouville inequality (a non zero integer has absolute value ≥ 1), the other supposes that a nice formula is known for the special value.

5.3.1. *Liouville inequality.* We are given an L-function which can be written

$$L(s) = \prod_{j=1}^{b}(1 - \beta_j q^{-s}) = Q(q^{-s})$$

with $\mathbb{Q}(T) \in \mathbb{Z}[T]$ and $|\beta_j| = q^{\frac{w}{2}}$ and w *even* (e.g. $L(s) = L(H^2(S), s)$ for a surface over $\mathbb{F}_q$ or $L(s) = L(H^1(X), s)$ for a curve or an abelian variety over $\mathbb{F}_q(B)$). Let ρ be the order of $L(s)$ at the integer $s = w/2$, we can factor $Q(T) = (1 - q^{\frac{w}{2}}T)^{\rho}Q_1(T)$, where $Q_1(T) := \prod_{j=1}^{b-\rho}(1 - \beta_j q^{-s})$ is in $\mathbb{Z}[T]$, with $Q_1(q^{-\frac{w}{2}}) \neq 0$.

LEMMA 5.3. *In the previous setting we have the lower bound:*

$$L^*\left(\frac{w}{2}\right) \geq q^{-b\frac{w}{4}}. \tag{5.7}$$

PROOF. We have

$$|L^*\left(\frac{w}{2}\right)| = (\log q)^{\rho}|Q_1(q^{-\frac{w}{2}})| \geq (\log q)^{\rho} q^{-\frac{w(b-\rho)}{2}} \geq q^{-b\frac{w}{2}}.$$

By using the functional equation $Q_1(T) = q^{(b-\rho)\frac{w}{2}}T^{b-\rho}Q_1(1/Tq^w)$ we can reduce the exponent by a factor 2. Indeed if $b - \rho$ is even, we may write $Q_1(T) = T^{(b-\rho)/2}R_1(T^{-1} + q^w T)$ with $R_1(T) \in \mathbb{Z}[T]$ and then

$$Q_1(q^{-w/2}) = q^{-w(b-\rho)/4}R_1(2q^{w/2}) \geq q^{-w(b-\rho)/4}.$$

If $b - \rho$ is odd, we have $Q_1(-q^{-w/2}) = 0$ hence $Q_1(T) = (1 + q^{w/2}T)Q_2(T)$ with now degree of $Q_2(T)$ even, hence $Q_1(q^{-w/2}) = 2Q_2(q^{-w/2})$ and we can apply the previous lower bound to $Q_2(q^{-w/2})$. □

In the case of hypersurfaces $X \subset \mathbb{P}^{n+1}$ with degree d, defined over $\mathbb{F}_q$ with n even, we consider $L(s) = L(H^n(X), s)$ and we obtain:

$$\frac{\log|L^*(\frac{n}{2})|}{p_g \log q} \geq -\frac{n}{4}\frac{b}{p_g},$$

where we recall that

$$b = \frac{(d-1)^{n+2} + (-1)^{n+1}}{d} \sim d^{n+1} \quad \text{and} \quad p_g = \frac{(d-1)\dots(d-n-1)}{(n+1)!} \sim \frac{d^{n+1}}{(n+1)!}.$$

Thus Liouville inequality implies:

$$\frac{\log|L^*(\frac{w+1}{2})|}{p_g \log q} \geq -\frac{n(n+1)!}{4} + o(1)\cdot$$

In particular, for surfaces $S/\mathbb{F}_q$, we get

$$\frac{\log|L^*(S,1)|}{p_g(S)\log q} \geq -3 + o(1). \tag{5.8}$$

5.3.2. *Applying special values.* The previous estimate (5.8) can be improved if we assume special value conjectures. For example if S is a surface in $\mathbb{P}^3$ that satisfies Artin-Tate conjecture, noticing the trivial $\sharp \mathrm{Br}(S/\mathbb{F}_q)\mathrm{Reg}(S/\mathbb{F}_q) \geq 1$, we conclude that

$$\frac{\log |L^*(S,1)|}{p_g \log q} \geq -1 + \frac{\log(q^{\delta(S)}/\sharp \mathrm{NS}(S)^2_{\mathrm{tor}})}{p_g \log q} \geq -1 + o(1)$$

This implies the lower bound $\liminf \mathfrak{Bs}(S) \geq 0$ quoted in section 3.1. As explained in section 3.2, the same lower bound is valid for abelian varieties over a global field, conditional to Birch & Swinnerton-Dyer conjecture.

5.3.3. *Is the lower bound* $\mathfrak{Bs}(X) \geq o(1)$ *optimal?* Our paper [**HiPa16**] contains a discussion of why the answer to the question should be "Yes", at least for abelian varieties over a global field. In particular, consider an elliptic curve given by the equation $y^2 = f(x)$, and the family $\mathcal{F}_E$ of twists E_D given by the equations $Dy^2 = f(x)$. Indeed the heuristic (loc. cit.) suggests:

CONJECTURE 5.4. [**HiPa16**] For the family of twists E_D of a fixed elliptic curve $E/\mathbb{Q}$ or $E/\mathbb{F}_q(T)$, as $|D|$ (where D a square free integer or a square free polynomial) goes to infinity:

$$0 = \liminf \mathfrak{Bs}(E_D) < \limsup \mathfrak{Bs}(E_D) = 1. \tag{5.9}$$

Over the function field $K = \mathbb{F}_q(B)$, the special case where E is defined over $\mathbb{F}_q$ is quite interesting, first because in this case finiteness of $\text{Ш}(E_D/K)$ [**Mi68**] and the BSD conjecture are known, second because another heuristic given in [**HiPa16**] points to conjecture 5.4.

5.3.4. *Central and near central point.* We have described analogies between the classical Brauer-Siegel theorem and similar problems in arithmetical geometrical context, we wish to conclude by stressing an important difference. The classical Brauer-Siegel theorem depends on estimates of the special value of $\zeta_F(s)$ at *near central point*, i.e. the point on the edge (to the right) of the critical strip. Thus assuming GRH or even a much weaker non vanishing property in a neighbourhood, we get a nice lower bound for the special value. Whereas most of the analogues we discussed depend on estimates of the special value at the *central point*, i.e. the center of symmetry of the functional equation. There the situation is different because, even assuming GRH, there exist zeroes very close to the value. This partly explains why the lower bound estimates are so much harder and why we expect a different behaviour.

As a simple illustratration we consider, for n *odd*, a hypersurface $X/\mathbb{F}_q$ in $\mathbb{P}^{n+1}$ and its associated L-function $L(H^n(X),s) = \prod_{j=1}^b (1-\beta_j q^{-s})$. We can estimate trivially the value at the integer $s = (n+1)/2$ (the integer at the near central point) by

$$(1-q^{-1/2})^b \leq L\left(H^n(X), \frac{n+1}{2}\right) \leq (1+q^{-1/2})^b.$$

But, invoking (5.4), we can do better and show that

$$\left|\log L\left(H^n(X), \frac{n+1}{2}\right)\right| \ll d^{\frac{1}{n}-\frac{2\delta}{n}} b^{1-\frac{1}{n}+\frac{2\delta}{n}} |\log \delta|$$

which, upon choosing $\delta = 1/\log d$, yields a bound $O(d^n \log\log d)$ and hence:

$$\lim_{d\to\infty} \frac{\log L\left(H^n(X), \frac{n+1}{2}\right)}{\log q^{p_g}} = 0.$$

References

[An04] Y. André, *Une introduction aux motifs (motifs purs, motifs mixtes, périodes)* (French, with English and French summaries), Panoramas et Synthèses [Panoramas and Syntheses], vol. 17, Société Mathématique de France, Paris, 2004. MR2115000

[Ba92] W. Bauer, *On the conjecture of Birch and Swinnerton-Dyer for abelian varieties over function fields in characteristic $p > 0$*, Invent. Math. **108** (1992), no. 2, 263–287, DOI 10.1007/BF02100606. MR1161093

[Bei85] A. Beilinson, *Higher regulators and values of L-functions*, Jour. Soviet. Math. **30** (1985), 2036–2070.

[Bel09] J. Bellaïche, *An introduction to the conjecture of Bloch and Kato*, Lectures at the Clay Mathematical Institute summer School, Honolulu, Hawaii, 2009, available at http://www.claymath.org/sites/default/files/bellaiche.pdf

[Bl84] S. Bloch, *Height pairings for algebraic cycles*, Proceedings of the Luminy conference on algebraic K-theory (Luminy, 1983), J. Pure Appl. Algebra **34** (1984), no. 2-3, 119–145, DOI 10.1016/0022-4049(84)90032-X. MR772054

[BlKa90] S. Bloch and K. Kato, *L-functions and Tamagawa numbers of motives*, The Grothendieck Festschrift, Vol. I, Progr. Math., vol. 86, Birkhäuser Boston, Boston, MA, 1990, pp. 333–400. MR1086888

[Br92] A. Brumer, *The average rank of elliptic curves. I*, Invent. Math. **109** (1992), no. 3, 445–472, DOI 10.1007/BF01232033. MR1176198

[De77] P. Deligne, *Valeurs de fonctions L et périodes d'intégrales* (French), Automorphic forms, representations and L-functions (Proc. Sympos. Pure Math., Oregon State Univ., Corvallis, Ore., 1977), Proc. Sympos. Pure Math., XXXIII, Amer. Math. Soc., Providence, R.I., 1979, pp. 313–346. With an appendix by N. Koblitz and A. Ogus. MR546622

[DrVl83] V. G. Drinfel′d and S. G. Vlăduţ, *The number of points of an algebraic curve* (Russian), Funktsional. Anal. i Prilozhen. **17** (1983), no. 1, 68–69. MR695100

[GoLu78] S. K. Gogia and I. S. Luthar, *The Brauer-Siegel theorem for algebraic function fields*, J. Reine Angew. Math. **299/300** (1978), 28–37. MR0485792

[Gor79] W. J. Gordon, *Linking the conjectures of Artin-Tate and Birch-Swinnerton-Dyer*, Compositio Math. **38** (1979), no. 2, 163–199. MR528839

[Gri16] R. Griffon, *Analogues du théorème de Brauer-Siegel pour quelques familles de courbes elliptiques*, PhD thesis, Université Paris Diderot, 2016.

[Gri18] R. Griffon, *A Brauer-Siegel theorem for Fermat surfaces over finite fields*, Journal of the London Mathematical Society 97 (2018), 523–549.

[Gri17b] R. Griffon, *Explicit L-function and a Brauer-Siegel theorem for Hessian elliptic curves*, to appear in Journal de théorie des nombres de Bordeaux, ArXiv:1709.02761

[Gri17c] R. Griffon, *Analogue of the Brauer-Siegel theorem for Legendre elliptic curves*, preprint arXiv:1706.07728

[Hi07] M. Hindry, *Why is it difficult to compute the Mordell-Weil group?*, Diophantine geometry, CRM Series, vol. 4, Ed. Norm., Pisa, 2007, pp. 197–219. MR2349656

[HiPaWa05] M. Hindry, A. Pacheco, and R. Wazir, *Fibrations et conjecture de Tate* (French, with English and French summaries), J. Number Theory **112** (2005), no. 2, 345–368, DOI 10.1016/j.jnt.2004.05.016. MR2141536

[HiPa16] M. Hindry and A. Pacheco, *An analogue of the Brauer-Siegel theorem for abelian varieties in positive characteristic*, Mosc. Math. J. **16** (2016), no. 1, 45–93. MR3470576

[In50] E. Inaba, *Number of divisor classes in algebraic function fields*, Proc. Japan Acad. **26** (1950), no. 7, 1–4. MR0046391

[JKS94] U. Jannsen, S. Kleiman, and J.-P. Serre (eds.), *Motives*, Proceedings of Symposia in Pure Mathematics, vol. 55, American Mathematical Society, Providence, RI, 1994. MR1265549

[KaTr03] K. Kato and F. Trihan, *On the conjectures of Birch and Swinnerton-Dyer in characteristic $p > 0$*, Invent. Math. **153** (2003), no. 3, 537–592, DOI 10.1007/s00222-003-0299-2. MR2000469

[Ka14] K. Kato, *Heights of motives*, Proc. Japan Acad. Ser. A Math. Sci. **90** (2014), no. 3, 49–53, DOI 10.3792/pjaa.90.49. MR3178484

[KuTs08] B. È. Kunyavskiĭ and M. A. Tsfasman, *Brauer-Siegel theorem for elliptic surfaces*, Int. Math. Res. Not. IMRN **8** (2008), Art. ID rnn009, 9, DOI 10.1093/imrn/rnn009. MR2428143

[La70] S. Lang, *Algebraic number theory*, Addison-Wesley Publishing Co., Inc., Reading, Mass.-London-Don Mills, Ont., 1970. MR0282947

[LeZy11] P. Lebacque and A. Zykin, *Asymptotic methods in number theory and algebraic geometry* (English, with English and French summaries), Actes de la Conférence "Théorie des Nombres et Applications", Publ. Math. Besançon Algèbre Théorie Nr., vol. 2011, Presses Univ. Franche-Comté, Besançon, 2011, pp. 47–73. MR2894268

[Li84] S. Lichtenbaum, *Values of zeta-functions at nonnegative integers*, Number theory, Noordwijkerhout 1983 (Noordwijkerhout, 1983), Lecture Notes in Math., vol. 1068, Springer, Berlin, 1984, pp. 127–138, DOI 10.1007/BFb0099447. MR756089

[Me86] J.-F. Mestre, *Formules explicites et minorations de conducteurs de variétés algébriques* (French), Compositio Math. **58** (1986), no. 2, 209–232. MR844410

[Mi68] J. S. Milne, *The Tate-Šafarevič group of a constant abelian variety*, Invent. Math. **6** (1968), 91–105, DOI 10.1007/BF01389836. MR0244264

[Mi75] J. S. Milne, *On a conjecture of Artin and Tate*, Ann. of Math. (2) **102** (1975), no. 3, 517–533, DOI 10.2307/1971042. MR0414558

[Mi81] J. S. Milne, *Comparison of the Brauer group with the Tate-Šafarevič group*, J. Fac. Sci. Univ. Tokyo Sect. IA Math. **28** (1981), no. 3, 735–743 (1982). MR656050

[Sc82] P. Schneider, *Zur Vermutung von Birch und Swinnerton-Dyer über globalen Funktionenkörpern* (German), Math. Ann. **260** (1982), no. 4, 495–510, DOI 10.1007/BF01457028. MR670197

[Se65] J.-P. Serre, *Zeta and L functions*, Arithmetical Algebraic Geometry (Proc. Conf. Purdue Univ., 1963), Harper & Row, New York, 1965, pp. 82–92. MR0194396

[Se69] J.-P. Serre, *Facteurs locaux des fonctions zêta des varietés algébriques (définitions et conjectures)* (French), Séminaire Delange-Pisot-Poitou. 11e année: 1969/70. Théorie des nombres. Fasc. 1: Exposés 1 à 15; Fasc. 2: Exposés 16 à 24, Secrétariat Math., Paris, 1970, pp. 15. MR3618526

[St74] H. M. Stark, *Some effective cases of the Brauer-Siegel theorem*, Invent. Math. **23** (1974), 135–152, DOI 10.1007/BF01405166. MR0342472

[Ta65] J. T. Tate, *Algebraic cycles and poles of zeta functions*, Arithmetical Algebraic Geometry (Proc. Conf. Purdue Univ., 1963), Harper & Row, New York, 1965, pp. 93–110. MR0225778

[Ta66] J. Tate, *On the conjectures of Birch and Swinnerton-Dyer and a geometric analog*, Séminaire Bourbaki, Vol. 9, Exp. No. 306, Soc. Math. France, Paris, 1995, pp. 415–440. MR1610977

[TsVl97] M. A. Tsfasman and S. G. Vlăduţ, *Infinite global fields and the generalized Brauer-Siegel theorem*, Mosc. Math. J. **2** (2002), no. 2, 329–402. Dedicated to Yuri I. Manin on the occasion of his 65th birthday. MR1944510

[Ul02] D. Ulmer, *Elliptic curves with large rank over function fields*, Ann. of Math. (2) **155** (2002), no. 1, 295–315, DOI 10.2307/3062158. MR1888802

[Ul18] D. Ulmer, *On the Brauer-Siegel ratio for abelin varieties over function fields*, Preprint June 2018, ArXiv:1806.01961v1.

[Zy09] A. I. Zykin, *Propriétés asymptotiques des corps globaux*, Thèse, Université de Marseille 2009. Asimptoticheskie svoĭcstva global'nykh poleĭ, thesis (Russian), Moscow 2010.

[Zy09a] A. I. Zykin, *The Brauer-Siegel theorem for families of elliptic surfaces over finite fields* (Russian), Mat. Zametki **86** (2009), no. 1, 148–150, DOI 10.1134/S000143460907013X; English transl., Math. Notes **86** (2009), no. 1-2, 140–142. MR2588645

[Zy09b] A. Zykin, *On the generalizations of the Brauer-Siegel theorem*, Arithmetic, geometry, cryptography and coding theory, Contemp. Math., vol. 487, Amer. Math. Soc., Providence, RI, 2009, pp. 195–206, DOI 10.1090/conm/487/09533. MR2555995

[Zy09c] A. I. Zykin, *Asymptotic properties of the Dedekind zeta function in families of number fields* (Russian), Uspekhi Mat. Nauk **64** (2009), no. 6(390), 175–176, DOI 10.1070/RM2009v064n06ABEH004657; English transl., Russian Math. Surveys **64** (2009), no. 6, 1145–1147. MR2640972

Institut de Mathématiques Jussieu – Paris Rive Gauche (IMG-PRG), 75252 Paris Cedex 05, France

Current address: UFR Mathématiques, Université Paris Diderot, Campus des Grands Moulins, Bâtiment Sophie Germain, Case 7012, 75205 Paris Cedex 13, France

Email address: marc.hindry@imj-prg.fr

Contemporary Mathematics
Volume **722**, 2019
https://doi.org/10.1090/conm/722/14533

The Belyi degree of a curve is computable

Ariyan Javanpeykar and John Voight

ABSTRACT. We exhibit an algorithm that, given input a curve X over a number field, computes as output the minimal degree of a Belyĭ map $X \to \mathbb{P}^1$. We discuss in detail the example of the Fermat curve of degree 4 and genus 3.

1. Introduction

Let $\overline{\mathbb{Q}} \subset \mathbb{C}$ be the algebraic closure of $\mathbb{Q}$ in $\mathbb{C}$. Let X be a smooth projective connected curve over $\overline{\mathbb{Q}}$; we call X just a curve. Belyĭ proved [**4, 5**] that there exists a finite morphism $\phi\colon X \to \mathbb{P}^1_{\overline{\mathbb{Q}}}$ unramified away from $\{0, 1, \infty\}$; we call such a map ϕ a Belyĭ map.

Grothendieck applied Belyĭ's theorem to show that the action of the absolute Galois group of $\mathbb{Q}$ on the set of dessins d'enfants is faithful [**27**, Theorem 4.7.7]. This observation began a flurry of activity [**24**]: for instance, the theory of dessins d'enfants was used to show that the action of the Galois group of $\mathbb{Q}$ on the set of connected components of the coarse moduli space of surfaces of general type is faithful [**2, 12**]. Indeed, the applications of Belyĭ's theorem are vast.

In this paper, we consider Belyĭ maps from the point of view of algorithmic number theory. We define the Belyĭ degree of X, denoted by $\mathrm{Beldeg}(X) \in \mathbb{Z}_{\geqslant 1}$, to be the minimal degree of a Belyĭ map $X \to \mathbb{P}^1_{\overline{\mathbb{Q}}}$. This integer appears naturally in Arakelov theory, the study of rational points on curves, and computational aspects of algebraic curves [**7, 14, 15, 25**]. It was defined and studied first by Liţcanu [**19**], whose work suggested that the Belyĭ degree behaves like a height.

The aim of this paper is to show that the Belyĭ degree is an effectively computable invariant of the curve X.

THEOREM 1.1. *There exists an algorithm that, given as input a curve X over $\overline{\mathbb{Q}}$, computes as output the Belyĭ degree* $\mathrm{Beldeg}(X)$.

The input curve X is specified by equations in projective space with coefficients in a number field. In fact, the resulting equations need only provide a birational model for X, as one can then effectively compute a smooth projective model birational to the given one.

In the proof of his theorem, Belyĭ provided an algorithm that, given as input a finite set of points $B \subset \mathbb{P}^1(\overline{\mathbb{Q}})$, computes a Belyĭ map $\phi\colon \mathbb{P}^1_{\mathbb{Q}} \to \mathbb{P}^1_{\mathbb{Q}}$ (defined over $\mathbb{Q}$)

2010 *Mathematics Subject Classification.* Primary 11G32, 11Y40.
Key words and phrases. Belyĭ map, Belyĭ degree, algorithm, effectively computable, Riemann-Roch space, moduli space of curves.

such that $\phi(B) \subseteq \{0,1,\infty\}$. Taking B to be the ramification set of any finite map $X \to \mathbb{P}^1_{\overline{\mathbb{Q}}}$, it follows that there is an algorithm that, given as input a curve X over $\overline{\mathbb{Q}}$, computes as output an *upper bound* for $\operatorname{Beldeg}(X)$. Khadjavi [**16**] has given an explicit such upper bound—see Proposition 2.10 for a precise statement. So at least one knows that the Belyĭ degree has a computable upper bound. However, neither of these results give a way to compute the Belyĭ degree: what one needs is the ability to test if a curve X has a Belyĭ map of a given degree d. Exhibiting such a test is the content of this paper, as follows.

A partition triple of d is a triple of partitions $\lambda = (\lambda_0, \lambda_1, \lambda_\infty)$ of d. The ramification type associates to each isomorphism class of Belyĭ map of degree d a partition triple λ of d.

THEOREM 1.2. *There exists an algorithm that, given as input a curve X over $\overline{\mathbb{Q}}$, an integer $d \geqslant 1$ and a partition triple λ of d, determines if there exists a Belyĭ map $\phi\colon X \to \mathbb{P}^1_{\overline{\mathbb{Q}}}$ of degree d with ramification type λ; and, if so, gives as output a model for such a map ϕ.*

Theorem 1.2 implies Theorem 1.1: for each $d \geqslant 1$, we loop over partition triples λ of d and we call the algorithm in Theorem 1.2; we terminate and return d when we find a map.

The plan of this paper is as follows. In section 2, we begin to study the Belyĭ degree and gather some of its basic properties. For instance, we observe that, for all odd $d \geqslant 1$, there is a curve of Belyĭ degree d. We also recall Khadjavi's effective version of Belyi's theorem. In section 3, we prove Theorem 1.2 by exhibiting equations for the space of Belyi maps on a curve with given degree and ramification type: see Proposition 3.16. These equations can be computed in practice, but unfortunately in general it may not be practical to detect if they have a solution over $\overline{\mathbb{Q}}$. In section 4, we sketch a second proof, which is much less practical but still proves the main result. Finally, in section 5 we discuss in detail the example of the Fermat curve $x^4 + y^4 = z^4$ of genus 3.

The theory of Belyĭ maps in characteristic $p > 0$ is quite different, and our main results rely fundamentally on the structure of the fundamental group of $\mathbb{C} \setminus \{0,1\}$, so we work over $\overline{\mathbb{Q}}$ throughout. However, certain intermediate results, including Lemma 4.1, hold over a general field.

Acknowledgements. This note grew out of questions asked to the authors by Yuri Bilu, Javier Fresán, David Holmes, and Jaap Top, and the authors are grateful for these comments. The authors also wish to thank Jacob Bond, Michael Musty, Sam Schiavone, and the anonymous referee for their feedback. Javanpeykar gratefully acknowledges support from SFB Transregio/45. Voight was supported by an NSF CAREER Award (DMS-1151047) and a Simons Collaboration Grant (550029).

2. The Belyi degree

In this section, we collect basic properties of the Belyĭ degree. Throughout, a curve X is a smooth projective connected variety of dimension 1 over $\overline{\mathbb{Q}}$; we denote its genus by $g = g(X)$. We write $\mathbb{P}^n$ and $\mathbb{A}^n$ for the schemes $\mathbb{P}^n_{\overline{\mathbb{Q}}}$ and $\mathbb{A}^n_{\overline{\mathbb{Q}}}$, respectively. A Belyĭ map on X is a finite morphism $X \to \mathbb{P}^1$ unramified away from $\{0,1,\infty\}$. Two Belyĭ maps $\phi\colon X \to \mathbb{P}^1$ and $\phi'\colon X' \to \mathbb{P}^1$ are isomorphic if there exists an isomorphism $i\colon X \xrightarrow{\sim} X'$ such that $\phi' \circ i = \phi$. For $d \geqslant 1$, define $\mathrm{Bel}_d(X)$ to be the set of isomorphism classes of Belyĭ maps of degree d on X, and let $\mathrm{Bel}(X) := \bigcup_d \mathrm{Bel}_d(X)$.

DEFINITION 2.1. The **Belyĭ degree** of X, denoted $\mathrm{Beldeg}(X) \in \mathbb{Z}_{\geqslant 1}$, is the minimal degree of a Belyĭ map on X.

In our notation, the Belyĭ degree of X is the smallest positive integer d such that $\mathrm{Bel}_d(X)$ is non-empty.

LEMMA 2.2. *Let $C \in \mathbb{R}_{\geqslant 1}$. Then the set of isomorphism classes of curves X with* $\mathrm{Beldeg}(X) \leqslant C$ *is finite.*

For an upper bound on the number of isomorphism classes of curves X with $\mathrm{Beldeg}(X) \leqslant C$ we refer to Liţcanu [**19**, Théorème 2.1].

PROOF. The monodromy representation provides a bijection between isomorphism classes of Belyĭ maps of degree d and permutation triples from S_d up to simultaneous conjugation; and there are only finitely many of the latter for each d. Said another way: the (topological) fundamental group of $\mathbb{P}^1(\mathbb{C}) \smallsetminus \{0, 1, \infty\}$ is finitely generated, and so there are only finitely many conjugacy classes of subgroups of bounded index. □

REMARK 2.3. One may also restrict to X over a number field $K \subseteq \overline{\mathbb{Q}}$ and ask for the minimal degree of a Belyĭ map defined over K: see Zapponi [**29**] for a discussion of this notion of relative Belyĭ degree.

Classical modular curves have their Belyĭ degree bounded above by the index of the corresponding modular group, as follows.

EXAMPLE 2.4. Let $\Gamma \leqslant \mathrm{PSL}_2(\mathbb{Z})$ be a finite index subgroup, and let $X(\Gamma) := \Gamma\backslash\mathbf{H}^{2*}$ where $\mathbf{H}^{2*}$ denotes the completed upper half-plane. Then $\mathrm{Beldeg}(X(\Gamma)) \leqslant [\mathrm{PSL}_2(\mathbb{Z}) : \Gamma]$, because the natural map $X(\Gamma) \to X(1) = \mathrm{PSL}_2(\mathbb{Z})\backslash\mathbf{H}^{2*} \xrightarrow{\sim} \mathbb{P}^1_{\mathbb{C}}$ descends to $\overline{\mathbb{Q}}$ and defines a Belyĭ map, where the latter isomorphism is the normalized modular j-invariant $j/1728$.

A lower bound on the Belyĭ degree may be given in terms of the genus, as we show now.

PROPOSITION 2.5. *For every curve X, the inequality* $\mathrm{Beldeg}(X) \geqslant 2g(X) + 1$ *holds.*

PROOF. By the Riemann–Hurwitz theorem, the degree of a map is minimized when its ramification is total, so for a Belyĭ map of degree d on X we have

$$2g - 2 \leqslant -2d + 3(d-1) = d - 3,$$

and therefore $d \geqslant 2g + 1$. □

As an application of Proposition 2.5, we now show that gonal maps on curves of positive genus are not Belyĭ maps.

COROLLARY 2.6. *Let X be a curve of gonality γ. A finite map $\phi\colon X \to \mathbb{P}^1$ with $\deg \phi = \gamma$ is a Belyĭ map only if ϕ is an isomorphism.*

PROOF. If $g(X) = 0$, then the result is clear. On the other hand, the gonality of X is bounded above by $\lceil g(X)/2 \rceil + 1$ by Brill–Noether theory [**1**, Chapter V], and the strict inequality $2g(X) + 1 > g(X)/2 + 1$ holds unless $g(X) = 0$, so the result follows from Proposition 2.5. □

EXAMPLE 2.7. Let $d = 2g + 1 \geqslant 1$ be odd, and let X be the curve defined by $y^2 - y = x^d$. Then X has genus g, and we verify that the map $y\colon X \to \mathbb{P}^1$ is a Belyĭ map of degree d. Therefore, the lower bound in Proposition 2.5 is sharp for every genus g.

REMARK 2.8. The bound in Proposition 2.5 gives a "topological" lower bound for the Belyĭ degree of X. One can also give "arithmetic" lower bounds as follows. Let p be a prime number, and let X be the elliptic curve given by the equation $y^2 = x(x-1)(x-p)$ over $\mathbb{Q}$. Then X has (bad) multiplicative reduction at p and this bad reduction persists over any extension field. It follows from work of Beckmann [**3**] that $\operatorname{Beldeg}(X) \geqslant p$ (see also Zapponi [**29**, Theorem 1.3]): if $\phi\colon X \to \mathbb{P}^1$ is a Belyĭ map of degree $d < p$, then the monodromy group G of ϕ has $p \nmid \#G$, and so ϕ and therefore X has potentially good reduction at p (in fact, obtained over an extension of $\mathbb{Q}$ unramified at p), a contradiction.

EXAMPLE 2.9. For every $n \geqslant 1$, the Belyĭ degree of the Fermat curve

$$X_n\colon x^n + y^n = z^n \subset \mathbb{P}^2$$

is bounded above by $\operatorname{Beldeg}(X_n) \leqslant n^2$, because there is a Belyĭ map

$$\begin{aligned} X_n &\to \mathbb{P}^1 \\ (x : y : z) &\mapsto (x^n : z^n) \end{aligned}$$

of degree n^2. On the other hand, we have $\operatorname{Beldeg}(X_n) \geqslant (n-1)(n-2) + 1 = n^2 - 3n + 3$ by Proposition 2.5.

For $n = 1, 2$, we have $X_n \simeq \mathbb{P}^1$ so $\operatorname{Beldeg}(X_1) = \operatorname{Beldeg}(X_2) = 1$. As observed by Zapponi [**29**, Example 1.2], for $n = 3$, the curve X_3 is a genus 1 curve with j-invariant 0, so isomorphic to $y^2 - y = x^3$, and $\operatorname{Beldeg}(X_3) = 3$ by Example 2.7.

We consider the case $n = 4$ in section 5, and show that $\operatorname{Beldeg}(X_4) = 8$ in Proposition 5.1.

We finish this section with an effective version of Belyĭ's theorem, due to Khadjavi [**16**]. (An effective version was also proven independently by Liţcanu [**19**, Théorème 4.3], with a weaker bound.) To give her result, we need the height of a finite subset of $\mathbb{P}^1(\overline{\mathbb{Q}})$. For K a number field and $a \in K$, we define the (exponential) height to be $H(a) := (\prod_v \max(1, \|\alpha\|_v))^{1/[K:\mathbb{Q}]}$, where the product runs over the set of absolute values indexed by the places v of K normalized so that the product formula holds [**16**, Section 2]. For a finite subset $B \subset \mathbb{P}^1(\overline{\mathbb{Q}})$, and K a number field over which the points B are defined, we define its (exponential) height by $H_B := \max\{H(\alpha) : \alpha \in B\}$, and we let N_B be the cardinality of the Galois orbit of B.

PROPOSITION 2.10 (Effective version of Belyĭ's theorem). *Let $B \subset \mathbb{P}^1(\overline{\mathbb{Q}})$ be a finite set. Write $N = N_B$. Then there exists a Belyĭ map $\phi\colon \mathbb{P}^1 \to \mathbb{P}^1$ such that $\phi(B) \subseteq \{0, 1, \infty\}$ and*

$$\deg \phi \leqslant (4NH_B)^{9N^3 2^{N-2} N!}.$$

PROOF. See Khadjavi [**16**, Theorem 1.1.c]. □

COROLLARY 2.11. *Let X be a curve, and let $\pi\colon X \to \mathbb{P}^1$ be a finite morphism with branch locus $B \subset \mathbb{P}^1(\overline{\mathbb{Q}})$. Write $N = N_B$. Then*

$$\operatorname{Beldeg}(X) \leqslant (4NH_B)^{9N^3 2^{N-2} N!} \deg \pi.$$

PROOF. Choose ϕ as in Proposition 2.10 and consider the composed morphism $\phi \circ \pi$. □

3. First proof of Theorem 1.2

Throughout this section, let K be a number field. We begin with two preliminary lemmas.

LEMMA 3.1. *There exists an algorithm that, given as input an affine variety $X \subset \mathbb{A}^n$ and $t \geqslant 1$, computes as output $N \geqslant 1$ and generators for an ideal $I \subseteq \overline{\mathbb{Q}}[x_1, \ldots, x_N]$ such that the zero locus of I is the variety obtained by removing all the diagonals from X^t/S_t.*

PROOF. Let $X = \operatorname{Spec} \overline{\mathbb{Q}}[x_1, \ldots, x_n]/I$. By (classical) invariant theory (see Sturmfels [**26**]), there is an algorithm to compute the coordinate ring of invariants $\left(\overline{\mathbb{Q}}[x_1, \ldots, x_n]/I\right)^{S_t}$. In other words, there is an algorithm which computes

$$X^t/S_t = \operatorname{Spec}\left(\left(\overline{\mathbb{Q}}[x_1, \ldots, x_n]/I\right)^{S_t}\right).$$

To conclude the proof, note that the complement of a divisor $D = Z(f)$ is again an affine variety, adding a coordinate z satisfying $zf - 1$. □

REMARK 3.2. We will use Lemma 3.1 below to parametrize extra ramification points, write equations in terms of these parameters, and check whether the system of equations has a solution over $\overline{\mathbb{Q}}$. For this purpose, we need not take the quotient by the symmetric group S_t, as the system of equations with unordered parameters has a solution over $\overline{\mathbb{Q}}$ if and only if the one with ordered parameters does.

Next, we show how to represent rational functions on X explicitly in terms of a Riemann–Roch basis.

LEMMA 3.3. *Let X be a curve over K of genus g, let $\mathscr{L}$ be an ample sheaf on X, and let d be a positive integer. Let*

$$t := \left\lceil \frac{d+g}{\deg \mathscr{L}} \right\rceil. \tag{3.4}$$

Then, for all $f \in \overline{\mathbb{Q}}(X)$ of degree d, there exist $a, b \in \mathrm{H}^0(X_{\overline{\mathbb{Q}}}, \mathscr{L}^{\otimes t})$ with $b \neq 0$ such that $f = a/b$.

PROOF. By definition, we have

$$t \deg \mathscr{L} - d + 1 - g \geqslant 1. \tag{3.5}$$

Let $\operatorname{div}_\infty f \geqslant 0$ be the divisor of poles of f. By Riemann–Roch,

$$\dim_{\overline{\mathbb{Q}}} \mathrm{H}^0(X_{\overline{\mathbb{Q}}}, \mathscr{L}^{\otimes t}(-\operatorname{div}_\infty f)) \geqslant t \deg \mathscr{L} - d + 1 - g \geqslant 1. \tag{3.6}$$

Let

$$b \in \mathrm{H}^0(X_{\overline{\mathbb{Q}}}, \mathscr{L}^{\otimes t}(-\operatorname{div}_\infty f)) \subseteq \mathrm{H}^0(X_{\overline{\mathbb{Q}}}, \mathscr{L}^{\otimes t})$$

be a nonzero element. Then $fb \in \mathrm{H}^0(X_{\overline{\mathbb{Q}}}, \mathscr{L}^{\otimes t})$. (In effect, we have "cancelled the poles" of f by the zeros of b, at the expense of possibly introducing new poles supported within $\mathscr{L}$.) Letting $fb = a \in \mathrm{H}^0(X, \mathscr{L}^{\otimes t})$ we have written $f = a/b$ as claimed. $\square$

The quantities in Lemma 3.3 can be effectively computed, as follows. Recall that a curve X over K is specified in bits by a set of defining equations in projective space with coefficients in K. (Starting with any birational model for X, we can effectively compute a smooth projective model.)

LEMMA 3.7. *Let $X \subset \mathbb{P}^n_K$ be a curve over K. Then the following quantities are effectively computable:*

(i) *The genus $g = g(X)$;*
(ii) *An effective divisor D on X over K and its degree.*
(iii) *Given a divisor D over K, a basis for the K-vector space $\mathrm{H}^0(X, \mathscr{O}_X(D))$.*

PROOF. For (a), to compute the genus we compute a Gröbner basis for the defining ideal I of X, compute its Hilbert polynomial, and recover the (arithmetic equals geometric) genus from the constant term. For (b), intersecting X with a hyperplane, we obtain an effective divisor D on X over K, and its degree is the leading term of the Hilbert polynomial computed in (a). For (c), it suffices to note that Riemann–Roch calculations can be done effectively: see e.g. Coates [**9**] or Hess [**13**]. $\square$

A ramification type for a positive integer d is a triple $\lambda = (\lambda_0, \lambda_1, \lambda_\infty)$ of partitions of d. For X a curve, d an integer, and λ a ramification type, let $\mathrm{Bel}_{d,\lambda}(X) \subseteq \mathrm{Bel}_d(X)$ be the subset of Belyĭ maps of degree d on X with ramification type λ. For the ramification type λ and $* \in \{0, 1, \infty\}$, let $\lambda_{*,1}, \dots, \lambda_{*,r_*}$ be the parts of λ (and r_* the number of parts), so

$$d = \lambda_{*,1} + \cdots + \lambda_{*,r_*}.$$

If $\phi\colon X \to \mathbb{P}^1$ is a Belyĭ map of degree d with ramification type λ, then the Riemann–Hurwitz formula is satisfied:

$$\begin{aligned} 2g - 2 &= -2d + \sum_{i=1}^{r_0} (\lambda_{0,i} - 1) + \sum_{i=1}^{r_1} (\lambda_{1,i} - 1) + \sum_{i=1}^{r_\infty} (\lambda_{\infty,i} - 1) \\ &= d - r_0 - r_1 - r_\infty. \end{aligned} \tag{3.8}$$

To prove our main theorem, we will show that one can compute equations whose vanishing locus over $\overline{\mathbb{Q}}$ is precisely the set $\mathrm{Bel}_{d,\lambda}(X)$ (see Proposition 3.16): we call such equations a model for $\mathrm{Bel}_{d,\lambda}(X)$.

On our way to prove Proposition 3.16, we first characterize Belyĭ maps of degree d with ramification type λ among rational functions on a curve written in terms of a Riemann–Roch basis. This characterization is technical but we will soon see that it is quite suitable for our algorithmic application.

PROPOSITION 3.9. *Let $d \geqslant 1$ be an integer and let $\lambda = (\lambda_0, \lambda_1, \lambda_\infty)$ be a ramification type for d with r_0, r_1, r_∞ parts, respectively. Let X be a curve over a number field K. Let g be the genus of X, and suppose that*

$$2g - 2 = d - r_0 - r_1 - r_\infty. \tag{3.10}$$

Let D_0 be an effective divisor of degree d_0, and let $\mathscr{L} = \mathscr{O}_X(D_0)$. Let $t \geqslant 1$ be the smallest positive integer such that $t \deg \mathscr{L} - d + 1 - g \geqslant 1$. Let $g_1, \dots, g_n \in K(X)$

be a basis for the K-vector space $\mathrm{H}^0(X, \mathscr{L}^{\otimes t})$. Let $0 \leqslant k, l \leqslant n$ be integers. Let m be minimal so that $g_k, g_l \in \mathrm{H}^0(X_{\overline{\mathbb{Q}}}, \mathscr{L}^{\otimes m}) \subseteq \mathrm{H}^0(X_{\overline{\mathbb{Q}}}, \mathscr{L}^{\otimes t})$. Let

$$\phi := \frac{a}{b} = \frac{a_1 g_1 + \ldots + a_k g_k}{b_1 g_1 + \ldots + b_l g_l}$$

be a nonconstant rational function with $a_1, \ldots, b_l \in \overline{\mathbb{Q}}$.

Then the rational function ϕ lies in $\mathrm{Bel}_{d,\lambda}(X)(\overline{\mathbb{Q}})$ if and only if there exists a partition $\mu = \mu_1 + \cdots + \mu_s$ of $md_0 - d$, distinct points

$$P_1, \ldots, P_{r_0}, Q_1, \ldots, Q_{r_1}, R_1, \ldots, R_{r_\infty} \in X(\overline{\mathbb{Q}})$$

and distinct points

$$Y_1, \ldots, Y_s \in X(\overline{\mathbb{Q}}),$$

allowing these two sets of points to meet, such that

$$\begin{aligned} \operatorname{div}(a) &= \sum_{i=1}^{r_0} \lambda_{0,i}[P_i] + \sum_{i=1}^{s} \mu_i[Y_i] - mD_0 \\ \operatorname{div}(a-b) &= \sum_{i=1}^{r_1} \lambda_{1,i}[Q_i] + \sum_{i=1}^{s} \mu_i[Y_i] - mD_0 \\ \operatorname{div}(b) &= \sum_{i=1}^{r_\infty} \lambda_{\infty,i}[R_i] + \sum_{i=1}^{s} \mu_i[Y_i] - mD_0. \end{aligned} \tag{3.11}$$

PROOF. We first prove the implication ($\Leftarrow$) of the proposition. Suppose ϕ satisfies the equations (3.11). Then

$$\operatorname{div} \phi = \operatorname{div}(a) - \operatorname{div}(b) = \sum_{i=1}^{r_0} \lambda_{0,i}[P_i] - \sum_{i=1}^{r_\infty} \lambda_{\infty,i}[R_i];$$

since the set of points $\{P_1, \ldots, P_{r_0}\}$ is disjoint from $\{R_1, \ldots, R_{r_\infty}\}$, we have $\deg \phi = d$. We see some ramification in $\phi\colon X \to \mathbb{P}^1$ above the points $0, 1, \infty$ according to the ramification type λ, specified by the equations (3.11); let ρ be the degree of the remaining ramification locus. We claim there can be no further ramification. Indeed, the Riemann–Hurwitz formula gives

$$\begin{aligned} 2g - 2 &= -2d + \sum_{i=0}^{r_0} (\lambda_{0,i} - 1) + \sum_{i=0}^{r_1} (\lambda_{1,i} - 1) + \sum_{i=0}^{r_\infty} (\lambda_{\infty,i} - 1) + \rho \\ &= d - r_0 - r_1 - r_\infty + \rho. \end{aligned} \tag{3.12}$$

On the other hand, we are given the equality 3.10, so $\rho = 0$. Therefore $\phi \in \mathrm{Bel}_{d,\lambda}(\overline{\mathbb{Q}})$.

We now prove the other implication ($\Rightarrow$). Suppose $\phi \in \mathrm{Bel}_{d,\lambda}(\overline{\mathbb{Q}})$. We have

$$\operatorname{div}(\phi) = \operatorname{div}(a) - \operatorname{div}(b) = \sum_{i=1}^{r_0} \lambda_{0,i}[P_i] - \sum_{i=1}^{r_\infty} \lambda_{\infty,i}[R_i] \tag{3.13}$$

and

$$\operatorname{div}(\phi - 1) = \operatorname{div}(a-b) - \operatorname{div}(b) = \sum_{i=1}^{r_1} \lambda_{1,i}[Q_i] - \sum_{i=1}^{r_\infty} \lambda_{\infty,i}[R_i] \tag{3.14}$$

for distinct points $P_1, \dots, P_{r_0}, Q_1, \dots, Q_{r_1}, R_1, \dots, R_{r_\infty} \in X(\overline{\mathbb{Q}})$. Moreover, since $a \in \mathrm{H}^0(X, \mathscr{L}^{\otimes t})$, we have

$$\operatorname{div}(a) = \sum_{i=1}^{r_0} \lambda_{0,i}[P_i] + E - mD_0$$

for some effective divisor E (not necessarily disjoint from D_0) with $\deg E = md_0 - d$; from (3.13) we obtain

$$\operatorname{div}(b) = \sum_{i=1}^{r_\infty} \lambda_{\infty,i}[R_i] + E - mD_0.$$

Writing out $E = \sum_{i=1}^{s} \mu_i[Y_i]$ with Y_i distinct as an effective divisor and arguing similarly for $\operatorname{div}(a - b)$, we conclude that the equations (3.11) hold. □

REMARK 3.15. $\mathrm{Bel}_{d,\lambda}(X)$ is a (non-positive dimensional) Hurwitz space: see for instance Bertin–Romagny [**6**, Section 6.6] (but also Mochizuki [**20**] and Romagny–Wewers [**23**]). Indeed, for a scheme S over $\overline{\mathbb{Q}}$, let $\underline{\mathrm{Bel}}_{d,\lambda,X}(S)$ be the groupoid whose objects are tuples $(\phi\colon Y \to \mathbb{P}^1_S,\ g\colon Y \to X_S)$, where Y is a smooth proper geometrically connected curve over S, the map $\phi\colon Y \to \mathbb{P}^1_S$ is a finite flat finitely-presented morphism of degree d ramified only over $0, 1, \infty$ with ramification type λ, and g is an isomorphism of S-schemes. This defines a (possibly empty) separated finite type Deligne–Mumford algebraic stack $\underline{\mathrm{Bel}}_{d,\lambda,X}$ over $\overline{\mathbb{Q}}$ which is usually referred to as a *Hurwitz stack*. Its coarse space, denoted by $\mathrm{Bel}_{d,\lambda,X}$, is usually referred to as a *Hurwitz space*. Since the set of $\overline{\mathbb{Q}}$-points $\mathrm{Bel}_{d,\lambda,X}(\overline{\mathbb{Q}})$ of its coarse space $\mathrm{Bel}_{d,\lambda,X}$ is naturally in bijection with $\mathrm{Bel}_{d,\lambda}(X)$, one could say that the following proposition says that there is an algorithm to compute a model for the Hurwitz space $\mathrm{Bel}_{d,\lambda,X}$.

We now prove the following key ingredient to our main result.

PROPOSITION 3.16. *There exists an algorithm that, given as input a curve X over $\overline{\mathbb{Q}}$, an integer d, and a ramification type λ of d, computes a model for $\mathrm{Bel}_{d,\lambda}(X)$.*

PROOF. Let K be a field of definition of X (containing the coefficients of the input model). Applying the algorithm in Lemma 3.7 to X over K, we compute the genus g of X.

Recall the Riemann–Hurwitz formula (3.8) for a Belyĭ map. If the Riemann–Hurwitz formula is not satisfied for d and the ramification type λ, there is no Belyĭ map of degree d with ramification type λ on X (indeed, on any curve of genus g), and the algorithm gives trivial output. So we may suppose that (3.8) holds.

Next, we compute an effective divisor D_0 on X with $\mathscr{L} := \mathscr{O}_X(D_0)$ and its degree $d_0 := \deg D_0$. Let

$$t := \left\lceil \frac{d+g}{\deg \mathscr{L}} \right\rceil$$

as in (3.4). By Lemma 3.7, we may compute a K-basis $g_1, \dots, g_n$ of $\mathrm{H}^0(X, \mathscr{L}^{\otimes t})$. Then by Lemma 3.3, if $\phi \in \overline{\mathbb{Q}}(X)$ is a degree d rational function on X, then there exist $a_1, \dots, a_n, b_1, \dots, b_n \in \overline{\mathbb{Q}}$ such that $a = \sum_{i=1}^{n} a_i g_i$ and $b = \sum_{i=1}^{n} b_i g_i$ satisfy $\phi = a/b$.

We now give algebraic conditions on the coefficients a_i, b_j that characterize the subset $\mathrm{Bel}_{d,\lambda}(X)$. There is a rescaling redundancy in the ratio a/b so we work affinely as follows. We loop over pairs $0 \leqslant k, \ell \leqslant n$ and consider functions

$$\phi = \frac{a}{b} = \frac{a_1 g_1 + \cdots + a_{k-1} g_{k-1} + a_k g_k}{b_1 g_1 + \cdots + b_{\ell-1} g_{\ell-1} + g_\ell} \tag{3.17}$$

with $a_k \neq 0$. Every function $\phi = a/b$ arises for a unique such k, ℓ. Let m be minimal so that $g_k, g_\ell \in \mathrm{H}^0(X_{\overline{\mathbb{Q}}}, \mathscr{L}^{\otimes m}) \subseteq \mathrm{H}^0(X_{\overline{\mathbb{Q}}}, \mathscr{L}^{\otimes t})$.

Note that Proposition 3.9 characterizes precisely when a rational function of the form (3.17) lies in $\mathrm{Bel}_{d,\lambda}(X)(\overline{\mathbb{Q}})$. Thus, by Proposition 3.9, we may finish by noting that the equations (3.11) can be written explicitly. To this end, we loop over the partitions μ and consider the configuration space of $r_0 + r_1 + r_\infty$ and s distinct points (but allowing the two sets to meet), which can be effectively computed by Lemma 3.1. Next, we write $D_0 = \sum_i \rho_i [D_{0i}]$ and loop over the possible cases where one of the points P_i, Q_i, R_i, Y_i is equal to one of the points D_{0i} or they are all distinct from D_{0i}. In each case, cancelling terms when they coincide, we impose the vanishing conditions on $a, a-b, b$ with multiple order vanishing defined by higher derivatives, in the usual way. For each such function, we have imposed that the divisor of zeros is at least as large in degree as the function itself, so there can be no further zeros, and therefore the equations (3.11) hold for any solution to this large system of equations. □

EXAMPLE 3.18. We specialize Proposition 3.16 to the case $X = \mathbb{P}^1$. (The Belyĭ degree of $\mathbb{P}^1$ is 1, but it is still instructive to see what the equations (3.11) look like in this case.) Let $X = \mathbb{P}^1$ with coordinate x, defined by $\mathrm{ord}_\infty x = -1$. We take $D_0 = (\infty)$. Then the basis of functions g_i is just $1, \dots, x^d$, and $f = a/b$ is a ratio of two polynomials of degree $\leqslant d$, at least one of which is degree exactly d. Having hit the degree on the nose, the "cancelling" divisor $E = \sum_{i=1}^{s} \mu_i [Y_i] = 0$ in the proof of Proposition 3.16 does not arise, and the equations for $a, b, a-b$ impose the required factorization properties of f. This method is sometimes called the *direct method* and has been frequently used (and adapted) in the computation of Belyĭ maps using Gröbner techniques [**25**, §2].

Given equations for the algebraic set $\mathrm{Bel}_{d,\lambda}(X)$, we now prove that there is an algorithm to check whether this set is empty or not.

LEMMA 3.19. *There exists an algorithm that, given as input an affine variety X over $\overline{\mathbb{Q}}$, computes as output whether $X(\overline{\mathbb{Q}})$ is empty or not.*

PROOF. Let I be an ideal defining the affine variety X (in some polynomial ring over $\overline{\mathbb{Q}}$). One can effectively compute a Gröbner basis for I [**11**, Chapter 15]. With a Gröbner basis at hand one can easily check whether 1 is in the ideal or not, and conclude by Hilbert's Nullstellensatz accordingly if $X(\overline{\mathbb{Q}})$ is empty or not. □

COROLLARY 3.20. *There exists an algorithm that, given as input a set S with a model computes as output whether S is empty or not.*

PROOF. Immediate from Lemma 3.19 and the definition of a model for a set S as being given by equations. □

We are now ready to give the first proof of the main result of this note.

FIRST PROOF OF THEOREM 1.2. Let X be a curve over $\overline{\mathbb{Q}}$. Let $d \geqslant 1$ be an integer, and let λ be a ramification type of d. To prove the theorem, it suffices to show that there is an algorithm which computes whether the set $\mathrm{Bel}_{d,\lambda}(X)$ of Belyĭ maps of degree d with ramification type λ is empty. We explain how to use the above results to do this.

By Proposition 3.16, we may (and do) compute a model for the set $\mathrm{Bel}_{d,\lambda}(X)$. By Corollary 3.20, we can check algorithmically whether this set is empty or not (by using the model we computed). This means that we can algorithmically check whether X has a Belyĭ map of degree d with ramification type λ. □

4. Second proof of Theorem 1.2

In this section, we sketch a second proof of Theorem 1.2. Instead of writing down equations for the Hurwitz space $\mathrm{Bel}_d(X)$, we enumerate all Belyĭ maps and effectively compute equations to check for isomorphism between curves. We saw this method already at work in Example 2.9.

Let X, Y be curves over $\overline{\mathbb{Q}}$. The functor $S \mapsto \mathrm{Isom}_S(X_S, Y_S)$ from the (opposite) category of schemes over $\overline{\mathbb{Q}}$ to the category of sets is representable [**10**, Theorem 1.11] by a finite étale $\overline{\mathbb{Q}}$-scheme $\underline{\mathrm{Isom}}(X,Y)$. Our next result shows that one can effectively compute a model for the (finite) set $\mathrm{Isom}(X,Y) = \underline{\mathrm{Isom}}(X,Y)(\overline{\mathbb{Q}})$ of isomorphisms from X to Y. Equivalently, one can effectively compute equations for the finite étale $\overline{\mathbb{Q}}$-scheme $\underline{\mathrm{Isom}}(X,Y)$.

LEMMA 4.1. *There exists an algorithm that, given as input curves X, Y over $\overline{\mathbb{Q}}$ with at least one of X or Y of genus at least 2, computes a model for the set* $\mathrm{Isom}(X,Y)$.

PROOF. We first compute the genera of X, Y (as in the proof of Lemma 3.3): if these are not equal, then we correctly return the empty set. Otherwise, we compute a canonical divisor K_X on X by a Riemann–Roch calculation [**13**] and the image of the pluricanonical map $\varphi\colon X \hookrightarrow \mathbb{P}^N$ associated to the complete linear series on the very ample divisor $3K_X$ via Gröbner bases. We repeat this with Y. An isomorphism $\mathrm{Isom}(X,Y)$ induces via its action on canonical divisors an element of $\mathrm{PGL}_{N-1}(\overline{\mathbb{Q}})$ mapping the canonically embedded curve X to Y, and vice versa, and so a model is provided by the equations that insist that a linear change of variables in $\mathbb{P}^N$ maps the ideal of X into the ideal of Y, which can again be achieved by Gröbner bases. □

COROLLARY 4.2. *There exists an algorithm that, given as input maps of curves $f\colon X \to \mathbb{P}^1$ and $h\colon Y \to \mathbb{P}^1$ over $\overline{\mathbb{Q}}$, computes as output whether there exists an isomorphism $\alpha\colon X \xrightarrow{\sim} Y$ such that $g = \alpha \circ f$ or not.*

Similarly, there exists an algorithm that, given as input curves X, Y over $\overline{\mathbb{Q}}$, computes as output whether $X \simeq Y$ or not.

As remarked by Ngo–Nguyen–van der Put–Top [**22**, Appendix], the existence of an algorithm which decides whether two curves are isomorphic over an algebraically closed field is well-known. We include the following proof for the sake of completeness.

PROOF OF COROLLARY 4.2. We compute the genera of X, Y and again if these are different we correctly return as output *no*. Otherwise, let g be the common genus.

If $g = 0$, we parametrize X and Y to get $X \simeq Y \simeq \mathbb{P}^1$ and then ask for $\alpha \in \mathrm{PGL}_2(\overline{\mathbb{Q}})$ to map f to g in a manner analogous to the proof of Lemma 4.1.

If $g = 1$, we loop over the preimages of $0 \in \mathbb{P}^1$ in X and Y as origins, we compute Weierstrass equations via Riemann–Roch, and return *no* if the j-invariants of X, Y are unequal. Otherwise, these j-invariants are equal and we compute an isomorphism $X \simeq Y$ of Weierstrass equations. The remaining isomorphisms are twists, and we conclude by checking if there is a twist α of the common Weierstrass equation that maps f to g.

If $g \geqslant 2$, we call the algorithm in Lemma 4.1: we obtain a finite set of isomorphisms, and for each $\alpha \in \operatorname{Isom}(X, Y)$ we check if $h = \alpha \circ f$.

The second statement is proven similarly, ignoring the map. □

We now give a second proof of our main result.

SECOND PROOF OF THEOREM 1.2. We first loop over integers $d \geqslant 1$ and all ramification types λ of d. For each λ, we count the number of permutation triples up to simultaneous conjugation with ramification type λ.

We then compute the set of Belyĭ maps of degree d with ramification type λ over $\overline{\mathbb{Q}}$ as follows. There are countably many number fields K, and they may be enumerated by a minimal polynomial of a primitive element. For each number field K, there are countably many curves X over K up to isomorphism over $\overline{\mathbb{Q}}$, and this set is computable: for $g = 0$ we have only $\mathbb{P}^1_K$, for $g = 1$ we can enumerate j-invariants, and for $g \geqslant 2$ we can enumerate candidate pluricanonical ideals (by Petri's theorem). Finally, for each curve X over K, there are countably many maps $f\colon X \to \mathbb{P}^1$, and these can be enumerated using Lemma 3.3. Diagonalizing, we can enumerate the entire countable set of such maps. For each such map f, using Gröbner bases we can compute the degree and ramification type of f, and in particular detect if f is a Belyĭ map of degree d with ramification type λ. Along the way in this (ghastly) enumeration, we can detect if two correctly identified Belyĭ maps are isomorphic using Corollary 4.2. Having counted the number of isomorphism classes of such maps, we know when to stop with the complete set of such maps.

Now, to see whether $\operatorname{Bel}_{d,\lambda}(X)$ is nonempty, we just check using Corollary 4.2 whether X is isomorphic to one of the source curves in the set of all Belyĭ maps of degree d and ramification type λ. □

5. The Fermat curve of degree four

In this section we prove the following proposition, promised in Example 2.9.

PROPOSITION 5.1. *The Belyĭ degree of the curve* $X\colon x^4 + y^4 = z^4$ *is equal to* 8.

PROOF. The curve X is a canonically embedded curve of genus 3. By Proposition 2.5, we have $\operatorname{Beldeg}(X) \geqslant 7$. On the other hand, X maps to the genus 1 curve with affine model $z^2 = x^4 + 1$ and j-invariant 1728, and this latter curve has a Belyĭ map of degree 4 taking the quotient by its automorphism group of order 4 as an elliptic curve, equipped with a point at infinity. Composing the two, we obtain a Belyĭ map of degree 8 on X defined by $(x : y : z) \mapsto x^2 + z^2$; therefore $\operatorname{Beldeg}(X) \leqslant 8$. So to show $\operatorname{Beldeg}(X) = 8$, it suffices to rule out the existence of a Belyĭ map of degree 7.

By enumeration of partitions and the Riemann–Hurwitz formula, we see that the only partition triple of 7 that gives rise to a Belyĭ map $\phi\colon X \to \mathbb{P}^1$ with X of genus 3 is $(7, 7, 7)$. By enumeration of permutation triples up to simultaneous conjugation, we compute that the Belyĭ maps of degree 7 and genus 3 have three possible monodromy groups: cyclic of order 7, the simple group $\mathrm{GL}_3(\mathbb{F}_2) \simeq \mathrm{PSL}_2(\mathbb{F}_7)$ of order 168, or the alternating group A_7. We rule these out by consideration of automorphism groups.

As in Lemma 4.1 but instead using the canonical embedding as K_X is already ample, we have $\mathrm{Aut}(X) \leqslant \mathrm{Aut}(\mathbb{P}^2) = \mathrm{PGL}_3(\overline{\mathbb{Q}})$, and a direct calculation yields that $\mathrm{Aut}(X) \simeq S_3 \ltimes (\mathbb{Z}/4\mathbb{Z})^2$ and $\#\mathrm{Aut}(X) = 96$. (For the automorphism group of the general Fermat curve X_n of degree $n \geqslant 4$, see Leopoldt [**18**] or Tzermias [**28**]: they prove that $\mathrm{Aut}(X_n) \simeq S_3 \ltimes (\mathbb{Z}/n\mathbb{Z})^2$.)

The cyclic case is a geometrically Galois map, but X does not have an automorphism of order 7, impossible. For the two noncyclic cases, computing the centralizers of the $2 + 23 = 25$ permutation triples up to simultaneous conjugation, we conclude that these Belyĭ maps have no automorphisms. An automorphism $\alpha \in \mathrm{Aut}(X)$ of order coprime to 7 cannot commute with a Belyĭ map of prime degree 7 because the quotient by α would be an intermediate curve. So if X had a Belyĭ map of degree 7, there would be 96 nonisomorphic such Belyĭ maps, but that is too many. □

REMARK 5.2. The above self-contained proof works because of the large automorphism group on the Fermat curve, and it seems difficult to make this strategy work for an arbitrary curve.

To illustrate how our algorithms work, we now show how they can be used to give two further proofs of Proposition 5.1.

EXAMPLE 5.3. We begin with the first algorithm exhibited in Proposition 3.16. We show that X has no Belyĭ map of degree 7 with explicit equations to illustrate our method; we finish the proof as above.

We take the divisor $D_0 = [D_{01}]$ where $D_{01} = (1 : 0 : 1) \in X(\mathbb{Q})$ and $\deg D_0 = d_0 = 1$. We write rational functions on X as ratios of polynomials in $\mathbb{Q}[x, y]$, writing x, y instead of $x/z, y/z$. According to (3.5), taking $\mathscr{L} = \mathscr{O}_X(D_0)$ we need $t - 7 + 1 - 3 \geqslant 1$, so we take $t = 10$. By a computation in MAGMA [**8**], the space $\mathrm{H}^0(X, \mathscr{L}^{\otimes 10})$ has dimension $n = 8$ and basis

$$
\begin{aligned}
g_1 &= 1 \\
g_2 &= \frac{x^3 + x^2 + x + 1}{y^3} \\
g_3 &= g_2/y \\
g_4 &= \frac{4(x^3 + x^2 + x + 1) - x^2y^4 - 2xy^4 - 3y^4}{4y^6} \\
g_5 &= g_5/y \\
g_6 &= g_6/y \\
g_7 &= \frac{16(x^3 + x^2 + x + 1) - 6x^3y^4 - 10x^2y^4 + xy^8 - 14xy^4 + 3y^8 - 18y^4}{6y^9} \\
g_8 &= \frac{32(x^3 + x^2 + x + 1) - 3x^2y^8 - 8x^2y^4 - 4xy^8 - 16xy^4 - 3y^8 - 24y^4}{32y^{10}}
\end{aligned}
\tag{5.4}
$$

We compute that $\mathrm{ord}_{D_0}\, g_i = 0, -3, -4, -6, -7, -8, -9, -10$.

The general case is where $a_8 b_8 \neq 0$, for which $k = \ell = 8$ and we may take

$$\phi = \frac{a}{b} = \frac{\sum_{i=1}^{8} a_i g_i}{\sum_{i=1}^{8} b_i g_i}$$

so we let $b_8 = 1$ and $m = t = 10$. As we already saw in Example 2.9, the only ramification type possible is $\lambda = (7, 7, 7)$, with $r_0 = r_1 = r_\infty = 1$ and $\lambda_0 = \lambda_1 = \lambda_\infty = 7$.

We have $md_0 - d = 10 - 7 = 3$, so we consider the partitions of 3. We start with the trivial partition $\mu = \mu_1 = 3$ with $s = 1$. Then the equations (3.11) read, dropping subscripts: we want distinct points $P, Q, R \in X(\overline{\mathbb{Q}})$ such that $\operatorname{div}(a) \geqslant 7[P] + 3[Y]$ and $\operatorname{div}(a - b) \geqslant 7[Q] + 3[Y]$ and $\operatorname{div}(b) \geqslant 7[R] + 3[Y]$.

Continuing in the general case, the points P, Q, R, Y, D_0 are all distinct, each such point belongs to the affine open with $z \neq 0$, and furthermore $x - x(Z)$ is a uniformizer at Z for each point $Z = P, \ldots, D_0$.

The conditions for the point P we write as follows: letting $P = (x_P : y_P : 1)$ with unknowns x_P, y_P, we add the equation $x_P^4 + y_P^4 = 1$ so that P lies on the curve X, and then (by Taylor expansion) to ensure $\operatorname{ord}_P a \geqslant 7$ we add the equations

$$\frac{\partial^j a}{\partial x^j}(x_P, y_P) = \sum_{i=1}^{8} a_i \frac{\partial^j g_i}{\partial x^j}(x_P, y_P) = 0 \tag{5.5}$$

for $j = 0, \ldots, 6$, and using implicit differentiation on the defining equation of X to obtain $\dfrac{\mathrm{d}y}{\mathrm{d}x} = -\dfrac{x^3}{y^3}$. For example, the case $j = 1$ (asserting that a vanishes to order at least 2 at P, assuming that $a(P) = 0$) is

$$\begin{aligned}
&(3x_P^6 y_P^7 + 3x_P^5 y_P^7 + 3x_P^4 y_P^7 + 3x_P^3 y_P^7 + 3x_P^2 y_P^{13} + 2x_P y_P^{13} + y_P^{13})a_2 \\
&+ (4x_P^6 y_P^6 + 4x_P^5 y_P^6 + 4x_P^4 y_P^6 + 4x_P^3 y_P^6 + 3x_P^2 y_P^{12} + 2x_P y_P^{12} + y_P^{12})a_3 \\
&+ (6x_P^6 y_P^4 - \tfrac{11}{2} x_P^5 y_P^8 + 6x_P^5 y_P^4 - 7x_P^4 y_P^8 + 6x_P^4 y_P^4 \\
&\quad - \tfrac{17}{2} x_P^3 y_P^8 + 6x_P^3 y_P^4 + 3x_P^2 y_P^{10} + 4x_P y_P^{10} + 2y_P^{10})a_4 \\
&+ (7x_P^6 y_P^3 - \tfrac{23}{4} x_P^5 y_P^7 + 7x_P^5 y_P^3 - \tfrac{15}{2} x_P^4 y_P^7 + 7x_P^4 y_P^3 \\
&\quad - \tfrac{37}{4} x_P^3 y_P^7 + 7x_P^3 y_P^3 + 3x_P^2 y_P^9 + 4x_P y_P^9 + 2y_P^9)a_5 \\
&+ (8x_P^6 y_P^2 - 6x_P^5 y_P^6 + 8x_P^5 y_P^2 - 8x_P^4 y_P^6 + 8x_P^4 y_P^2 \\
&\quad - 10x_P^3 y_P^6 + 8x_P^3 y_P^2 + 3x_P^2 y_P^8 + 4x_P y_P^8 + 2y_P^8)a_6 \\
&+ (5x_P^6 y_P^5 - 24x_P^6 y_P + 11x_P^5 y_P^5 - 24x_P^5 y_P - \tfrac{19}{2} x_P^4 y_P^9 + 17x_P^4 y_P^5 - 24x_P^4 y_P \\
&\quad - \tfrac{25}{2} x_P^3 y_P^9 + 23x_P^3 y_P^5 - 24x_P^3 y_P + 6x_P^2 y_P^7 + 4x_P y_P^7 + 3y_P^7)a_7 \\
&+ (10x_P^6 - \tfrac{143}{16} x_P^5 y_P^8 - \tfrac{13}{2} x_P^5 y_P^4 + 10x_P^5 - \tfrac{37}{4} x_P^4 y_P^8 - 9x_P^4 y_P^4 + 10x_P^4 \\
&\quad - \tfrac{143}{16} x_P^3 y_P^8 - \tfrac{23}{2} x_P^3 y_P^4 + 10x_P^3 + 3x_P^2 y_P^6 + 6x_P y_P^6 + 3y_P^6)a_8 \\
&= 0.
\end{aligned} \tag{5.6}$$

The equations for the points Q, R are the same, with $a - b$ and b in place of a, and again for Y but with a *and* b in place of a. We must also impose the conditions that the points are distinct and that $a_8 \neq 0$: for example, to say $P \neq Q$ we introduce the variable z_{PQ} and the equation

$$((x_P - x_Q)z_{PQ} - 1)((y_P - y_Q)z_{PQ} - 1) = 0. \tag{5.7}$$

In this general case, we end up with $8+7+2\cdot 4+10=33$ variables

$$a_1,\ldots,a_8,b_1,\ldots,b_7,x_{P_1},y_{P_1},x_{Q_1},y_{Q_1},x_{R_1},y_{R_1},x_{Y_1},y_{Y_1},z_{PQ},\ldots,z_{RD_0} \tag{5.8}$$

and $8\cdot 3+7+10=41$ equations.

Moving on from the general case, we consider also the case where x does not yield a uniformizer for one of the points; that one of the points lies along the line $z=0$; or that some of the points coincide. After this, we have completed the case $k=\ell=8$, and consider more degenerate cases (k,ℓ).

Finally, we repeat the entire process again with the partitions $\mu=2+1$ and $\mu=1+1+1$.

We conclude by a version of the second proof of our main result, explained in section 4.

Example 5.9. We compute each Belyĭ map of degree 7 and genus 3 and show that no source curve is isomorphic to X.

As above, there are three cases to consider. The first cyclic case is the map in Example 2.7 above, followed by its post-composition by automorphisms of $\mathbb{P}^1$ permuting $\{0,1,\infty\}$. But the curve $y^2-y=x^7$ has an automorphism of order 7, and X does not.

The genus 3 Belyĭ maps of degree 7 in the noncyclic case with 2 permutation triples up to conjugation was computed by Klug–Musty–Schiavone–Voight [**17**, Example 5.27]: using the algorithm in Lemma 4.1 we find that X is not isomorphic to either source curve. Alternatively, these two curves are minimally defined over $\mathbb{Q}(\sqrt{-7})$ (and are conjugate under $\mathrm{Gal}(\mathbb{Q}(\sqrt{-7})\,|\,\mathbb{Q})$), whereas X can be defined over $\mathbb{Q}$.

In the third case, we apply the same argument, appealing to the exhaustive computation of Belyĭ maps of small degree by Musty–Schiavone–Sijsling–Voight [**21**] and again checking for isomorphism.

References

[1] E. Arbarello, M. Cornalba, P. A. Griffiths, and J. Harris, *Geometry of algebraic curves. Vol. I*, Grundlehren der Mathematischen Wissenschaften [Fundamental Principles of Mathematical Sciences], vol. 267, Springer-Verlag, New York, 1985. MR770932

[2] Ingrid Bauer, Fabrizio Catanese, and Fritz Grunewald, *Faithful actions of the absolute Galois group on connected components of moduli spaces*, Invent. Math. **199** (2015), no. 3, 859–888, DOI 10.1007/s00222-014-0531-2. MR3314516

[3] Sybilla Beckmann, *Ramified primes in the field of moduli of branched coverings of curves*, J. Algebra **125** (1989), no. 1, 236–255, DOI 10.1016/0021-8693(89)90303-7. MR1012673

[4] G. V. Belyĭ, *Galois extensions of a maximal cyclotomic field* (Russian), Izv. Akad. Nauk SSSR Ser. Mat. **43** (1979), no. 2, 267–276, 479. MR534593

[5] G. V. Belyĭ, *A new proof of the three-point theorem* (Russian, with Russian summary), Mat. Sb. **193** (2002), no. 3, 21–24, DOI 10.1070/SM2002v193n03ABEH000633; English transl., Sb. Math. **193** (2002), no. 3-4, 329–332. MR1913596

[6] José Bertin and Matthieu Romagny, *Champs de Hurwitz* (French, with English and French summaries), Mém. Soc. Math. Fr. (N.S.) **125-126** (2011), 219. MR2920693

[7] Yuri F. Bilu and Marco Strambi, *Quantitative Riemann existence theorem over a number field*, Acta Arith. **145** (2010), no. 4, 319–339, DOI 10.4064/aa145-4-2. MR2738151

[8] Wieb Bosma, John Cannon, and Catherine Playoust, *The Magma algebra system. I. The user language*, J. Symbolic Comput. **24** (1997), no. 3-4, 235–265, DOI 10.1006/jsco.1996.0125. Computational algebra and number theory (London, 1993). MR1484478

[9] J. Coates, *Construction of rational functions on a curve*, Proc. Cambridge Philos. Soc. **68** (1970), 105–123. MR0258831

[10] P. Deligne and D. Mumford, *The irreducibility of the space of curves of given genus*, Inst. Hautes Études Sci. Publ. Math. **36** (1969), 75–109. MR0262240

[11] David Eisenbud, *Commutative algebra*, Graduate Texts in Mathematics, vol. 150, Springer-Verlag, New York, 1995. With a view toward algebraic geometry. MR1322960

[12] G. González-Diez and D. Torres-Teigell, *Non-homeomorphic Galois conjugate Beauville structures on* PSL$(2,p)$, Adv. Math. **229** (2012), no. 6, 3096–3122, DOI 10.1016/j.aim.2012.02.014. MR2900436

[13] F. Hess, *Computing Riemann-Roch spaces in algebraic function fields and related topics*, J. Symbolic Comput. **33** (2002), no. 4, 425–445, DOI 10.1006/jsco.2001.0513. MR1890579

[14] Ariyan Javanpeykar, *Polynomial bounds for Arakelov invariants of Belyi curves*, Algebra Number Theory **8** (2014), no. 1, 89–140, DOI 10.2140/ant.2014.8.89. With an appendix by Peter Bruin. MR3207580

[15] Ariyan Javanpeykar and Rafael von Känel, *Szpiro's small points conjecture for cyclic covers*, Doc. Math. **19** (2014), 1085–1103. MR3272921

[16] Lily S. Khadjavi, *An effective version of Belyi's theorem*, J. Number Theory **96** (2002), no. 1, 22–47. MR1931191

[17] Michael Klug, Michael Musty, Sam Schiavone, and John Voight, *Numerical calculation of three-point branched covers of the projective line*, LMS J. Comput. Math. **17** (2014), no. 1, 379–430, DOI 10.1112/S1461157014000084. MR3356040

[18] Heinrich-Wolfgang Leopoldt, *Über die Automorphismengruppe des Fermatkörpers* (German, with English summary), J. Number Theory **56** (1996), no. 2, 256–282, DOI 10.1006/jnth.1996.0017. MR1373551

[19] Răzvan Liţcanu, *Propriétés du degré des morphismes de Belyi* (French, with English summary), Monatsh. Math. **142** (2004), no. 4, 327–340, DOI 10.1007/s00605-003-0142-2. MR2085047

[20] Shinichi Mochizuki, *The geometry of the compactification of the Hurwitz scheme*, Publ. Res. Inst. Math. Sci. **31** (1995), no. 3, 355–441, DOI 10.2977/prims/1195164048. MR1355945

[21] Michael Musty, Sam Schiavone, Jeroen Sijsling, and John Voight, *A database of Belyi maps*, `arXiv:1805.07751`, 2018.

[22] L. X. Chau Ngo, K. A. Nguyen, M. van der Put, and J. Top, *Equivalence of differential equations of order one*, J. Symbolic Comput. **71** (2015), 47–59, DOI 10.1016/j.jsc.2014.09.041. MR3345314

[23] Matthieu Romagny and Stefan Wewers, *Hurwitz spaces* (English, with English and French summaries), Groupes de Galois arithmétiques et différentiels, Sémin. Congr., vol. 13, Soc. Math. France, Paris, 2006, pp. 313–341. MR2316356

[24] Leila Schneps (ed.), *The Grothendieck theory of dessins d'enfants*, London Mathematical Society Lecture Note Series, vol. 200, Cambridge University Press, Cambridge, 1994. Papers from the Conference on Dessins d'Enfant held in Luminy, April 19–24, 1993. MR1305390

[25] J. Sijsling and J. Voight, *On computing Belyi maps* (English, with English and French summaries), Numéro consacré au trimestre "Méthodes arithmétiques et applications", automne 2013, Publ. Math. Besançon Algèbre Théorie Nr., vol. 2014/1, Presses Univ. Franche-Comté, Besançon, 2014, pp. 73–131. MR3362631

[26] Bernd Sturmfels, *Algorithms in invariant theory*, Texts and Monographs in Symbolic Computation, Springer-Verlag, Vienna, 1993. MR1255980

[27] Tamás Szamuely, *Galois groups and fundamental groups*, Cambridge Studies in Advanced Mathematics, vol. 117, Cambridge University Press, Cambridge, 2009. MR2548205

[28] Pavlos Tzermias, *The group of automorphisms of the Fermat curve*, J. Number Theory **53** (1995), no. 1, 173–178, DOI 10.1006/jnth.1995.1085. MR1344839

[29] Leonardo Zapponi, *On the Belyi degree(s) of a curve defined over a number field*, `arXiv:0904.0967`, 2009.

MATHEMATICAL INSTITUTE, JOHANNES-GUTENBERG UNIVERSITY, MAINZ, GERMANY
Email address: `peykar@uni-mainz.de`

DEPARTMENT OF MATHEMATICS, DARTMOUTH COLLEGE, 6188 KEMENY HALL, HANOVER, NEW HAMPSHIRE 03755
Email address: `jvoight@gmail.com`

Contemporary Mathematics
Volume **722**, 2019
https://doi.org/10.1090/conm/722/14535

Weight enumerators of Reed-Muller codes from cubic curves and their duals

Nathan Kaplan

ABSTRACT. Let $\mathbb{F}_q$ be a finite field of characteristic not equal to 2 or 3. We compute the weight enumerators of some projective and affine Reed-Muller codes of order 3 over $\mathbb{F}_q$. These weight enumerators answer enumerative questions about plane cubic curves. We apply the MacWilliams theorem to give formulas for coefficients of the weight enumerator of the duals of these codes. We see how traces of Hecke operators acting on spaces of cusp forms for $\mathrm{SL}_2(\mathbb{Z})$ play a role in these formulas.

1. Introduction

Reed-Muller codes are some of the most famous and well studied examples of *evaluation codes*. Let $\mathcal{P} = \{P_1, \ldots, P_N\}$ be a subset of points of the affine space $\mathbb{F}_q^n$ and let $\mathcal{V}$ be a finite subspace of polynomials in $\mathbb{F}_q[x_1, \ldots, x_n]$. Consider the evaluation map:

$$\begin{array}{rcl} \mathrm{ev}_{\mathcal{P}} \colon \mathcal{V} & \to & \mathbb{F}_q^N \\ f & \to & (f(P_1), \ldots, f(P_N)) \,. \end{array}$$

The *evaluation code* $\mathrm{ev}(\mathcal{V}, \mathcal{P})$ is the image of this map. The *affine Reed-Muller Code* of order k and length q^n, denoted $\mathrm{RM}_q^A(k,n)$, comes from choosing $\mathcal{V}$ to be $\mathbb{F}_q[x_1, \ldots, x_n]_{\leq k}$, the vector space of polynomials of degree at most k, and $\mathcal{P}$ to be the set of all points in the affine space $\mathbb{F}_q^n$. When the evaluation map is injective, which is certainly the case for $k < q$, $\mathrm{RM}_q^A(k,n) \subset \mathbb{F}_q^{q^n}$ is a $\binom{n+k}{k}$-dimensional linear code.

Let $\mathcal{P} = \{P_1, \ldots, P_N\}$ be a subset of points in the projective space $\mathbb{P}^n(\mathbb{F}_q)$. It does not make sense to evaluate an element of $\mathbb{F}_q[x_0, \ldots, x_n]$ at a projective point, so we make a choice of affine representative for each, giving a set $\mathcal{P}' = \{P_1', \ldots, P_N'\}$ with each $P_i' \in \mathbb{F}_q^{n+1}$. Let $\mathcal{V}$ be a subspace of $\mathbb{F}_q[x_0, \ldots, x_n]_k$, the set of homogeneous polynomials of degree k (including the zero polynomial). We now define an evaluation code as in the previous paragraph. When $\mathcal{P}$ consists of all points in $\mathbb{P}^n(\mathbb{F}_q)$ and $\mathcal{V}$ is all of $\mathbb{F}_q[x_0, \ldots, x_n]_k$, this construction defines the *projective Reed-Muller Code* of order k and length $N = |\mathbb{P}^n(\mathbb{F}_q)| = (q^{n+1}-1)/(q-1)$, denoted $\mathrm{RM}_q^P(k,n)$. When the evaluation map is injective, which is certainly the

The author is supported by NSA Young Investigator Grant H98230-16-10305.

case when $k < q$, $\mathrm{RM}_q^P(k,n) \subseteq \mathbb{F}_q^N$, is a $\binom{n+k}{k}$-dimensional linear code. In this paper we write $C_{n,k}$ for $\mathrm{RM}_q^P(k,n)$ and $C_{n,k}^A$ for $\mathrm{RM}_q^A(k,n)$.

REMARK 1. (1) This definition of Reed-Muller codes depends on an ordering of the points, and in the projective case, on a choice of affine representatives. These Reed-Muller codes are not uniquely defined, but satisfy a strong form of equivalence [**14**, Section 1.7].

An $n \times n$ *monomial matrix* is a matrix of the form $M = DP$ where D is an $n \times n$ invertible diagonal matrix and P is a permutation matrix. A linear code C with generator matrix G is *monomially equivalent* to a linear code C' of the same length n if there exists an $n \times n$ monomial matrix M such that $G' = GM$ is a generator matrix for C'. Informally, two codes are monomially equivalent if you can get from one to the other by permuting coordinates and then scaling each coordinate by an element of $\mathbb{F}_q^*$. The Reed-Muller codes introduced above are not uniquely defined, but are well defined up to monomial equivalence.

(2) The definition of the projective Reed-Muller code given by Lachaud in [**19**] is phrased differently, but corresponds to making the standard choice where the affine representative $(x_0', \dots, x_n')$ for $[x_0 : \cdots : x_n]$ satisfies $x_i' = 1$ for the smallest i such that $x_i \neq 0$.

Let $C \subseteq \mathbb{F}_q^N$ be a code and choose a subset $S \subseteq \{1, 2, \dots, N\}$ of size m. The *punctured code* $C' \subseteq \mathbb{F}_q^{N-m}$ comes from taking each codeword of C and erasing the coordinates in positions in S. The affine Reed-Muller code $C_{n,k}^A$ is the projective Reed-Muller code $C_{n,k}$ punctured at the positions corresponding to the $\mathbb{F}_q$-points of any $\mathbb{F}_q$-rational hyperplane. Since $\mathrm{PGL}_{n+1}(\mathbb{F}_q)$ acts transitively on hyperplanes, the corresponding punctured codes are monomially equivalent.

Properties of $C_{n,k}$ and $C_{n,k}^A$ answer questions about the set of all degree k hypersurfaces in n-dimensional affine and projective space over $\mathbb{F}_q$.

DEFINITION 1. For elements of $\mathbb{F}_q^N$, $x = (x_1, \dots, x_N)$ and $y = (y_1, \dots, y_N)$, the *Hamming distance* of x and y is

$$d(x,y) = \#\{i \mid x_i \neq y_i\}.$$

The *Hamming weight* of x is its Hamming distance from the all zero vector,

$$\mathrm{wt}(x) = d(x, \mathbf{0}) = \#\{i \mid x_i \neq 0\}.$$

The Hamming weight enumerator of a code $C \subseteq \mathbb{F}_q^N$ is a homogeneous polynomial in two variables that keeps track of the number of codewords of C of each weight. More formally,

$$W_C(X,Y) = \sum_{c \in C} X^{N-\mathrm{wt}(c)} Y^{\mathrm{wt}(c)} = \sum_{i=0}^{N} A_i X^{N-i} Y^i,$$

where $A_i = \#\{c \in C \mid \mathrm{wt}(c) = i\}$.

We see that the following questions are equivalent.

QUESTION 1. Let $N = |\mathbb{P}^n(\mathbb{F}_q)| = (q^{n+1}-1)/(q-1)$. Suppose that $k < q$, so $C_{n,k} \subseteq \mathbb{F}_q^N$ is a $\binom{n+k}{k}$-dimensional linear code.

(1) How many homogeneous polynomials of degree k, $f \in \mathbb{F}_q[x_0, \dots, x_n]_k$ are such that the hypersurface $\{f = 0\} \subseteq \mathbb{P}^n$ has exactly $N - i$ $\mathbb{F}_q$-rational points?

(2) What is the weight i coefficient A_i of $W_{C_{n,k}}(X,Y)$?

We can ask analogous questions for the affine Reed-Muller codes.

Many authors have studied minimum distances and other invariants of affine and projective Reed-Muller codes. For example, see [**4**, **10**, **12**, **19**, **26**]. There are not many examples for which the weight enumerators of Reed-Muller codes have been computed explicitly. For $k = 1$ these codes come from hyperplanes and their weight enumerators are easy to compute. See [**16**] for more refined information on these codes. Aubry considers codes from projective quadric hypersurfaces in [**1**]. Elkies computes the weight enumerators of the duals of these codes in the unpublished preprint [**7**]. Elkies also computes the weight enumerator of the code of projective plane cubic curves $C_{2,3}$ and the weight enumerator of the code of cubic surfaces $C_{3,3}$. Knowledge of $W_{C_{2,3}}(X,Y)$ plays an important role in the analogous computation for $C_{3,3}$ since cones over plane cubic curves arise as singular cubic hypersurfaces in $\mathbb{P}^3$. For $n = 1$ and any k, affine Reed-Muller codes are Reed-Solomon codes and projective Reed-Muller codes are (doubly) extended, or projective, Reed-Solomon codes. In these cases the weight enumerators are well understood.

Results for small k lead to corresponding results for k large by considering dual codes.

DEFINITION 2. For $x, y \in \mathbb{F}_q^N$, $x = (x_1, \ldots, x_N)$ and $y = (y_1, \ldots, y_N)$, let

$$\langle x, y\rangle = \sum_{i=1}^{N} x_i y_i \in \mathbb{F}_q.$$

The *dual code* $C^\perp$ of a linear code $C \subset \mathbb{F}_q^N$ is defined by

$$\{y \in \mathbb{F}_q^N \mid \langle x, y\rangle = 0 \ \forall x \in C\}.$$

The dual of a Reed-Muller code is also a Reed-Muller code. For details, see [**21**, Theorem 6.11.3].

The MacWilliams theorem states that the weight enumerator of a linear code C determines the weight enumerator of $C^\perp$.

THEOREM 1 (MacWilliams). *Let $C \subseteq \mathbb{F}_q^N$ be a linear code. Then*

$$W_{C^\perp}(X,Y) = \frac{1}{|C|} W_C(X + (q-1)Y, X - Y).$$

For a proof, see for example [**21**, Theorem 3.5.3].

Theorem 1 implies that we can give an expression for $W_{C_{n,n-k-1}}(X,Y)$ in terms of $W_{C_{n,k}}(X,Y)$. However, this does not make it clear what sorts of inputs are necessary to give formulas for the coefficients of $W_{C_{n,n-k-1}}(X,Y)$. Applying the MacWilliams theorem in this way, Elkies gives closed formulas for the coefficients of $W_{C_{n,n-3}}(X,Y)$ [**7**].

We have two main goals in this paper:

(1) Compute $W_{C_{2,3}}(X,Y)$ following the strategy of Elkies [**7**], and use it to compute the significantly more complicated $W_{C_{2,3}^A}(X,Y)$.

(2) Apply the MacWilliams theorem to give formulas for low-weight coefficients of $W_{C_{2,3}^\perp}(X,Y)$ and $W_{(C_{2,3}^A)^\perp}(X,Y)$. Computing these expressions leads to evaluating sums that have been considered by Birch [**2**] and Ihara [**15**] and gives a connection between traces of Hecke operators acting on spaces of cusp forms for $\mathrm{SL}_2(\mathbb{Z})$ and coefficients of weight enumerators of Reed-Muller codes.

1.1. Comparison with previous work. The results of this paper are similar in spirit to those of the author's paper with Petrow [**17**]. In that paper, we study a refined weight enumerator of $C_{1,4}$, the *quadratic residue weight enumerator*, that keeps track of not just the number of coordinates of a codeword that are zero or nonzero, but also the number of nonzero coordinates that are squares in $\mathbb{F}_q^*$. This leads to statements about the number of homogeneous quartic polynomials $f \in \mathbb{F}_q[x,z]_4$ for which $y^2 = f(x,z)$ is an elliptic curve with a given number of $\mathbb{F}_q$-points and for which $f(x,z)$ has a specified number of $\mathbb{F}_q$-rational roots. Formulas for low-weight coefficients of the quadratic residue weight enumerator of $C_{1,4}^\perp$ involve traces of Hecke operators acting on spaces of cusp forms for the congruence subgroups $\mathrm{SL}_2(\mathbb{Z}), \Gamma_0(2)$, and $\Gamma_0(4)$. In this paper we see that elliptic curves with prescribed 3-torsion play a role in computing $W_{C_{2,3}^A}(X,Y)$. If we isolate the contribution to $W_{C_{2,3}^\perp}(X,Y)$ from elliptic curves with specified 3-torsion, we get formulas involving traces of Hecke operators acting on spaces of cusp forms for the congruence subgroups $\mathrm{SL}_2(\mathbb{Z}), \Gamma_0(3)$, and $\Gamma(3)$.

A smooth projective plane cubic curve C over $\mathbb{F}_q$ has genus 1 and every such curve has an $\mathbb{F}_q$-rational point, so C defines an elliptic curve. Several authors have studied families of codes coming from elliptic curves over finite fields where traces of Hecke operators play a role in formulas for coefficients of the weight enumerators. For example, weight enumerators of Zetterberg and Melas codes are studied in [**9, 25**], and the Eichler-Selberg trace formula for $\Gamma_1(4)$ plays an important role in the proofs. In [**8**], the authors study families of codes related to supersingular elliptic curves in characteristic 2 and to certain Reed-Muller codes. Other results of this type are described in the survey of Schoof [**24**].

In our study of affine plane cubic curves, we see that 3-torsion of elliptic curves over $\mathbb{F}_q$ plays an important role. We see formulas that are reminiscent of Schoof's formulas for the number of projective equivalence classes of plane cubic curves [**23**].

1.2. Outline of the paper. Our strategy is to separate the weight enumerators into the contribution from singular cubics and the contribution from smooth cubics. That is, we write

$$W_{C_{2,3}}(X,Y) = W_{C_{2,3}}^{\text{singular}}(X,Y) + W_{C_{2,3}}^{\text{smooth}}(X,Y).$$

Applying the MacWilliams theorem we see that for $q \geq 3$,

$$q^{10} W_{C_{2,3}^\perp}(X,Y) = W_{C_{2,3}}^{\text{singular}}(X+(q-1)Y, X-Y) + W_{C_{2,3}}^{\text{smooth}}(X+(q-1)Y, X-Y).$$

We compute the contribution to the weight enumerator of $C_{2,3}^\perp$ from singular cubics and from smooth cubics separately. We follow a similar strategy for affine cubics, but the details are more complicated.

In the next section we recall the rational point count distribution for singular plane cubic curves in $\mathbb{P}^2(\mathbb{F}_q)$. In Section 3, we recall some results about elliptic curves over finite fields and compute $W_{C_{2,3}}^{\text{smooth}}(X,Y)$. In Section 4, we combine formulas of Birch and Ihara with the MacWilliams theorem to give formulas for low-weight coefficients of $W_{C_{2,3}^\perp}(X,Y)$. We analyze singular affine cubics in Section 5, and smooth affine cubics in Section 6. This involves a discussion of rational inflection points on smooth cubic curves. In the final section, we discuss formulas for low-weight coefficients of $W_{(C_{2,3}^A)^\perp}(X,Y)$. We recall some generalizations of the results of Birch and Ihara due to the author and Petrow that allow us to understand the contribution to these weight enumerators from elliptic curves with prescribed 3-torsion.

REMARK 2. (1) We have done extensive calculations in the computer algebra system Sage verifying the results of this paper. It would not have been possible to write down the formulas we obtain here, see for example Theorem 3, without explicit computer verification.
(2) In Section 5 there are issues related to singular irreducible cubic curves that arise in the computation of $W_{C_{2,3}^{A}}(X,Y)$ that are different in characteristic 3. It is likely that with some additional effort, this case can be addressed. While some behavior for cubic curves in characteristic 2 is special [**13**, Chapter 11], our results seem to hold here. Since we have not verified our results computationally in this case, we assume that the characteristic of $\mathbb{F}_q$ is not 2 in the statement of our main theorems.

2. Singular projective plane cubic curves

There is a short list of isomorphism classes of singular projective plane cubic curves in $\mathbb{P}^2(\mathbb{F}_q)$. For an extensive discussion of the classification of such curves, including normal forms for projective equivalence classes, see [**13**, Chapter 11]. We recall the following chart from the paper of Elkies [**7**], which is related to [**13**, Theorem 11.3.11]

LEMMA 1. *Every homogeneous cubic $f \in \mathbb{F}_q[x_1,x_2,x_3]_3$ such that $\{f=0\}$ is not a smooth plane cubic is one of the* 15 *types listed in the following table together with the number of $f \in C_{2,3}$ of that type and the weight of every such f.*

Shape of $\{f=0\}$	*number of such* $f \in C_{2,3}$	$\mathrm{wt}(f)$
$\mathbb{P}^2(\mathbb{F}_q)$	1	0
triple line	q^3-1	q^2
line and double line	$(q^3-1)(q^2+q)$	q^2-q
3 *concurrent lines:*		
all rational	$(q^3-1)(q^3-q)/6$	q^2-2q
1 *rational,* 2 *conjugate by* $\mathrm{Gal}(\mathbb{F}_{q^2}/\mathbb{F}_q)$	$(q^3-1)(q^3-q)/2$	q^2
3 *conjugate by* $\mathrm{Gal}(\mathbb{F}_{q^3}/\mathbb{F}_q)$	$(q^3-1)(q^3-q)/3$	q^2+q
3 *non-concurrent lines:*		
all rational	$(q^3-1)(q^4+q^3)/6$	$(q-1)^2$
1 *rational,* 2 *conjugate by* $\mathrm{Gal}(\mathbb{F}_{q^2}/\mathbb{F}_q)$	$(q^3-1)(q^4-q^3)/2$	q^2-1
3 *conjugate by* $\mathrm{Gal}(\mathbb{F}_{q^3}/\mathbb{F}_q)$	$(q-1)^2(q^5-q^3)/3$	q^2+q+1
conic and tangent line	$(q^5-q^2)(q^2-1)$	q^2-q
conic and line meeting:		
in 2 *rational points*	$(q^6-q^3)(q^2-1)/2$	q^2-q+1
in 2 *conjugate points*	$(q^6-q^3)(q-1)^2/2$	q^2-q-1
cubic with a cusp	$(q^3-1)(q^3-q)q^2$	q^2
cubic with a node:		
with rational slopes	$(q^3-1)(q^3-q)(q^3-q^2)/2$	q^2+1
with slopes conjugate by $\mathrm{Gal}(\mathbb{F}_{q^2}/\mathbb{F}_q)$	$(q^3-1)(q^3-q)(q^3-q^2)/2$	q^2-1
TOTAL	$q^9+q^8-q^6-q^5+q^4$	

We can state this result in terms of the contribution to $W_{C_{2,3}}(X,Y)$ from homogeneous cubic polynomials f for which $\{f=0\}$ is not a smooth cubic curve.

LEMMA 2. *We have*

$$W^{sing}_{C_{2,3}}(X,Y) = X^{q^2+q+1} + \frac{(q^3-1)(q^3-q)}{6}X^{3q+1}Y^{q^2-2q}$$
$$+ \frac{(q^3-1)(q^4+q^3)}{6}X^{3q}Y^{q^2-2q+1} + \frac{(q^3-1)(q^3-q^2)(q^2-q)}{2}X^{2q+2}Y^{q^2-q-1}$$
$$+ (q^3-1)(q^2+q)(q^2-q+1)X^{2q+1}Y^{q^2-q} + \frac{(q^6-q^3)(q^2-1)}{2}X^{2q}Y^{q^2-q+1}$$
$$+ \frac{(q^3-1)(q^6-q^5)}{2}X^{q+2}Y^{q^2-1} + \frac{(q^3-1)(2q^5-q^3-q+2)}{2}X^{q+1}Y^{q^2}$$
$$+ \frac{(q^3-1)(q^3-q)(q^3-q^2)}{2}X^{q}Y^{q^2+1} + \frac{(q^3-1)(q^3-q)}{3}XY^{q^2+q}$$
$$+ \frac{(q-1)(q^3-q)(q^3-q^2)}{3}Y^{q^2+q+1}.$$

The coefficients of $W^{\text{sing}}_{C_{2,3}}(X,Y)$ are polynomials in q, so for each j, the contribution to the $X^{q^2+q+1-j}Y^j$ coefficient of $W_{C_{2,3}^{\perp}}(X,Y)$ from these singular cubics is a polynomial in q. For example, the $X^{q^2+q}Y$ coefficient of $W^{\text{sing}}_{C_{2,3}}(X+(q-1)Y, X-Y)$ is $(q^3-1)(q^3-q)(q^4-q^3)$.

3. Smooth projective plane cubic curves

In order to determine $W^{\text{smooth}}_{C_{2,3}}(X,Y)$ we answer two questions.

QUESTION 2. (1) Given an isomorphism class of an elliptic curve E defined over $\mathbb{F}_q$, how many elements of $C_{2,3}$ define a smooth plane cubic isomorphic to E? (2) How many isomorphism classes of an elliptic curve $E/\mathbb{F}_q$ have $\#E(\mathbb{F}_q)=q+1-t$?

The first of these questions is answered by Elkies in [**7**].

LEMMA 3. *For every elliptic curve $E/\mathbb{F}_q$, the number of polynomials $f \in \mathbb{F}_q[x,y,z]_3$ such that the zero-locus $\{f=0\}$ is isomorphic to E equals $\#\mathrm{GL}_3(\mathbb{F}_q)/\#\mathrm{Aut}_{\mathbb{F}_q}(E)$.*

We give the proof from [**7**] for completeness, including some of the presentation in [**23**, Section 5]. This argument will play a role in our analysis of smooth affine cubics.

PROOF. The number of polynomials $f \in \mathbb{F}_q[x,y,z]_3$ such that $\{f=0\} \cong E$ is $q-1$ times the number of smooth projective plane cubic curves C with $C \cong E$.

Let $i\colon E \hookrightarrow \mathbb{P}^2$ be a closed immersion defined over $\mathbb{F}_q$ with image $i(E) = C$. There is a one-to-one correspondence between $E(\mathbb{F}_q)$ and $C(\mathbb{F}_q)$. The sheaf $i^*\mathcal{O}(1)$ is a very ample invertible sheaf $\mathcal{L}(D)$. We have that D is a divisor of degree 3 defined over $\mathbb{F}_q$, and only its class is determined by i.

An isomorphism between C and E requires a choice of a point $P_0 \in C(\mathbb{F}_q)$ that is sent to the identity O of the group law on E. There are $\#E(\mathbb{F}_q) = q+1-t$ choices for P_0.

We claim that the number of pairs (C, P_0) where C is a smooth projective plane cubic defined over $\mathbb{F}_q$ with $C \cong E$ and $P_0 \in C(\mathbb{F}_q)$ is $(q+1-t)\#\mathrm{PGL}_3(\mathbb{F}_q)/\#\mathrm{Aut}_{\mathbb{F}_q}(E)$. Proving this claim completes the proof of the lemma. A plane cubic curve C gives not only a genus one curve E, but a degree 3 divisor class on E and a choice of basis for the global sections of that divisor class (x,y,z) up to $\mathbb{F}_q^*$ scaling.

Starting from E, there are $\#E(\mathbb{F}_q) = q+1-t$ divisor classes of E of degree 3 defined over $\mathbb{F}_q$. We claim that for each one, we get $\#\mathrm{PGL}_3(\mathbb{F}_q)/\#\mathrm{Aut}_{\mathbb{F}_q}(E)$ embeddings $i\colon E \hookrightarrow \mathbb{P}^2$ defined over $\mathbb{F}_q$. This will complete the proof.

By Riemann-Roch, a degree 3 divisor has a 3-dimensional space of global sections. Choosing a basis for the space of sections gives a closed immersion $i\colon E \hookrightarrow \mathbb{P}^2$ defined over $\mathbb{F}_q$. Two choices yield the same embedding, including the same image P_0 of the identity element $O \in E(\mathbb{F}_q)$, if and only if they are related by an automorphism of E and an $\mathbb{F}_q^*$ scaling. □

We turn to the second part of Question 2, following the presentation in [**18**]. Let E be an elliptic curve defined over $\mathbb{F}_q$. When we mention an elliptic curve E we always implicitly mean the isomorphism class of E. With this convention in mind, let $\mathcal{C} = \{E/\mathbb{F}_q\}$, the set of $\mathbb{F}_q$-isomorphism classes of elliptic curves defined over $\mathbb{F}_q$. We have

$$\sum_{E\in\mathcal{C}} \frac{1}{\#\mathrm{Aut}_{\mathbb{F}_q}(E)} = q,$$

so the finite set $\mathcal{C}$ is a probability space where a singleton $\{E\}$ occurs with probability

$$\mathbb{P}_q(\{E\}) = \frac{1}{q\#\mathrm{Aut}_{\mathbb{F}_q}(E)}.$$

Let $t_E \in \mathbb{Z}$ denote the trace of the Frobenius endomorphism associated to E. We have $t_E = q+1-\#E(\mathbb{F}_q)$ and by Hasse's theorem $t_E^2 \leq 4q$. For an integer t, let $\mathcal{C}(t)$ be the subset of $\mathcal{C}$ for which $t_E = t$. Using this terminology, Lemma 3 gives an expression for $W^{\mathrm{smooth}}_{C_{2,3}}(X,Y)$.

PROPOSITION 1. *Let $q \geq 3$. Then*

$$W^{smooth}_{C_{2,3}}(X,Y) = (q^3-1)(q^3-q)(q^3-q^2)q \sum_{t^2\leq 4q} \mathbb{P}_q(\mathcal{C}(t))X^{q+1-t}Y^{q^2+t}.$$

We recall results due to Deuring [**5**], Waterhouse [**27**], and Schoof [**23**], that express $\mathbb{P}_q(\mathcal{C}(t))$ in terms of class numbers of orders in imaginary quadratic fields. For $d < 0$ with $d \equiv 0,1 \pmod 4$, let $h(d)$ be the class number of the unique quadratic order of discriminant d. Let

$$h_w(d) \overset{\mathrm{def}}{=} \begin{cases} h(d)/3, & \text{if } d=-3,\\ h(d)/2, & \text{if } d=-4,\\ h(d) & \text{if } d<0,\ d\equiv 0,1 \pmod 4, \text{ and } d\neq -3,-4,\\ 0 & \text{otherwise,}\end{cases} \tag{1}$$

and for $\Delta \equiv 0,1 \pmod 4$ let

$$H(\Delta) \overset{\mathrm{def}}{=} \sum_{d^2\mid\Delta} h_w\left(\frac{\Delta}{d^2}\right) \tag{2}$$

be the *Hurwitz-Kronecker class number*. For $a \in \mathbb{Z}$ and n a positive integer, the *Kronecker symbol* $\left(\frac{a}{n}\right)$ is defined to be the completely multiplicative function of n

such that if p is an odd prime $\left(\frac{a}{p}\right)$ is the quadratic residue symbol, and if $p = 2$,

$$\left(\frac{a}{2}\right) \stackrel{\text{def}}{=} \begin{cases} 0 & \text{if } 2 \mid a, \\ 1 & \text{if } a \equiv \pm 1 \pmod 8, \\ -1 & \text{if } a \equiv \pm 5 \pmod 8. \end{cases} \tag{3}$$

The following is a weighted version of [**23**, Theorem 4.6].

LEMMA 4. *Let $t \in \mathbb{Z}$. Suppose $q = p^v$ where p is prime and $v \geq 1$. Then if q is not a square*

$$\begin{aligned} \mathbb{P}_q(\mathcal{C}(t)) =& \frac{1}{2q} H(t^2 - 4q) && \textit{if } t^2 < 4q \textit{ and } p \nmid t, \\ =& \frac{1}{2q} H(-4p) && \textit{if } t = 0, \\ =& \frac{1}{4q} && \textit{if } t^2 = 2q \textit{ and } p = 2, \\ =& \frac{1}{6q} && \textit{if } t^2 = 3q \textit{ and } p = 3, \end{aligned}$$

and if q is a square

$$\begin{aligned} \mathbb{P}_q(\mathcal{C}(t)) =& \frac{1}{2q} H(t^2 - 4q) && \textit{if } t^2 < 4q \textit{ and } p \nmid t, \\ =& \frac{1}{4q}\left(1 - \left(\frac{-4}{p}\right)\right) && \textit{if } t = 0, \\ =& \frac{1}{6q}\left(1 - \left(\frac{-3}{p}\right)\right) && \textit{if } t^2 = q, \\ =& \frac{p-1}{24q} && \textit{if } t^2 = 4q, \end{aligned}$$

and $\mathbb{P}_q(\mathcal{C}(t)) = 0$ in all other cases.

Combining Proposition 1 and Lemma 4 gives an expression for $W^{\text{smooth}}_{C_{2,3}}(X, Y)$ in terms of class numbers of orders in imaginary quadratic fields.

We close this section with a result about $\mathbb{F}_q$-rational inflection lines of smooth plane projective cubic curves that we apply in Section 6 when we compute $W^{\text{smooth}}_{C^A_{2,3}}(X, Y)$.

DEFINITION 3. A non-singular point P of an absolutely irreducible cubic curve C is an *inflection point* if the tangent line L at P has contact order 3 with C. In this case L is called an *inflection line*. In particular, L does not intersect any other points of C.

Let $\mathbb{I}(C)$ denote the number of $\mathbb{F}_q$-rational inflection lines of C. For an extensive discussion the possibilities for $\mathbb{I}(C)$, see [**13**, Chapter 11].

PROPOSITION 2. *Let E be an elliptic curve defined over $\mathbb{F}_q$.*

(1) *If $E(\mathbb{F}_q)[3]$ is trivial, then every smooth projective plane cubic C with $C \cong E$ has $\mathbb{I}(C) = 1$.*
(2) *If $E(\mathbb{F}_q)[3] \cong \mathbb{Z}/3\mathbb{Z}$, then $1/3$ of the smooth projective plane cubics C with $C \cong E$ have $\mathbb{I}(C) = 3$, and the remaining $2/3$ have $\mathbb{I}(C) = 0$.*

(3) *If* $E(\mathbb{F}_q)[3] \cong \mathbb{Z}/3\mathbb{Z} \times \mathbb{Z}/3\mathbb{Z}$, *then* $1/9$ *of the smooth projective plane cubics* C *with* $C \cong E$ *have* $\mathbb{I}(C) = 9$, *and the remaining* $8/9$ *have* $\mathbb{I}(C) = 0$.

We recall some material from [**3**, Section 4], which cites [**20**], to give a description of the inflection points in terms of the geometry of C. Let $P * Q$ denote the third intersection of the line PQ with C. When $Q = P$, the line PQ is the tangent at P and $P * P = P_t$ is the *tangential* of P. If P is an inflection point, $P_t = P$. Choose a point $O \in C(\mathbb{F}_q)$ to be the identity of the group law on $C(\mathbb{F}_q)$ and recall that the group operation is defined by

$$P \oplus Q = (P * Q) * O.$$

Let N denote O_t, the tangential of O. Then P is an inflection point if and only if $3P = N$.

We give a proof of Proposition 2 similar to the proof of [**17**, Lemma 3].

PROOF. Recall from the proof of Lemma 3 that C defines not only an elliptic curve E, but also a degree 3 divisor of E defined over $\mathbb{F}_q$. There is a one-to-one correspondence between $C(\mathbb{F}_q)$ and $E(\mathbb{F}_q)$. As we vary over all cubics C with $C \cong E$ we get each degree 3 divisor class of E defined over $\mathbb{F}_q$ the same number of times.

This divisor is linearly equivalent to a unique one of the form $2O + P$ where O is the identity element of the group law on E and $P \in E(\mathbb{F}_q)$. A point $Q \in E(\overline{\mathbb{F}}_q)$ gives an inflection point of the cubic if and only if $3Q \sim 2O + P$, or equivalently, $3Q = P$ in the group law on E.

We vary over all choices of P and consider how many Q occur as points with $3Q = P$.

(1) If $\#E(\mathbb{F}_q) \not\equiv 0 \pmod 3$, then the map $Q \to 3Q$ is an isomorphism and every P gives exactly one such Q.

(2) If $\#E(\mathbb{F}_q) \equiv 0 \pmod 3$ then there are two possibilities for the group structure of $E(\mathbb{F}_q)[3]$.

(a) If $E(\mathbb{F}_q)[3] \cong \mathbb{Z}/3\mathbb{Z}$, then $2/3$ of the points of $E(\mathbb{F}_q)$ have 0 preimages under the map $Q \to 3Q$, and $1/3$ have exactly 3.

(b) If $E(\mathbb{F}_q)[3] \cong \mathbb{Z}/3\mathbb{Z} \times \mathbb{Z}/3\mathbb{Z}$ then $8/9$ of the points of $E(\mathbb{F}_q)$ have 0 preimages under the map $Q \to 3Q$, and $1/9$ have exactly 9.

□

4. Low-weight coefficients of $W_{C_{2,3}^{\perp}}(X, Y)$

In this section we give formulas for low-weight coefficients of the weight enumerator of $C_{2,3}^{\perp}$. By the MacWilliams theorem, we have

$$q^{10} W_{C_{2,3}^{\perp}}(X, Y) = W_{C_{2,3}}^{\text{singular}}(X + (q-1), X - Y) + W_{C_{2,3}}^{\text{smooth}}(X + (q-1)Y, X - Y).$$

Since the coefficients of $W_{C_{2,3}}^{\text{singular}}(X, Y)$ are polynomials in q, the $X^{q^2+q+1-j}Y^j$ coefficient of $W_{C_{2,3}}^{\text{singular}}(X + (q-1)Y, X - Y)$ is a polynomial in q for any j. For the rest of this section we focus on $W_{C_{2,3}}^{\text{smooth}}(X + (q-1)Y, X - Y)$.

Applying the binomial theorem, we see that the $X^{q^2+q+1-j}Y^j$ coefficient of

$$W_{C_{2,3}}^{\text{smooth}}(X + (q-1)Y, X - Y) =$$
$$(q^3 - 1)(q^3 - q)(q^3 - q^2)q \sum_{t^2 \le 4q} \mathbb{P}_q(\mathcal{C}(t))(X + (q-1)Y)^{q+1-t}(X - Y)^{q^2+t}$$

is a linear combination of expressions

$$\sum_{t^2\leq 4q} \mathbb{P}_q(\mathcal{C}(t))t^k$$

for $k \in \{0, 1, \ldots, j\}$, with coefficients that are polynomials in q. Since $\mathbb{P}_q(\mathcal{C}(t)) = \mathbb{P}_q(\mathcal{C}(-t))$ for any t, we see that when k is odd

$$\sum_{t^2\leq 4q} \mathbb{P}_q(\mathcal{C}(t))t^k = 0.$$

Birch gives formulas for these types of expressions [**2**, equation (4)]. For a nonnegative integer R, let

$$\mathbb{E}_q(t_E^{2R}) = \frac{1}{q}\sum_{E\in\mathcal{C}} \frac{t_E^{2R}}{\#\mathrm{Aut}_{\mathbb{F}_q}(E)} = \sum_{t^2\leq 4q} \mathbb{P}_q(\mathcal{C}(t))t^{2R}.$$

THEOREM 2 (Birch). *For prime $p \geq 5$ we have*

$$\begin{aligned}
p\mathbb{E}_p(1) =& p\\
p\mathbb{E}_p(t_E^2) =& p^2 - 1\\
p\mathbb{E}_p(t_E^4) =& 2p^3 - 3p - 1\\
p\mathbb{E}_p(t_E^6) =& 5p^4 - 9p^2 - 5p - 1\\
p\mathbb{E}_p(t_E^8) =& 14p^5 - 28p^3 - 20p^2 - 7p - 1\\
p\mathbb{E}_p(t_E^{10}) =& 42p^6 - 90p^4 - 75p^3 - 35p^2 - 9p - 1 - \tau(p),
\end{aligned}$$

where $\tau(p)$ is Ramanujan's τ-function.

Birch only gives such formulas for prime fields, but the extension to all finite fields is now well known and is implicit in work of Ihara [**15**]. In the case where $q = p^v$ for $v \geq 2$, the $q\mathbb{E}_q(t_E^{10})$ formula also involves $\tau(q/p^2)$. Formulas for $\mathbb{E}_q(t_E^{2R})$ involve traces of the Hecke operators T_q and T_{q/p^2} acting on spaces of cusp forms of weight at most $2R+2$ for $\mathrm{SL}_2(\mathbb{Z})$. See [**18**, Theorem 2] for a precise statement.

Applying these computations to the study of $W_{C_{2,3}^\perp}(X,Y)$ gives the following formulas.

THEOREM 3. *Let $\mathbb{F}_q$ be a finite field of characteristic not equal to 2 or 3. We have*

$$\begin{aligned}
W_{C_{2,3}^\perp}(X,Y) = X^{q^2+q+1} + q(q+1)(q-1)^2(q-2)(q^2+q+1)\Bigg[\frac{1}{5!}(q-3)X^{q^2+q-4}Y^5\\
+\frac{1}{6!}(q-5)(q-4)(q-3)X^{q^2+q-5}Y^6 + \frac{1}{7!}(q-5)(q-4)(q-3)(q^2-6q+15)X^{q^2+q-6}Y^7\\
+\frac{1}{8!}(q-3)(2q^6-3q^5+79q^4-797q^3+2829q^2-5110q+4200)X^{q^2+q-7}Y^8\\
+\frac{1}{9!}\Big(q^{10}+3q^9-16q^8-585q^7+4262q^6-7310q^5-24393q^4+138512q^3\\
-293174q^2+333900q-176400\Big)X^{q^2+q-8}Y^9\Bigg] + O(Y^{10}).
\end{aligned}$$

When $p \geq 5$ is prime, the $X^{p^2+p-9}Y^{10}$ coefficient of $W_{C_{2,3}^{\perp}}(X,Y)$ is

$$\frac{p(p+1)(p-1)^2(p^2+p+1)}{10!}\Bigg(\Big(p^{14}-43p^{12}+117p^{11}-2327p^{10}+40444p^9-287841p^8+1088452p^7 -2263884p^6+1782811p^5+3312614p^4-12006000p^3+17345160p^2-13807584p+5080320\Big) -(p-1)p^2\tau(p)\Bigg).$$

REMARK 3. When q is a prime power the $X^{q^2+q-9}Y^{10}$ coefficient of $W_{C_{2,3}^{\perp}}(X,Y)$ also contains a term involving $\tau(q/p^2)$. In general, the formula for the $X^{q^2+q+1-j}Y^j$ coefficient of $W_{C_{2,3}^{\perp}}(X,Y)$ involves traces of the Hecke operators T_q and T_{q/p^2} acting on spaces of cusp forms for $\mathrm{SL}_2(\mathbb{Z})$ of weight at most $j+2$.

The coefficients computed in Theorem 3 count something. A weight 5 codeword of $C_{2,3}^{\perp}$ corresponds to a tuple of points $p_1,\ldots,p_5 \in \mathbb{P}^2(\mathbb{F}_q)$ along with elements $a_1,\ldots,a_5 \in \mathbb{F}_q^*$ such that

$$a_1f(p_1') + a_2f(p_2') + a_3f(p_3') + a_4f(p_4') + a_5f(p_5') = 0$$

for all homogeneous cubics $f \in \mathbb{F}_q[x,y,z]_3$, where p_i' denotes an affine representative of p_i. It is not difficult to show that if $p_1,\ldots,p_5$ are not collinear, there is a cubic vanishing at 4 of these points, but not all 5. In fact, the weight 5 coefficient is exactly $q-1$ times the number of sets of 5 collinear points in $\mathbb{P}^2(\mathbb{F}_q)$. The points corresponding to a codeword of weight 8 may be either collinear or lie on a smooth conic, and for larger weights more types of point configurations are possible. In general, low-weight dual coefficients of $C_{n,k}^{\perp}$ count collections of points in $\mathbb{P}^n(\mathbb{F}_q)$ that fail to impose independent conditions on degree k hypersurfaces. The study of these collections is the subject of *Interpolation Problems in Algebraic Geometry*. For more information, see [**6, 11**].

5. Singular affine plane cubic curves

In this section we adapt our computation of $W_{C_{2,3}}^{\text{singular}}(X,Y)$ to compute $W_{C_{2,3}^A}^{\text{singular}}(X,Y)$. We divide the singular affine cubics into two groups: those that are absolutely irreducible and those that are not. We divide the set of reducible cubics into two further groups: those that contain an $\mathbb{F}_q$-rational affine line and those that do not.

5.1. Absolutely irreducible singular affine plane cubic curves. We give a result parallel to Proposition 2 for rational inflection lines of singular absolutely irreducible projective cubics. As detailed in Lemma 1 there are three isomorphism classes of such cubics:

(1) Cuspidal cubics, which have $q+1$ $\mathbb{F}_q$-points;
(2) Split nodal cubics, which have q $\mathbb{F}_q$-points;
(3) Non-split nodal cubics, which have $q+2$ $\mathbb{F}_q$-points.

As in Section 3, we write $\mathbb{I}(C)$ for the number of $\mathbb{F}_q$-rational inflection lines at nonsingular points of C. In each case, the set of *nonsingular* $\mathbb{F}_q$-rational points of C forms a finite abelian group and $\mathbb{I}(C)$ depends on the structure of this group. We assume that the characteristic of $\mathbb{F}_q$ is not 3, because in characteristic 3 all $\mathbb{F}_q$-points of the cuspidal cubic $\{zy^2 - x^3 = 0\}$ are inflection points. Outside of this exceptional case, every absolutely irreducible singular plane cubic curve has exactly 3 collinear inflection points over $\overline{\mathbb{F}}_q$ [**3**, Theorem 5.1]. When the group of nonsingular points has order divisible by 3, there are two possibilities for $\mathbb{I}(C)$. When this group has order not divisible by 3, $\mathbb{I}(C) = 1$. We summarize these statements below; see [**3**] and [**13**, Chapter 11] for further discussion.

PROPOSITION 3. *Suppose $\mathbb{F}_q$ is a finite field of characteristic not equal to 3.*

(1) *If $q \equiv 1 \pmod 3$ every cuspidal cubic and every non-split nodal cubic has $\mathbb{I}(C) = 1$. Of the $(q^3-1)(q^3-q)(q^3-q^2)/2$ homogeneous cubic polynomials defining split nodal cubics, $2/3$ have $\mathbb{I}(C) = 0$ and $1/3$ have $\mathbb{I}(C) = 3$.*

(2) *If $q \equiv 2 \pmod 3$ every cuspidal cubic and every split nodal cubic has $\mathbb{I}(C) = 1$. Of the $(q^3-1)(q^3-q)(q^3-q^2)/2$ homogeneous cubic polynomials defining non-split nodal cubics, $2/3$ have $\mathbb{I}(C) = 0$ and $1/3$ have $\mathbb{I}(C) = 3$.*

Let C be a projective plane cubic curve. We get an affine plane cubic by considering $C \setminus \{z = 0\}$, the part of C away from the *line at infinity.* The number of $\mathbb{F}_q$-points on this affine curve is

$$\#C(\mathbb{F}_q) - \#(C \cap \{z = 0\})(\mathbb{F}_q).$$

Our strategy for computing the contribution to $W_{C_{2,3}^A}(X, Y)$ from a particular type of a cubic C will be to determine the distribution of rational point counts for $C \cap L$ as we vary over all $\mathbb{F}_q$-rational lines L. Since $\mathrm{PGL}_3(\mathbb{F}_q)$ acts transitively on lines, we can determine the contribution from curves isomorphic to C as an average involving these point counts.

Let C be a singular absolutely irreducible projective plane cubic curve with $\#C(\mathbb{F}_q) = q + 1 - t$. Let $\mathbb{L}_i(C)$ be the number of lines in $\mathbb{P}^2(\mathbb{F}_q)$ such that $\#(C \cap L)(\mathbb{F}_q) = i$. We have

$$\mathbb{L}_0(C) + \mathbb{L}_1(C) + \mathbb{L}_2(C) + \mathbb{L}_3(C) = q^2 + q + 1. \tag{4}$$

Every $\mathbb{F}_q$-rational line L through the singular point of C intersects C in at most one other $\mathbb{F}_q$-point. Exactly $q - t$ of these lines have $\#(C \cap L)(\mathbb{F}_q) = 2$. The remaining $1+t$ of these lines have $\#(C \cap L)(\mathbb{F}_q) = 1$. There are $q-t$ tangent lines at nonsingular points in $C(\mathbb{F}_q)$. The $\mathbb{I}(C)$ $\mathbb{F}_q$-rational inflection lines have $\#(C \cap L)(\mathbb{F}_q) = 1$, and the remaining $q - t - \mathbb{I}(C)$ tangent lines have $\#(C \cap L)(\mathbb{F}_q) = 2$. Therefore,

$$\mathbb{L}_2(C) = 2q - 2t - \mathbb{I}(C). \tag{5}$$

Every other line passing through 2 distinct points of $C(\mathbb{F}_q)$ also passes through a third. Therefore,

$$\mathbb{L}_3(C) = \frac{\binom{q-t}{2} - (q - t - \mathbb{I}(C))}{3}. \tag{6}$$

Since there are $q + 1$ $\mathbb{F}_q$-rational lines passing through each point of $C(\mathbb{F}_q)$, we see that

$$\mathbb{L}_1(C) + 2\mathbb{L}_2(C) + 3\mathbb{L}_3(C) = (q + 1)(q + 1 - t). \tag{7}$$

Equations (4), (5), (6), and (7) determine $\mathbb{L}_0(C), \mathbb{L}_1(C), \mathbb{L}_2(C)$, and $\mathbb{L}_3(C)$.

PROPOSITION 4. *Let $\mathbb{F}_q$ be a finite field of characteristic not equal to 3. Let $W_{C_{2,3}^A}^{\text{sing. irred.}}(X, Y)$ denote the contribution to $W_{C_{2,3}^A}(X, Y)$ from singular affine cubics irreducible over $\overline{\mathbb{F}}_q$.*

Let

$$W^{cusp}_{C^A_{2,3}}(X,Y) = (q-1)(q^3-q)q^2\Bigg(\frac{(q+1)(q-1)}{3}X^{q+1}Y^{q^2-q-1} + \frac{q^2-q+4}{2}X^qY^{q^2-q} + (2q-1)X^{q-1}Y^{q^2-q+1} + \frac{(q-1)(q-2)}{6}X^{q-2}Y^{q^2-q+2}\Bigg),$$

$$W^{split}_{C^A_{2,3}}(X,Y) = \frac{(q-1)(q^3-q)(q^3-q^2)}{2}\Bigg(\frac{q(q+1)}{3}X^qY^{q^2-q} + \frac{q^2-q+6}{2}X^{q-1}Y^{q^2-q+1} + (2q-3)X^{q-2}Y^{q^2-q+2} + \frac{(q-2)(q-3)}{6}X^{q-3}Y^{q^2-q+3}\Bigg),$$

and let

$$W^{non\text{-}split}_{C^A_{2,3}}(X,Y) = \frac{(q-1)(q^3-q)(q^3-q^2)}{2}\Bigg(\frac{q(q-1)}{3}X^{q+2}Y^{q^2-q-2} + \frac{q(q-1)}{2}X^{q+1}Y^{q^2-q-1} + (2q+1)X^qY^{q^2-q} + \frac{q(q-1)}{6}X^{q-1}Y^{q^2-q+1}\Bigg).$$

We have

$$W^{sing.\ irred.}_{C^A_{2,3}}(X,Y) = W^{cusp}_{C^A_{2,3}}(X,Y) + W^{split}_{C^A_{2,3}}(X,Y) + W^{non\text{-}split}_{C^A_{2,3}}(X,Y).$$

PROOF. Proposition 3 and the discussion of the $\mathbb{L}_i(C)$ following it lead to the formula for $W^{\text{cusp}}_{C^A_{2,3}}(X,Y)$ for all $q \not\equiv 0 \pmod 3$. This also gives the formula for $W^{\text{split}}_{C^A_{2,3}}(X,Y)$ when $q \equiv 2 \pmod 3$. To see that the same formula holds when $q \equiv 1 \pmod 3$, simplify the expression

$$W^{\text{split}}_{C^A_{2,3}}(X,Y) = \frac{(q-1)(q^3-q)(q^3-q^2)}{2}\Bigg(\left(\frac{2}{3}\cdot\frac{q^2+q+1}{3} + \frac{1}{3}\cdot\frac{(q+2)(q-1)}{3}\right)X^qY^{q^2-q} + \left(\frac{2}{3}\cdot\frac{q^2-q+4}{2} + \frac{1}{3}\cdot\frac{q^2-q+10}{2}\right)X^{q-1}Y^{q^2-q+1} + \left(\frac{2}{3}\cdot(2q-2) + \frac{1}{3}\cdot(2q-5)\right)X^{q-2}Y^{q^2-q+2} + \left(\frac{2}{3}\cdot\frac{(q-1)(q-4)}{6} + \frac{1}{3}\cdot\frac{q^2-5q+10}{6}\right)X^{q-3}Y^{q^2-q+3}\Bigg).$$

A similar simplification shows that the expression for $W^{\text{non-split}}_{C^A_{2,3}}(X,Y)$ holds when $q \equiv 1 \pmod 3$ and when $q \equiv 2 \pmod 3$. □

5.2. Reducible affine cubic curves. We divide the affine cubics reducible over $\overline{\mathbb{F}}_q$ into two groups: cubics that contain an $\mathbb{F}_q$-rational affine line and those that do not. We begin with the second type. There is a short list of such cubics.

(1) **Three lines conjugate by** $\text{Gal}(\mathbb{F}_{q^3}/\mathbb{F}_q)$: These three lines can either be concurrent or not. In the concurrent case, the $\mathbb{F}_q$-rational intersection point of these lines can either be in the affine plane, or on the line at infinity. The contribution to $W_{C^A_{2,3}}(X,Y)$ from such cubics is

$$\frac{(q-1)q^2(q^3-q)}{3}XY^{q^2-1} + \frac{(q-1)(q+1)(q^3-q)}{3}Y^{q^2} + \frac{(q-1)^2(q^5-q^3)}{3}Y^{q^2}.$$

(2) **Triple line at infinity**: Such cubics contribute $(q-1)Y^{q^2}$ to $W_{C^A_{2,3}}(X,Y)$.

(3) **Line at infinity together with a smooth conic**: The contribution to $W_{C_{2,3}^A}(X,Y)$ from such cubics is

$$(q-1)^2(q+1)q^2X^qY^{q^2-q}+\frac{(q-1)^2q^3(q+1)}{2}X^{q-1}Y^{q^2-q+1}+\frac{(q-1)^3q^3}{2}X^{q+1}Y^{q^2-q-1}.$$

(4) **Line at infinity together with two lines conjugate by** $\mathrm{Gal}(\mathbb{F}_{q^2}/\mathbb{F}_q)$: The $\mathbb{F}_q$-rational intersection point of these Galois-conjugate lines can either lie in the affine plane or on the line at infinity. The contribution to $W_{C_{2,3}^A}(X,Y)$ from such cubics is

$$(q-1)q^2\frac{q^2-q}{2}XY^{q^2-1}+(q-1)(q+1)\frac{q^2-q}{2}Y^{q^2}.$$

We now analyze affine plane cubics that contain an $\mathbb{F}_q$-rational line. We begin with the contribution to the weight enumerator from cubics that contain the particular line $\{x=0\}$. The contribution to $W_{C_{2,3}^A}(X,Y)$ from such cubics is

$$\begin{aligned}&(q-1)\frac{2q^3-q^2-q+6}{2}X^qY^{q^2-q}+\frac{q^2(q-1)^3}{2}X^{q+1}Y^{q^2-q-1}\\&+\frac{(q-2)q^2(q-1)^3}{4}X^{2q-3}Y^{q^2-2q+3}+2q^2(q-1)^3X^{2q-2}Y^{q^2-2q+2}\\&+\frac{(q-1)q^3(q^2-2q+7)}{2}X^{2q-1}Y^{q^2-2q+1}+(q-1)^2(q^3-q^2+3)X^{2q}Y^{q^2-2q}\\&+\frac{(q-2)q^2(q-1)^3}{4}X^{2q+1}Y^{q^2-2q-1}+\frac{q^2(q-1)^3}{2}X^{3q-3}Y^{q^2-3q+3}\\&+2(q-1)^2q^2X^{3q-2}Y^{q^2-3q+2}+\frac{(q-2)(q-1)^2}{2}X^{3q}Y^{q^2-3q}.\end{aligned}$$

This computation is elementary but intricate, and we omit the details.

The contribution to $W_{C_{2,3}^A}(X,Y)$ from cubics that contain at least one $\mathbb{F}_q$-rational affine line is the number of lines in the affine plane, q^2+q, times the polynomial above, minus the contribution from cubics that contain exactly 2 $\mathbb{F}_q$-rational lines, minus twice the contribution from cubics that contain exactly 3 $\mathbb{F}_q$-rational lines. There are not many types of affine cubics that contain two or three $\mathbb{F}_q$-rational lines, and we omit this calculation. This completes the determination of $W^{\mathrm{singular}}_{C_{2,3}^A}(X,Y)$.

6. Smooth affine plane cubic curves

We begin with an argument parallel to the one given in Section 5 about intersections of $\mathbb{F}_q$-rational lines and irreducible singular cubic curves. Let C be a smooth projective plane cubic curve with $\#C(\mathbb{F}_q)=q+1-t$ and $\mathbb{I}(C)$ $\mathbb{F}_q$-rational inflection lines. Recall from Proposition 2 that for an elliptic curve E, the number of smooth cubics C such that $C\cong E$ and $\mathbb{I}(C)=k$ depends on $E(\mathbb{F}_q)[3]$. Let $\mathbb{L}_i(C)$ be the number of $\mathbb{F}_q$-rational lines L with $\#(C\cap L)(\mathbb{F}_q)=i$. We follow the argument given for singular cubics in Section 5 and see that:

$$\begin{aligned}&\mathbb{L}_0(C)+\mathbb{L}_1(C)+\mathbb{L}_2(C)+\mathbb{L}_3(C)=q^2+q+1,\\&\mathbb{L}_1(C)+2\mathbb{L}_2(C)+3\mathbb{L}_3(C)=(q+1)(q+1-t),\\&\mathbb{L}_2(C)=q+1-t-\mathbb{I}(C),\\&\mathbb{L}_3(C)=\frac{1}{3}\left(\binom{q+1-t}{2}-\mathbb{L}_2(C)\right).\end{aligned}$$

These equations determine $\mathbb{L}_i(C)$ given $\mathbb{I}(C)$ and t.

PROPOSITION 5. *Let E be an elliptic curve defined over $\mathbb{F}_q$ with $\#E(\mathbb{F}_q) = q+1-t$. The number of pairs (C, L) such that C is a projective plane cubic curve with $C \cong E$ and L an $\mathbb{F}_q$-rational line such that $\#(C \cap L)(\mathbb{F}_q) = i$ is $\#\mathrm{PGL}_3(\mathbb{F}_q)/\#\mathrm{Aut}_{\mathbb{F}_q}(E)$ times the polynomial given below:*

$$\frac{q^2+qt+t^2-q+t}{3} \quad \text{if } i=0; \qquad \frac{q^2-t^2+q+t+2}{2} \quad \text{if } i=1;$$
$$q-t \quad \text{if } i=2; \qquad \frac{(q-t)(q-t-1)}{6} \quad \text{if } i=3.$$

PROOF. By Lemma 3, there are $\#\mathrm{PGL}_3(\mathbb{F}_q)/\#\mathrm{Aut}_{\mathbb{F}_q}(E)$ cubic curves C with $C \cong E$. If $E(\mathbb{F}_q)[3]$ is trivial, then every smooth cubic C with $C \cong E$ has $\mathbb{I}(C) = 1$. If $\mathbb{I}(C) = 1$,

$$\mathbb{L}_0(C) = \frac{q^2+qt+t^2-q+t}{3}, \qquad \mathbb{L}_1(C) = \frac{q^2-t^2+q+t+2}{2},$$
$$\mathbb{L}_2(C) = q-t, \qquad \mathbb{L}_3(C) = \frac{(q-t)(q-t-1)}{6},$$

matching the polynomials given in the statement of the proposition.

We can similarly solve for $\mathbb{L}_i(C)$ for the other possible values of $\mathbb{I}(C)$.

(1) When $E(\mathbb{F}_q)[3] \cong \mathbb{Z}/3\mathbb{Z}$ Proposition 2 says that $1/3$ of all cubics C with $C \cong E$ have $\mathbb{I}(C) = 3$ and $2/3$ have $\mathbb{I}(C) = 0$. Adding $1/3$ times the value of $\mathbb{L}_i(C)$ for $\mathbb{I}(C) = 3$ and $2/3$ times the value of $\mathbb{L}_i(C)$ for $\mathbb{I}(C) = 0$ gives the value of $\mathbb{L}_i(C)$ stated in the proposition.

(2) When $E(\mathbb{F}_q)[3] \cong \mathbb{Z}/3\mathbb{Z}\times\mathbb{Z}/3\mathbb{Z}$ Proposition 2 says that $1/9$ of all cubics C with $C \cong E$ have $\mathbb{I}(C) = 9$ and $8/9$ have $\mathbb{I}(C) = 0$. Adding $1/9$ times the value of $\mathbb{L}_i(C)$ for $\mathbb{I}(C) = 9$ and $8/9$ times the value of $\mathbb{L}_i(C)$ for $\mathbb{I}(C) = 0$ gives the value of $\mathbb{L}_i(C)$ stated in the proposition.

□

THEOREM 4. *Let $\mathbb{F}_q$ be a finite field of characteristic not equal to 3. Let*

$$\begin{aligned} W^\alpha(X,Y,t) = {} & \tfrac{q^2+qt+t^2-q+t}{3} X^{q+1-t} Y^{q^2-q-1+t} + \tfrac{q^2-t^2+q+t+2}{2} X^{q-t} Y^{q^2-q+t} \\ & + (q-t) X^{q-1-t} Y^{q^2-q+1+t} + \tfrac{(q-t)(q-t-1)}{6} X^{q-2-t} Y^{q^2-q+2+t}. \end{aligned}$$

We have

$$W^{smooth}_{C^A_{2,3}}(X,Y) = q(q-1)(q^3-q)(q^3-q^2) \sum_{\substack{t^2 \le 4q \\ t \not\equiv 0 \pmod 3}} \mathbb{P}_q(\mathcal{C}(t)) W^\alpha(X,Y,t).$$

PROOF. Let E be an elliptic curve defined over $\mathbb{F}_q$ with $\#E(\mathbb{F}_q) = q+1-t$. Proposition 5 gives the distribution of $\#C(\mathbb{F}_q) - \#(C \cap L)(\mathbb{F}_q)$ as we vary over all smooth cubic curves C with $C \cong E$ and all $\mathbb{F}_q$-rational lines L. Since $\mathrm{PGL}_3(\mathbb{F}_q)$ acts transitively on $\mathbb{F}_q$-rational lines, this is q^2+q+1 times the distribution of values of $\#C(\mathbb{F}_q) - \#(C \cap \{z=0\})(\mathbb{F}_q)$. There are $q-1$ nonzero polynomials defining each smooth cubic curve C. Therefore, such an elliptic curve E contributes

$$(q-1)\frac{\#\mathrm{PGL}_3(\mathbb{F}_q)}{(q^2+q+1)\#\mathrm{Aut}_{\mathbb{F}_q}(E)} W^\alpha(X,Y,t)$$

to the weight enumerator $W^{\text{smooth}}_{C^A_{2,3}}(X,Y)$. Varying over all $E \in \mathcal{C}$ completes the proof. □

7. Low-weight coefficients of $W_{(C^A_{2,3})^\perp}(X,Y)$

We apply Theorem 1 to the expression for $W_{C^A_{2,3}}(X,Y)$ from Theorem 4 in order to prove formulas analogous to those of Theorem 3. As in the projective case, this leads to the expressions considered in Theorem 2.

THEOREM 5. *Let $\mathbb{F}_q$ be a finite field of characteristic not equal to 2 or 3. We have*

$$W_{(C^A_{2,3})^\perp}(X,Y) = X^{q^2} + (q-2)(q-1)^2q^2(q+1)$$
$$\left[\frac{1}{5!}(q-4)(q-3)X^{q^2-5}Y^5 + \frac{1}{6!}(q-5)^2(q-4)(q-3)X^{q^2-6}Y^6\right.$$
$$+\frac{1}{7!}(q-6)(q-5)(q-4)(q-3)(q^2-6q+15)X^{q^2-7}Y^7$$
$$+\frac{1}{8!}(q-3)(2q^7-17q^6+121q^5-1161q^4+7127q^3-23212q^2+39340q-29400)X^{q^2-8}Y^8$$
$$+\frac{1}{9!}\Big(q^{11}-5q^{10}-12q^9-485q^8+8788q^7-53642q^6+142167q^5-30540q^4$$
$$\left.-818744q^3+2249352q^2-2731680q+1411200\Big)X^{q^2-9}Y^9\right]$$
$$+O(Y^{10}).$$

When $p \geq 5$ is prime, the $X^{p^2-10}Y^{10}$ coefficient of $W_{(C^A_{2,3})^\perp}(X,Y)$ is

$$\frac{(p-1)^2p^2(p+1)}{10!}\Big((p^{15}-9p^{14}-7p^{13}+384p^{12}-4514p^{11}+68191p^{10}-706065p^9$$
$$+4482991p^8-18172206p^7+47512147p^6-75728017p^5+54600840p^4+36872568p^3$$
$$-125756064p^2+120294720p-45722880)$$
$$-p(p-1)(p^2-9p+36)\tau(p)\Big).$$

REMARK 4. (1) As in the discussion following Theorem 3, weight k codewords in $(C^A_{2,3})^\perp$ come from special configurations of k points in affine space failing to impose independent conditions on homogeneous cubic polynomials. For example, the weight 5 coefficient of $W_{(C^A_{2,3})^\perp}(X,Y)$ is $q-1$ times the number of collections of 5 collinear points in the affine space $\mathbb{F}_q^2$.
(2) Also as in the discussion following Theorem 3, when q is a prime power, the $X^{q^2-10}Y^{10}$ coefficient of $W_{(C^A_{2,3})^\perp}(X,Y)$ includes a term involving $\tau(q/p^2)$, but we do not compute it here. In general, the weight k coefficeint of $W_{(C^A_{2,3})^\perp}(X,Y)$ will involve traces of the Hecke operators T_q and T_{q/p^2} acting on spaces of cusp forms for $\mathrm{SL}_2(\mathbb{Z})$ of weight at most $k+2$.

7.1. The weight enumerator of the dual code and curves with prescribed 3-torsion. Proposition 2 demonstrates how elliptic curves with prescribed 3-torsion play a role in enumerative questions about affine plane cubic curves. In Proposition 5 we saw that when we average over all cubic curves C isomorphic to a particular elliptic curve E, the effect of $E(\mathbb{F}_q)[3]$ disappears. We do not need to consider cubic curves in groups based on the 3-torsion of the associated elliptic curve, but recent work of the author and Petrow shows that we could divide things in this way and still obtain explicit formulas.

We consider set of smooth projective plane cubic curves C based on $E(\mathbb{F}_q)[3]$ for the elliptic curve E satisfying $C \cong E$. By inclusion-exclusion, we can do this by dividing these cubics into those for which $E(\mathbb{F}_q)[3]$ has a subgroup isomorphic to $\mathbb{Z}/3\mathbb{Z}$ and those for which $E(\mathbb{F}_q)[3] \cong \mathbb{Z}/3\mathbb{Z} \times \mathbb{Z}/3\mathbb{Z}$. Following the terminology from [**18**], let $\mathcal{C}(A_{3,3}t)$ denote the set of isomorphism classes of elliptic curves $E \in \mathcal{C}$ with $E(\mathbb{F}_q)[3] \cong \mathbb{Z}/3\mathbb{Z} \times \mathbb{Z}/3\mathbb{Z}$ and $\#E(\mathbb{F}_q) = q + 1 - t$. The following result is a special case of a weighted version of [**23**, Theorem 4.9].

LEMMA 5. *Suppose that p is the characteristic of $\mathbb{F}_q$ and that $t \in \mathbb{Z}$ satisfies $t^2 \leq 4q$. Then*

$$\begin{aligned}\mathbb{P}_q(\mathcal{C}(A_{3,3}t)) &= \frac{1}{2q} H\left(\frac{t^2-4q}{9}\right) \text{ if } q \equiv 1 \pmod 3,\ p \nmid t, \text{ and } t \equiv q+1 \pmod 9;\\ &= \mathbb{P}_q(\mathcal{C}(2\sqrt{q})) \text{ if } q \text{ is a square } p \neq 3,\ t = 2\sqrt{q}, \text{ and } \sqrt{q} \equiv 1 \pmod 3;\\ &= \mathbb{P}_q(\mathcal{C}(-2\sqrt{q})) \text{ if } q \text{ is a square } p \neq 3,\ t = -2\sqrt{q}, \text{ and } \sqrt{q} \equiv -1 \pmod 3;\\ &= 0 \text{ otherwise.}\end{aligned}$$

Applying the MacWilliams theorem, using the binomial theorem to isolate a particular coefficient of $W_{(C_{2,3}^A)^\perp}(X, Y)$ leads to expressions of the following type:

$$\begin{aligned}\mathbb{E}_q(t_E^R \Phi_{\mathbb{Z}/3\mathbb{Z}}) &:= \sum_{\substack{t^2 \leq 4q \\ q+1-t \equiv 0 \pmod 3}} q\mathbb{P}_q(\mathcal{C}(t))t^R;\\ \mathbb{E}_q(t_E^R \Phi_{\mathbb{Z}/3\mathbb{Z}\times\mathbb{Z}/3\mathbb{Z}}) &:= \sum_{\substack{t^2 \leq 4q \\ q+1-t \equiv 0 \pmod 9}} q\mathbb{P}_q(\mathcal{C}(A_{3,3}t))t^R.\end{aligned}$$

The author and Petrow give formulas for exactly these types of expressions in Theorem 3 of [**18**]. Stating the full result would require introducing too much additional notation, so we refer to [**18**] for details. The case where R is even and $q = p$ is prime is addressed in [**18**, Examples 1 and 2]. We see that $\mathbb{E}_p(t_E^{2R}\Phi_{\mathbb{Z}/3\mathbb{Z}})$ can be expressed in terms of the traces of the Hecke operator T_p acting on spaces of cusp forms for $\Gamma_1(3)$ of weight at most $2R+2$, and that $\mathbb{E}_p(t_E^{2R}\Phi_{\mathbb{Z}/3\mathbb{Z}\times\mathbb{Z}/3\mathbb{Z}})$ can be expressed in terms of traces of T_p acting on spaces of cusp forms for $\Gamma(3)$ of weight at most $2R+2$. Formulas for these quantities when q is a prime power, or when R is odd, are more intricate. See [**18**, Theorem 3] for details.

We give an example of a formula we get from applying the MacWilliams theorem to a subset of smooth projective plane cubic curves that satisfy additional constraints on the 3-torsion of the associated elliptic curve.

THEOREM 6. (1) *Let q be a prime with $q \equiv 1 \pmod 3$. The $X^{q^2+q-1}Y^2$ coefficient of*

$$\sum_{\substack{t^2 \leq 4q \\ q+1-t \equiv 0 \pmod 9}} \mathbb{P}_q(\mathcal{C}(A_{3,3}t))(X + (q-1)Y)^{q+1-t}(X-Y)^{q^2+t}$$

is

$$\frac{-(q+1)(q-1)^3q^4(q^2+q+1)}{48}\Big((7q+3) + q\mathrm{Tr}(\mathrm{T_q}|\mathrm{S_4}(\Gamma(3)))\Big),$$

where $\mathrm{Tr}(\mathrm{T_q}|\mathrm{S}_4(\Gamma(3)))$ *denotes the trace of the* T_q *Hecke operator acting on the space of holomorphic weight* 4 *cusp forms for* $\Gamma(3)$.

(2) *Let* q *be a prime with* $q \equiv 2 \pmod 3$. *The* $X^{q^2+q-3}Y^4$ *coefficient of*

$$\sum_{\substack{t^2 \leq 4q \\ q+1-t \equiv 0 \pmod 3}} \mathbb{P}_q(\mathcal{C}(t))(X+(q-1)Y)^{q+1-t}(X-Y)^{q^2+t}$$

is

$$\frac{-(q+1)(q-1)^3q^4(q^2+q+1)}{48}\Big((q+1)(q^5-7q^4+20q^3-26q^2+13q+2)+q^3\mathrm{Tr}(\mathrm{T_q}|\mathrm{S}_6(\Gamma_0(3)))\Big),$$

where $\mathrm{Tr}(\mathrm{T_q}|\mathrm{S}_6(\Gamma_0(3)))$ *denotes the trace of the* T_q *Hecke operator acting on the space of holomorphic weight* 6 *cusp forms for* $\Gamma_0(3)$.

8. Acknowledgments

Part of this project grew out of the PhD thesis of the author. He thanks Noam Elkies for his extensive guidance and for many helpful conversations. The author thanks the referee for several very helpful suggestions. He also thanks Joseph Gunther for helpful discussions.

References

[1] Y. Aubry, *Reed-Muller codes associated to projective algebraic varieties*, Coding theory and algebraic geometry (Luminy, 1991), Lecture Notes in Math., vol. 1518, Springer, Berlin, 1992, pp. 4–17, DOI 10.1007/BFb0087988. MR1186411

[2] B. J. Birch, *How the number of points of an elliptic curve over a fixed prime field varies*, J. London Math. Soc. **43** (1968), 57–60, DOI 10.1112/jlms/s1-43.1.57. MR0230682

[3] A. A. Bruen, J. W. P. Hirschfeld, and D. L. Wehlau, *Cubic curves, finite geometry and cryptography*, Acta Appl. Math. **115** (2011), no. 3, 265–278, DOI 10.1007/s10440-011-9620-z. MR2823118

[4] P. Delsarte, J.-M. Goethals, and F. J. MacWilliams, *On generalized Reed-Muller codes and their relatives*, Information and Control **16** (1970), 403–442. MR0274186

[5] M. Deuring, *Die Typen der Multiplikatorenringe elliptischer Funktionenkörper* (German), Abh. Math. Sem. Hansischen Univ. **14** (1941), 197–272. MR0005125

[6] D. Eisenbud, M. Green, and J. Harris, *Cayley-Bacharach theorems and conjectures*, Bull. Amer. Math. Soc. (N.S.) **33** (1996), no. 3, 295–324, DOI 10.1090/S0273-0979-96-00666-0. MR1376653

[7] N. D. Elkies, *Linear codes and algebraic geometry in higher dimensions*, Preprint, 2006.

[8] G. van der Geer and M. van der Vlugt, *Reed-Muller codes and supersingular curves. I*, Compositio Math. **84** (1992), no. 3, 333–367. MR1189892

[9] G. van der Geer, R. Schoof, and M. van der Vlugt, *Weight formulas for ternary Melas codes*, Math. Comp. **58** (1992), no. 198, 781–792, DOI 10.2307/2153217. MR1122080

[10] G. van der Geer and M. van der Vlugt, *Generalized Reed-Muller codes and curves with many points*, J. Number Theory **72** (1998), no. 2, 257–268, DOI 10.1006/jnth.1998.2277. MR1651693

[11] J. Harris, *Interpolation*, Current developments in algebraic geometry, Math. Sci. Res. Inst. Publ., vol. 59, Cambridge Univ. Press, Cambridge, 2012, pp. 165–176. MR2931869

[12] P. Heijnen and R. Pellikaan, *Generalized Hamming weights of q-ary Reed-Muller codes*, IEEE Trans. Inform. Theory **44** (1998), no. 1, 181–196, DOI 10.1109/18.651015. MR1486657

[13] J. W. P. Hirschfeld, *Projective geometries over finite fields*, 2nd ed., Oxford Mathematical Monographs, The Clarendon Press, Oxford University Press, New York, 1998. MR1612570

[14] W. C. Huffman and V. Pless, *Fundamentals of error-correcting codes*, Cambridge University Press, Cambridge, 2003. MR1996953

[15] Y. Ihara, *Hecke Polynomials as congruence ζ functions in elliptic modular case*, Ann. of Math. (2) **85** (1967), 267–295, DOI 10.2307/1970442. MR0207655

[16] R. P. M. J. Jurrius, *Weight enumeration of codes from finite spaces*, Des. Codes Cryptogr. **63** (2012), no. 3, 321–330, DOI 10.1007/s10623-011-9557-2. MR2892517

[17] N. Kaplan and I. Petrow, *Traces of Hecke operators and refined weight enumerators of Reed-Solomon codes*, Trans. Amer. Math. Soc. **370** (2018), no. 4, 2537–2561, DOI 10.1090/tran/7089. MR3748576

[18] N. Kaplan and I. Petrow, *Elliptic curves over a finite field and the trace formula*, Proc. Lond. Math. Soc. (3) **115** (2017), no. 6, 1317–1372, DOI 10.1112/plms.12069. MR3741853

[19] G. Lachaud, *The parameters of projective Reed-Muller codes* (English, with French summary), Discrete Math. **81** (1990), no. 2, 217–221, DOI 10.1016/0012-365X(90)90155-B. MR1054981

[20] F. Lang, *Geometry and group structures of some cubics*, Forum Geom. **2** (2002), 135–146. MR1940110

[21] J. H. van Lint, *Introduction to coding theory*, 3rd ed., Graduate Texts in Mathematics, vol. 86, Springer-Verlag, Berlin, 1999. MR1664228

[22] F. J. MacWilliams and N. J. A. Sloane, *The theory of error-correcting codes. I*, North-Holland Publishing Co., Amsterdam-New York-Oxford, 1977. North-Holland Mathematical Library, Vol. 16. MR0465509

[23] R. Schoof, *Nonsingular plane cubic curves over finite fields*, J. Combin. Theory Ser. A **46** (1987), no. 2, 183–211, DOI 10.1016/0097-3165(87)90003-3. MR914657

[24] R. Schoof, *Families of curves and weight distributions of codes*, Bull. Amer. Math. Soc. (N.S.) **32** (1995), no. 2, 171–183, DOI 10.1090/S0273-0979-1995-00586-0. MR1302786

[25] R. Schoof and M. van der Vlugt, *Hecke operators and the weight distributions of certain codes*, J. Combin. Theory Ser. A **57** (1991), no. 2, 163–186, DOI 10.1016/0097-3165(91)90016-A. MR1111555

[26] A. B. Sørensen, *Projective Reed-Muller codes*, IEEE Trans. Inform. Theory **37** (1991), no. 6, 1567–1576, DOI 10.1109/18.104317. MR1134296

[27] W. C. Waterhouse, *Abelian varieties over finite fields*, Ann. Sci. École Norm. Sup. (4) **2** (1969), 521–560. MR0265369

Department of Mathematics, University of California, Irvine, California 92697

Email address: nckaplan@math.uci.edu

Contemporary Mathematics
Volume **722**, 2019
https://doi.org/10.1090/conm/722/14536

The distribution of the trace in the compact group of type $\mathbf{G}_2$

Gilles Lachaud

Our friend and colleague Gilles Lachaud died on February 21, 2018, before he had a chance to correct the handful of typographical errors in this paper noted by the referee. We have taken the liberty of making these small changes, and of reformatting the paper to fit the style of the volume. —The editors.

ABSTRACT. We study the distribution of the trace of the seven-dimensional representation of the exceptional compact simple Lie group G of type G_2. The interest for this distribution comes from its relevance to the equidistribution of several families of exponential sums involving a seven degree binomial phase, established by N. Katz. Firstly, we give a construction of the algebraic group of type G_2 and of its Lie algebra, whose G is the compact form. We then define the Steinberg map of G, defined from the traces of the fundamental representations, inducing a homeomorphism from the alcove of G (the simplex parametrizing conjugacy classes) to a compact set in the affine space. By combining Weyl's integration formula and the Steinberg map, we obtain an explicit expression for the probability density function of the distribution of the trace function on G in terms of Gauss hypergeometric function, and some other special functions. This answers a question raised by J.-P. Serre and N. Katz.

CONTENTS

2010 *Mathematics Subject Classification.* Primary 11L07, 14D05, 20G41, 22E45, 60B15, 60B20; Secondary 11K36, 33D80.

Key words and phrases. Compact Lie groups of type G_2, distribution of the trace of matrices, random matrices, Weyl's integration formula, Steinberg map, equidistribution, generalized Sato–Tate conjecture.

1. Introduction

There is, up to isomorphism, only one compact simple Lie group of type G_2, and we denote it by $\mathbf{UG}_2$. This exceptional group enjoys a unique class of irreducible continuous representations π of degree 7. We are interested here in the probability distribution of the trace τ of π. The map

$$g \mapsto \tau(g) = \operatorname{Tr} \pi(g)$$

is a continuous central function on $\mathbf{UG}_2$, whose values lie in the compact interval $I := [-2, 7]$. The *distribution* or *law* of τ is the measure μ_τ on I which is the image by τ of the mass one Haar measure dg on $\mathbf{UG}_2$. That is, for any continuous real function $\varphi \in \mathscr{C}(I)$, we impose the integration formula

$$\int_I \varphi(x)\, \mu_\tau(x) = \int_{\mathbf{UG}_2} \varphi \circ \boldsymbol{\tau}(g)\; dg\,. \tag{1.1}$$

Alternately, if $x \in \mathbb{R}$, then

$$\int_{-\infty}^{x} \mu_\tau = \text{volume}\,\{g \in \mathbf{UG}_2 \mid \operatorname{Tr} \pi(g) \leq x\}\,.$$

The distribution μ_τ has a *probability density function* f_τ with respect to the Lebesgue measure: this is a positive continuous function defined on $[-2, 7]$ such that $\mu_\tau(x) = f_\tau(x)dx$. This density is real analytic outside the point $x = -1$, and the main result of this article is to obtain an explicit expression for this function. The same question for the symplectic group $\mathbf{USp}_{2g}$ is solved in [**19**].

The distribution laws of trace of linear representations of Lie groups present an intrinsic interest, but they take also place in several arithmetical topics, like Sato–Tate laws, see Serre [**24**]. They also occur in equidistribution theory of the number of points of families of curves over finite fields, or families of exponential sums, see Katz [**13**]. Several examples with $\mathbf{USp}_{2g}$ are given in [**19**]. The distribution law of the trace τ of $\mathbf{UG}_2$ appears in several occurrences; in particular, it governs the behaviour of certain exponential sums, as described in section 2.

Apart from the introductory sections, there are three parts in this paper. The first one (sections 3 and 4) leads to a complete description of the complex algebraic group $\mathbf{G}_2$, which can be defined as the isotropy subgroup of an orthogonal group of type B_3: this is performed in section 3. This definition of $\mathbf{G}_2$ allows a simple description of the Lie algebra $\mathfrak{g}_2$ of $\mathbf{G}_2$, of the corresponding root system in the dual of a suitable torus T, and provides easily a Chevalley system. There are two Lie algebras which are real forms of type G_2: we define the compact form $\mathfrak{ug}_2$ in section 4, leading to the Lie group $\mathbf{UG}_2$, which is the unique compact form of $\mathbf{G}_2$.

The goal of the second part (sections 5 to 10) is to set up an environment suitable for the calculation of the integrals (1.1) and the density function f_τ, and for this we need several tools. We make use of the Steinberg map, defined for the algebraic group $\mathbf{G}_2$ in section 5. In general, if $\mathbf{G}$ is a semi-simple simply connected algebraic group of rank n, with maximal torus $\mathbf{T}$, the *Steinberg map*

$$\tau \colon \mathbf{G} \longrightarrow \mathbb{A}^n$$

is defined by $\tau(x) = (\tau_1(x), \ldots, \tau_r(x))$. This map is surjective and its fibres are unions of conjugacy classes, with each fibre containing a unique semisimple class. This map is described in section 5 if $\mathbf{G} = \mathbf{G}_2$. If we restrict the Steinberg map to the compact form **UG** of **G**, we get a map

$$\tau\colon \mathbf{UG} \longrightarrow \Sigma$$

providing a homeomorphism of the space of conjugacy classes $\mathrm{Cl}(\mathbf{UG})$ onto a compact subset Σ of $\mathbb{R}^n$, and a homeomorphism

$$\psi\colon A_0 \longrightarrow \Sigma$$

of the fundamental alcove A_0 of **UG** onto Σ. This is performed in the particular case of $\mathbf{UG}_2$ in section 6 and 7. The integral (1.1) can be reduced to an integral over the maximal torus, by means of Weyl's integration formula, as explained in Section 8.

In the third and last part (sections 9 to 11), we provide an expression for the density distribution of the trace and its moments. The picture of Σ in the plane is drawn in section 9, and already gives an intuitive idea of this distribution. A combination of Weyl's integration formula with the Steinberg map leads in the main section 10 to a new expression of the initial integral as an integral over the set Σ. We then get an explicit expression of the density distribution in terms of Gauss' hypergeometric function (and other special functions). Other expressions of this density are given, in terms of Legendre functions, Legendre's elliptic integrals, or Meijer G-function. Finally several formulas for the moments of the trace are discussed in section 11.

I would like to thank Jean-Pierre Serre for enlightening conversations and for the handover of his unpublished correspondence with N. M. Katz of 2002–2008 on the topic described here; the results of the present article are an answer to a question appearing in these letters [**22**], [**14**].

2. Exponential sums

Let p be a prime number different from 2 and 7. We consider the one parameter family of exponential sums introduced by N. Katz [**15**]:

$$S(t) = \sum_{x\in\mathbb{F}_p^\times} \chi_2(x) \exp\frac{2i\pi(x^7+tx)}{p}\,,$$

where $\chi_2(x) = \left(\frac{x}{p}\right)$ is the quadratic character of $\mathbb{F}_p^\times$, where $t \in \mathbb{F}_p$ is a parameter, and where the phase of this sum is the polynomial $x^7 + tx$. These sums appear in the trace of the Frobenius in the cohomology of the complete nonsingular model of the affine space curve with equation

$$z^2 = x\,, \qquad y^p - y = x^7 - tx\,,$$

see [**15**, p. 258], see also [**17**, §4.7]. N. Katz determined the exact shape of the equidistribution law for these sums, and we recall the result here. These sums have to be normalised as follows.

The *quadratic Gauss sum* is

$$G(\chi_2) = \sum_{x\in\mathbb{F}_p^\times} \left(\frac{x}{p}\right) \exp\frac{2i\pi x}{p}\,.$$

Hence,

$$G(\chi_2) = \begin{cases} \sqrt{p} & \text{if } p \equiv 1 \pmod 4 \\ i\sqrt{p} & \text{if } p \equiv 3 \pmod 4 . \end{cases}$$

We normalize the sums by putting

$$\widetilde{S}(t) = \left(\frac{p}{7}\right) \frac{S(t)}{G(\chi_2)} .$$

Then $\widetilde{S}(t)$ is real and belongs to $[-2, 7]$, since

$$\widetilde{S}(t) = 1 + a_1 + a_2 + a_1 a_2 + \frac{1}{a_1} + \frac{1}{a_2} + \frac{1}{a_1 a_2}$$

with a_1, a_2 on the unit circle; this is the typical form of the trace of an element of $\mathbf{UG}_2$, see (6.1). Now Katz proved in [**15**, Th. 5.3] that, as $p \to \infty$, the p real numbers $\widetilde{S}(t)$, with $t \in \mathbb{F}_p$, becomes equidistributed in $[-2, 7]$ for the measure μ_τ, that is:

THEOREM 2.1 (Katz). *The monodromy group of $\widetilde{S}(t)$ is equal to $\mathbf{UG}_2$. Hence,*

$$\frac{\left|\left\{t \in \mathbb{F}_p \mid \widetilde{S}(t) \le x\right\}\right|}{p} = \operatorname{vol}\{g \in \mathbf{UG}_2 \mid \tau_1(g) \le x\} + O\left(p^{-1/2}\right),$$

as $p \to \infty$. □

In other words [**15**, Cor. 5.9], the sums

$$\widetilde{S}(t) = \begin{cases} \left(\frac{p}{7}\right) p^{-1/2} \sum_{x \in \mathbb{F}_p^\times} \left(\frac{x}{p}\right) \cos \frac{2\pi(x^7 + tx)}{p} & \text{if } p \equiv 1 \pmod 4, \\ \left(\frac{p}{7}\right) p^{-1/2} \sum_{x \in \mathbb{F}_p^\times} \left(\frac{x}{p}\right) \sin \frac{2\pi(x^7 + tx)}{p} & \text{if } p \equiv 3 \pmod 4 \end{cases}$$

follow the same equidistribution law. These results are generalised in [**16**].

Figure 1 shows the bar chart of the distribution of these sums for $p = 1019$, and compare the distribution of these sums to the density function of the trace τ in $\mathbf{UG}_2$.

3. The group $\mathbf{G}_2$ and its Lie algebra

The complex simple Lie algebra $\mathfrak{g}_2$ was discovered by Killing in 1887, and Engel gave in 1900 a description of the corresponding complex Lie group $\mathbf{G}_2$, which is, as he says, "as elegant as one can wish for"; see Agricola [**3**] for these historical references. We shall use his description, but as a preamble we define an orthogonal Lie algebra. We define on $V = \mathbb{C}^7$ the nondegenerate symmetric bilinear form

$$\Psi(x, y) = -\frac{x_1 y_1}{2} + x_2 y_5 + x_5 y_2 + x_3 y_6 + x_6 y_3 + x_4 y_7 + x_7 y_4 .$$

Hence, $\Psi(x, y) = {}^t x.S.y$, with the 7×7 symmetric matrix

$$S = \begin{pmatrix} -\frac{1}{2} & 0 & 0 \\ 0 & 0 & \mathbf{I}_3 \\ 0 & \mathbf{I}_3 & 0 \end{pmatrix} . \tag{3.1}$$

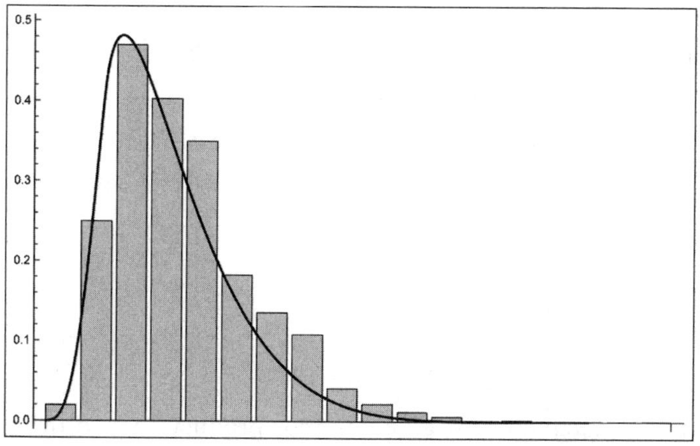

FIGURE 1. Bar chart of the values, and density of the trace.

The form Ψ has maximum index of isotropy 3, signature $(3,4)$ when restricted to $\mathbb{R}^7$, discriminant $1/2$, and associated quadratic form

$$Q(x) = \Psi(x,x) = -\frac{x_1^2}{2} + 2x_2x_5 + 2x_3x_6 + 2x_4x_7 .$$

We denote by

$$\mathbf{O}(\Psi) = \left\{ g \in \mathbf{GL}_7(\mathbb{C}) \mid {}^t g.S.g = S \right\}$$

the *orthogonal group* of Ψ, which is the totality of invertible matrices leaving Q invariant. Then $\mathbf{O}(\Psi) = \{\pm 1\} \times \mathbf{SO}(\Psi)$, where

$$\mathbf{SO}(\Psi) = \mathbf{O}(\Psi) \cap \mathbf{SL}_7(\mathbb{C})$$

is the *special orthogonal group* of Ψ. If X is a complex square matrix of order 7, then

$$\Psi(x, X.y) = \Psi(X^\dagger.x, y)\,, \quad \text{where} \quad X^\dagger = S^{-1}.{}^tX.S\,,$$
$$\Psi(x, X.y) = -\Psi(X.x, y) \quad \text{if and only if} \quad S.X + {}^tX.S = 0\,.$$

The Lie algebra of $\mathbf{SO}(\Psi)$ is the *orthogonal Lie algebra associated to* S, namely,

$$\mathfrak{o}(\Psi) = \left\{ X \in \mathfrak{sl}_7(\mathbb{C}) \mid S.X + {}^tX.S = 0 \right\}.$$

This is a 21-dimensional simple Lie subalgebra of $\mathfrak{sl}_7(\mathbb{C})$, of type B_3, whose elements are, comp. [**7**, Ch. 8, §13, no. 2]:

$$X = \left(\begin{array}{c|ccc|ccc} 0 & 2d & 2e & 2f & 2a & 2b & 2c \\ \hline a & & & & 0 & m_1 & m_3 \\ b & & A & & -m_1 & 0 & m_2 \\ c & & & & -m_3 & -m_2 & 0 \\ \hline d & 0 & n_1 & n_3 & & & \\ e & -n_1 & 0 & n_2 & & -{}^tA & \\ f & -n_3 & -n_2 & 0 & & & \end{array} \right), \quad A = \begin{pmatrix} x_1 & u_1 & u_3 \\ v_1 & x_2 & u_2 \\ v_3 & v_2 & x_3 \end{pmatrix} \in \mathfrak{gl}_3(\mathbb{C})\,.$$

We give now the construction of $\mathbf{G}_2$ and its Lie algebra along the lines of Engel, following Sato and Kimura [**21**, pp. 20 and 85], [**18**, pp. 50 and 243]. We denote by π_1 the identity representation of $\mathbf{GL}_7(\mathbb{C})$ on V, such that $\pi_1(g)v = g.v$

if $g \in \mathbf{GL}_7(\mathbb{C})$ and $v \in V$. The third exterior power $\pi = \wedge^3 \pi_1$ is a representation of $\mathbf{GL}_7(\mathbb{C})$ on $\wedge^3 V$. Let $(e_1, \dots, e_7)$ be the canonical basis of V. We put

$$\mathsf{w} = e_1 \wedge e_2 \wedge e_5 + e_1 \wedge e_3 \wedge e_6 + e_1 \wedge e_4 \wedge e_7 + e_2 \wedge e_3 \wedge e_4 + e_5 \wedge e_6 \wedge e_7 .$$

The algebraic group $\mathbf{G}_2$ is the isotropy subgroup of $\mathbf{SL}_7(\mathbb{C})$ at w:

$$\mathbf{G}_2 = \{ g \in \mathbf{SL}_7(\mathbb{C}) \mid \pi(g)\mathsf{w} = \mathsf{w} \} .$$

PROPOSITION 3.1. *The group* $\mathbf{G}_2$ *is a subgroup of* $\mathbf{SO}(\Psi)$.

PROOF. Let $\lambda = e_1 \wedge \cdots \wedge e_7$ be the canonical element of $\wedge^7 V$. We define a bilinear form B_{w} on V^* by the relation

$$B_{\mathsf{w}}(a, b)\, \lambda = \frac{1}{3} (a \lrcorner \mathsf{w}) \wedge (b \lrcorner \mathsf{w}) \wedge \mathsf{w} , \tag{3.2}$$

where $a, b \in V^*$, and $a \lrcorner \mathsf{w} \in \wedge^2 V$ is the interior product defined by

$$\langle \mathsf{c}, a \lrcorner \mathsf{w} \rangle = \langle a \wedge \mathsf{c}, \mathsf{w} \rangle , \quad \mathsf{c} \in \wedge^2 V.$$

If $g \in \mathbf{GL}_7(\mathbb{C})$, $\mathsf{x} \in \wedge V$ and $\mathsf{a} \in \wedge V^*$, then

$$\mathsf{a} \lrcorner g\mathsf{x} = g({}^t g \mathsf{a} \lrcorner \mathsf{x}) ,$$

see [**5**, Eq. 56, p. 599]. This implies, for $a, b \in V^*$,

$$({}^t g a \lrcorner \mathsf{w}) \wedge ({}^t g b \lrcorner \mathsf{w}) \wedge \mathsf{w} = g^{-1}((a \lrcorner g\mathsf{w}) \wedge (b \lrcorner g\mathsf{w}) \wedge g\mathsf{w}) .$$

Hence, if $g \in \mathbf{G}_2$, then

$$B_{\mathsf{w}}({}^t g a, {}^t g b) = B_{\mathsf{w}}(a, b) . \tag{3.3}$$

A direct computation shows that the matrix of B_{w} is inverse to S:

$$B_{\mathsf{w}}(a, b) = {}^t a . S^{-1} . b .$$

If $g \in \mathbf{G}_2$, we have $g.S^{-1}.{}^t g = S^{-1}$ by (3.3). Then ${}^t g^{-1}.S.g^{-1} = S$, which means that $g^{-1} \in \mathbf{SO}(\Psi)$. □

The infinitesimal representation $d\pi$ of $\mathfrak{gl}_7(\mathbb{C})$ on $\wedge^3 V$ corresponding to π is given by

$$d\pi(X).(e_i \wedge e_j \wedge e_k) = X.e_i \wedge e_j \wedge e_k + e_i \wedge X.e_j \wedge e_k + e_i \wedge e_j \wedge X.e_k$$

for $X \in \mathfrak{gl}_7(\mathbb{C})$. Hence, the Lie algebra $\mathfrak{g}_2$ of $\mathbf{G}_2$ is the *isotropy subalgebra* of w:

$$\mathfrak{g}_2 = \{ X \in \mathfrak{gl}_7(\mathbb{C}) \mid d\pi(X)\mathsf{w} = 0 \} .$$

Since $\mathfrak{g}_2 \subset \mathfrak{o}(\Psi)$, it suffices to determine which matrices of $\mathfrak{o}(\Psi)$ satisfy the condition $d\pi(X)\mathsf{w} = 0$ in order to obtain an extensional description of $\mathfrak{g}_2$. We get

$$x_3 = -x_1 - x_2, \quad m_1 = f, \quad m_2 = d, \quad m_3 = -e, \quad n_1 = -c, \quad n_2 = -a, \quad n_3 = b,$$

and $\mathfrak{g}_2$ is the 14-dimensional Lie algebra of matrices

$$X = \left(\begin{array}{c|ccc|ccc} 0 & 2d & 2e & 2f & 2a & 2b & 2c \\ \hline a & & & & 0 & f & -e \\ b & & A & & -f & 0 & d \\ c & & & & e & -d & 0 \\ \hline d & 0 & -c & b & & & \\ e & c & 0 & -a & & -{}^t A & \\ f & -b & a & 0 & & & \end{array} \right), \quad A = \begin{pmatrix} x_1 & u_1 & u_3 \\ v_1 & x_2 & u_2 \\ v_3 & v_2 & x_3 \end{pmatrix} \in \mathfrak{sl}_3(\mathbb{C}) . \tag{3.4}$$

REMARKS 3.2.

(i) One finds a similar description of $\mathfrak{g}_2$ in Humphreys [**11**, p. 103].
(ii) As observed by Élie Cartan, the group $\mathbf{G}_2$ can be realised as the automorphism group of a complex octonion algebra, see [**27**, Ch. 2]. The Lie algebra $\mathfrak{g}_2$ then appears as a derivation algebra.

The Killing form B of $\mathfrak{g}_2$ is computed as usual from the matrix of the adjoint representation. One finds

$$B(X,X') = 4\operatorname{Tr}(X.X') = 24(ad' + a'd + be' + b'e + cf' + c'f) + 8\operatorname{Tr}(A.A') .$$

Moreover the Killing form of $\mathfrak{sl}_3(\mathbb{C})$ is equal to $6\operatorname{Tr}(A.A')$, and with notation as in (3.4):

$$\operatorname{Tr}(A.A') = 2x_1x_1' + 2x_2x_2' + x_1x_2' + x_1'x_2 + u_1v_1' + u_1'v_1 + u_2v_2' + u_2'v_2 + u_3v_3' + u_3'v_3.$$

Since B is nondegenerate, the algebra $\mathfrak{g}_2$ is semisimple. A splitting Cartan subalgebra $\mathfrak{h}$ of $\mathfrak{g}_2$ is obtained by taking a Cartan subalgebra of the subalgebra $\mathfrak{sl}_3$. We take for $\mathfrak{h}$ the 2-dimensional subalgebra of diagonal elements of $\mathfrak{g}_2$, with basis

$$h_1 = H(1,0)\,, \quad h_2 = H(0,1)\,, \tag{3.5}$$

where

$$H(x_1,x_2) = \begin{pmatrix} 0 & 0 & 0 & 0 & 0 & 0 & 0 \\ 0 & x_1 & 0 & 0 & 0 & 0 & 0 \\ 0 & 0 & x_2 & 0 & 0 & 0 & 0 \\ 0 & 0 & 0 & -x_1-x_2 & 0 & 0 & 0 \\ 0 & 0 & 0 & 0 & -x_1 & 0 & 0 \\ 0 & 0 & 0 & 0 & 0 & -x_2 & 0 \\ 0 & 0 & 0 & 0 & 0 & 0 & x_1+x_2 \end{pmatrix} . \tag{3.6}$$

PROPOSITION 3.3. *Let $(\varepsilon_1, \varepsilon_2)$ be the dual basis in $\mathfrak{h}^*$ of the basis (h_1, h_2) of $\mathfrak{h}$. The root system of $(\mathfrak{g}_2, \mathfrak{h})$ is*

$$\Phi = \{\pm\varepsilon_1, \pm\varepsilon_2, \pm(\varepsilon_1+\varepsilon_2), \pm(\varepsilon_1-\varepsilon_2), \pm(2\varepsilon_1+\varepsilon_2), \pm(\varepsilon_1+2\varepsilon_2)\} .$$

Let $\alpha_1 = \varepsilon_2$ and $\alpha_2 = \varepsilon_1 - \varepsilon_2$. Then $\Delta = \{\alpha_1, \alpha_2\}$ is a base of Φ, and the coordinates of the positive roots in the basis $(\varepsilon_1, \varepsilon_2)$ are

$$\alpha_1 = (0,1)\,, \qquad \alpha_2 = (1,-1)\,, \qquad \alpha_3 = (1,0)\,,$$
$$\alpha_4 = (1,1)\,, \qquad \alpha_5 = (1,2)\,, \qquad \alpha_6 = (2,1)\,.$$

□

We have $|\Phi| = 12$. Moreover

$$\alpha_3 = \alpha_1 + \alpha_2\,, \qquad \alpha_4 = \alpha_1 + \alpha_3 = 2\alpha_1 + \alpha_2\,,$$
$$\alpha_5 = \alpha_1 + \alpha_4 = 3\alpha_1 + \alpha_2\,, \qquad \alpha_6 = \alpha_2 + \alpha_5 = 3\alpha_1 + 2\alpha_2\,,$$

the half-sum of positive roots is

$$\rho = 5\alpha_1 + 3\alpha_2\,,$$

and the highest root is $\widetilde{\alpha} = \alpha_6$. For $1 \le i \le 6$, we put $\alpha_{-i} = -\alpha_i$; these are the negative roots. If $1 \le |i| \le 6$, the one-dimensional eigenspace $\mathfrak{g}^{(i)}$ corresponding to the root α_i can be described explicitly, as follows: let $X \in \mathfrak{g}_2$ be given by (3.4), and write

$$X = \gamma_1 h_1 + \gamma_2 h_2 \\ + bX_1 + x_1 X_2 + aX_3 + fX_4 + x_2 X_5 + x_3 X_6 \\ - eX_{-1} - y_1 X_{-2} - dX_{-3} - cX_{-4} - y_2 X_{-5} - y_3 X_{-6}\,. \tag{3.7}$$

This defines 14 matrices $h_1, h_2, X_1, \ldots, X_6, X_{-1}, \ldots, X_{-6}$ constituting a basis B of $\mathfrak{g}_2$. These matrices are numbered in such a way that if $H \in \mathfrak{h}$, then

$$[H, X_i] = \alpha_i(H) X_i\,, \quad 1 \leq |i| \leq 6 \tag{3.8}$$

and X_i is a basis of $\mathfrak{g}^{(i)}$. The positive elements of the coroot system $\Phi^\vee \subset \mathfrak{h}$ are

$$H_1 = H(-1,2)\,, \qquad H_2 = H(1,-1)\,, \qquad H_3 = H(2,-1)\,,$$
$$H_4 = H(1,1)\,, \qquad H_5 = H(0,1)\,, \qquad H_6 = H(1,0)\,,$$

since the choice of signs in the basis B implies

$$[X_i, X_{-i}] = -H_i\,, \quad 1 \leq i \leq 6\,. \tag{3.9}$$

Let

$$C = \begin{pmatrix} \frac{1}{2} & 0 & 0 \\ 0 & \mathbf{I}_3 & 0 \\ 0 & 0 & \mathbf{I}_3 \end{pmatrix}. \tag{3.10}$$

Then the automorphism $\vartheta\colon \mathfrak{g}_2 \to \mathfrak{g}_2$ given by

$$\vartheta(X) = -C^{-1}.{}^tX.C$$

is equal to -1 on $\mathfrak{h}$ and takes X_α to $X_{-\alpha}$ for all $\alpha \in \Phi$ (here, an automorphism involving the transposition must contain a conjugation, since $\mathfrak{g}_2$ is not self-adjoint). We have proved:

PROPOSITION 3.4. *The family $(X_\alpha)_{\alpha\in\Phi}$ is a Chevalley system for $(\mathfrak{g}, \mathfrak{h})$, with the involution ϑ.* □

For the definition of a Chevalley system, see [**7**, VIII.2.4, Def. 3]. Assume $k = \mathbb{R}$. The bilinear form on $\mathfrak{h}^*$ defined by

$$B_\Phi(\lambda, \mu) = \sum_{\alpha\in\Phi} \lambda(H_\alpha)\mu(H_\alpha)$$

is symmetric, positive definite, nondegenerate, and $B_\Phi(\lambda, \mu) = 8(\lambda \mid \mu)$, with the scalar product

$$(\lambda \mid \mu) = 2\lambda_1\mu_1 + 2\lambda_2\mu_2 - \lambda_1\mu_2 - \lambda_2\mu_1$$

and norm given by $\|\lambda\|^2 = 2(\lambda_1^2 + \lambda_2^2 - \lambda_1\lambda_2)$. There are roots of two different lengths:

$$\text{short roots:} \quad \|\alpha_1\| = \|\alpha_3\| = \|\alpha_4\| = \sqrt{2}\,;$$
$$\text{long roots:} \quad \|\alpha_2\| = \|\alpha_5\| = \|\alpha_6\| = \sqrt{6}\,.$$

The root system Φ is depicted in Figure 2. Notice that

$$(\alpha_1 \mid \alpha_6) = (\alpha_2 \mid \alpha_4) = (\alpha_3 \mid \alpha_5) = 0\,.$$

PROPOSITION 3.5. *The algebra $\mathfrak{g}_2$ is the complex simple Lie algebra with root system of type G_2.*

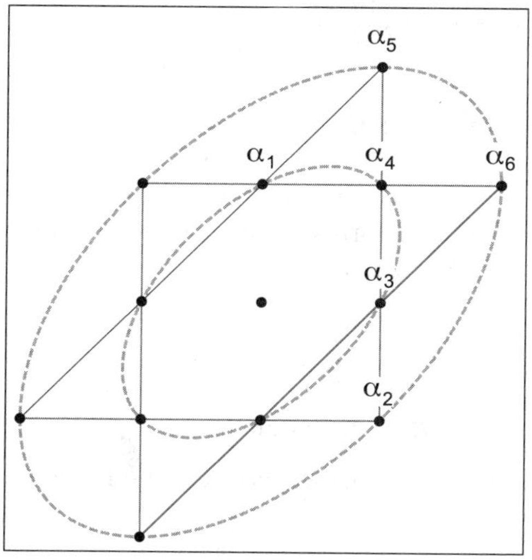

FIGURE 2. The root system Φ.

PROOF. The root system Φ of $\mathfrak{g}_2$ is of rank 2 with a base (α_1, α_2) such that

$$\widehat{(\alpha_1, \alpha_2)} = \frac{5\pi}{6}, \quad \text{i.e.} \quad (\alpha_1 \mid \alpha_2) = -\frac{\sqrt{3}}{2} \|\alpha_1\| \|\alpha_2\|, \quad \text{and} \quad \|\alpha_2\| = \sqrt{3}\, \|\alpha_1\| .$$

The Dynkin graph of Φ is therefore

$$\circ\!\!\Rrightarrow\!\!\circ$$

This proves that Φ is of type G_2, according to the classification of irreducible reduced root systems [**6**, Ch. 6, §4, Th. 13]. □

The Weyl group W of Φ is isomorphic to the dihedral group $D_6 = S_3 \times C_2$ of order 12. The matrices

$$r = \begin{bmatrix} 0 & -1 \\ 1 & -1 \end{bmatrix}, \quad s = \begin{bmatrix} 0 & 1 \\ 1 & 0 \end{bmatrix}$$

generate a normal subgroup N of W of order 6 isomorphic to the symmetric group S_3, and W is the direct product in $\mathbf{GL}_2(\mathbb{Z})$ of N and $C_2 = \{0, -\mathbf{I}_2\}$.

The fundamental weights ϖ_1 and ϖ_2 are orthogonal to α_2 and α_1, hence, proportional to α_4 and α_6. Since $\varpi_1 + \varpi_2 = \rho = 5\alpha_1 + 3\alpha_2$, we have

$$\varpi_1 = \alpha_4 = 2\alpha_1 + \alpha_2\,, \quad \varpi_2 = \alpha_6 = 3\alpha_1 + 2\alpha_2\,. \tag{3.11}$$

The highest root $\widetilde{\alpha}$ is equal to α_6, and the root and weight lattices coincide.

By the *formula of Hermann Weyl*, the dimension of the space of the representation π_ω of $\mathfrak{g}_2$ of highest weight $\omega = p\varpi_1 + q\varpi_2$ is

$$\deg \pi_\omega = \frac{1}{5!}(p+1)(q+1)(p+q+2)(p+2q+3)(p+3q+4).(2p+3q+5)\,. \tag{3.12}$$

See [**7**, Ch. 8, §9, no 2] and [**9**, p. 414]. Speiser [**26**] obtained the following formula:

$$\deg \pi_\omega = \frac{\sqrt{3}}{1280} \operatorname{Im} z_\omega^6\,,$$

where $z_\omega = (p+1)\varpi_1 + i(q+1)\varpi_2$.

REMARK 3.6. The root system G_2 is frequently defined with the base $\Delta' = \{\alpha_1', \alpha_2'\}$, where

$$\alpha_1' = (1, 0), \quad \alpha_2' = \left(-\frac{3}{2}, \frac{\sqrt{3}}{2}\right).$$

This base is used in [9] for the construction of the Lie algebra $\mathfrak{g}_2$, and enjoys a distinctive feature, namely, the corresponding bilinear form is the Euclidean scalar product. Observe that $P.\Delta' = \Delta$, with

$$P = \begin{bmatrix} 0 & 2/\sqrt{3} \\ 1 & 1/\sqrt{3} \end{bmatrix}.$$

4. Real forms

The complex Lie algebra $\mathfrak{g}_2$ has exactly two real forms [10, Ch. 10, §6]. The *split real form* of $\mathfrak{g}_2$ is

$$\mathfrak{g}_2(\mathbb{R}) = \mathfrak{g}_2 \cap \mathbf{M}_7(\mathbb{R}),$$

with Cartan subalgebra $\mathfrak{h}(\mathbb{R}) = \mathfrak{h} \cap \mathbf{M}_7(\mathbb{R})$. Then

$$\mathfrak{g}_2 = \mathfrak{g}_2(\mathbb{R}) \oplus i\mathfrak{g}_2(\mathbb{R}),$$

and the conjugation of $\mathfrak{g}_2$ relative to $\mathfrak{g}_2(\mathbb{R})$ is the usual conjugation $X \mapsto \overline{X}$ in $\mathbf{M}_7(\mathbb{C})$. The other one is the *compact real form* $\mathfrak{ug}_2$ of $\mathfrak{g}_2$. In order to describe it, we first define a nondegenerate positive hermitian form on $V = \mathbb{C}^7$, where C is defined in (3.10):

$$H(x, y) = {}^t x.C.\overline{y} = \frac{x_1\overline{y_1}}{2} + \sum_{i=2}^{7} x_i\overline{y_i},$$

and if $M \in \mathbf{M}_7(\mathbb{C})$, we denote by

$$M^* = C^{-1}.{}^t\overline{M}.C$$

the adjoint of M relative to H, such that $H(M^*x, y) = H(x, My)$. The *special unitary group* associated to H is

$$\mathbf{SU}(H) = \{g \in \mathbf{SL}_7(\mathbb{C}) \mid g^*.g = \mathbf{I}_7\};$$

this is a simple compact Lie group, whose Lie algebra is

$$\mathfrak{su}(H) = \{X \in \mathfrak{sl}_7(\mathbb{C}) \mid X^* = -X\}.$$

This algebra is isomorphic to $\mathfrak{su}(7, \mathbb{C})$ [7, Ch. 9, §3, no. 4].

PROPOSITION 4.1. *With the preceding notation:*

(i) *The algebra*

$$\mathfrak{ug}_2 = \mathfrak{g}_2 \cap \mathfrak{su}(H) = \{X \in \mathfrak{g}_2 \mid X^* = -X\}$$

is the compact real form of $\mathfrak{g}_2$, and $\iota(X) = -X^*$ is the conjugation of $\mathfrak{g}_2$ relative to $\mathfrak{ug}_2$. Moreover $\mathfrak{t} = i\mathfrak{h}(\mathbb{R})$ is a Cartan subalgebra of $\mathfrak{ug}_2$.

(ii) *The real Lie group*

$$\mathbf{UG}_2 = \mathbf{G}_2 \cap \mathbf{SU}(H) = \{g \in \mathbf{G}_2 \mid g^*.g = \mathbf{I}_7\}$$

is a maximal compact subgroup of $\mathbf{G}_2$, with Lie algebra $\mathfrak{ug}_2$.

PROOF. Recall that the involution of $\mathfrak{g}_2$ defined by the Chevalley system of Proposition 3.4 is given by $\vartheta(X) = -C.{}^tX.C^{-1}$. Notice that

$$\vartheta(\overline{X}) = -X^*.$$

This proves [**7**, Ch. 9, §3, Prop. 2] that the algebra $\mathfrak{ug}_2$ of fixed points of the involution ι is the compact real form of $\mathfrak{g}_2$. From this (i) and (ii) follow. □

We obtain a description of $\mathfrak{ug}_2$ by applying Proposition 4.1(i). The matrix (3.4) is fixed by τ if and only if

$$d = -\bar{a}, \quad e = -\bar{b}, \quad f = -\bar{c}, \quad y_1 = \overline{x_1}, \quad y_2 = \overline{x_2}, \quad y_3 = \overline{x_3}, \quad \operatorname{Re}\lambda_1 = \operatorname{Re}\lambda_2 = 0,$$

hence, the elements of $\mathfrak{ug}_2$ are the matrices

$$X = \left(\begin{array}{c|ccc|ccc} 0 & -2\bar{a} & -2\bar{b} & -2\bar{c} & 2a & 2b & 2c \\ \hline a & & & & 0 & -\bar{c} & b \\ b & & A & & \bar{c} & 0 & -\bar{a} \\ c & & & & -\bar{b} & \bar{a} & 0 \\ \hline -\bar{a} & 0 & -c & b & & & \\ -\bar{b} & c & 0 & -a & & -{}^tA & \\ -\bar{c} & -b & a & 0 & & & \end{array}\right), \quad A = \begin{pmatrix} i\theta_1 & x_1 & x_3 \\ \overline{x_1} & i\theta_2 & x_2 \\ \overline{x_3} & \overline{x_2} & i\theta_3 \end{pmatrix} \in \mathfrak{su}_3,$$

with $x_i \in \mathbb{C}$, $\theta_i \in \mathbb{R}$, and $\theta_1 + \theta_2 + \theta_3 = 0$.

5. The Steinberg map of $\mathbf{G}_2$

The group $\mathbf{G}_2$ is the only connected algebraic group with Lie algebra $\mathfrak{g}_2$, since the weight lattice Λ and the root lattice of a root system of type G_2 are equal. The group $\mathbf{G}_2$ is simple, and simply connected. The subgroup $\mathbf{T}$ of diagonal matrices

$$\mathbf{t}(a_1, a_2) = \begin{pmatrix} 1 & & & & & & \\ & a_1 & & & & 0 & \\ & & a_2 & & & & \\ & & & (a_1a_2)^{-1} & & & \\ & & & & a_1^{-1} & & \\ & 0 & & & & a_2^{-1} & \\ & & & & & & a_1a_2 \end{pmatrix}, \quad (a_1, a_2) \in \mathbb{G}_m^2,$$

is a maximal torus of $\mathbf{G}_2$, with Lie algebra $\mathfrak{h}$ as in (3.6). Since $\mathbf{G}_2$ is simply connected, the semisimple conjugacy classes of $\mathbf{G}_2$ are in natural correspondence with the elements of $\mathbf{T}/W$. Denote by $\mathsf{X}(\mathbf{T})$ the commutative group of morphisms from $\mathbf{T}$ to $\mathbf{GL}_1$. If $\lambda \in \Lambda$, we define

$$\chi_\lambda(\exp H) = e^{\langle \lambda, H\rangle} \quad \text{if} \quad H \in \mathfrak{h}.$$

The map $\lambda \mapsto \chi_\lambda$ provides a bijection

$$\Lambda \xrightarrow{\ \sim\ } \mathsf{X}(\mathbf{T}),$$

mapping the root system $\Phi(\mathfrak{g}_2, \mathfrak{h})$ to the set $\Phi(\mathbf{G}_2, \mathbf{T})$ of roots of $(\mathbf{G}_2, \mathbf{T})$. Conversely, if $\chi \in \mathsf{X}(\mathbf{T})$, we denote by $d\chi \in \mathfrak{h}^*$ the differential of χ taken at the group unit. For $1 \le j \le 6$, write $\chi_j = \chi_{\alpha_j}$. The positive roots of $(\mathbf{G}_2, \mathbf{T})$ are given, for $t = \mathbf{t}(a_1, a_2)$, by

$$\chi_1(t) = (0,1) = a_2\,, \qquad \chi_2(t) = (1,-1) = a_1 a_2^{-1}, \qquad \chi_3(t) = (1,0) = a_1\,,$$
$$\chi_4(t) = (1,1) = a_1 a_2\,, \qquad \chi_5(t) = (1,\ \ 2) = a_1 a_2^2\,, \qquad \chi_6(t) = (2,1) = a_1^2 a_2\,.$$

By (3.12), the fundamental representations of $\mathbf{G}_2$ are:

— the identity representation π_1, of degree 7. The fundamental representation $d\pi_1$ of $\mathfrak{g}_2$ has highest weight ϖ_1;

— the adjoint representation π_2, of degree 14. The fundamental representation $d\pi_2$ of $\mathfrak{g}_2$ has highest weight ϖ_2.

Call τ_1, τ_2 the fundamental characters of $\mathbf{G}_2$, and ψ_1, ψ_2 their restriction to $\mathbf{T}$. If $t \in \mathbf{T}$, then

$$\psi_1(t) = \operatorname{Tr} \pi_1(t) = 1 + \sum_{\alpha \in \Phi_S} \chi_\alpha(t)\,,$$
$$\psi_2(t) = \operatorname{Tr} \pi_2(t) = 2 + \sum_{\alpha \in \Phi} \chi_\alpha(t)\,,$$

where $\Phi_S = \{\pm\alpha_1, \pm\alpha_3, \pm\alpha_4\}$ is the set of short roots. Then

$$\text{(5.1)}\quad \psi_1 \circ \mathbf{t}(a_1, a_2) = 1 + a_1 + a_2 + a_1 a_2 + \frac{1}{a_1} + \frac{1}{a_2} + \frac{1}{a_1 a_2}\,,$$

$$\text{(5.2)}\quad \psi_2 \circ \mathbf{t}(a_1, a_2) = 1 + \psi_1 \circ \mathbf{t}(a_1, a_2) + a_1^2 a_2 + a_1 a_2^2 + \frac{a_1}{a_2} + \frac{a_2}{a_1} + \frac{1}{a_1 a_2^2} + \frac{1}{a_1^2 a_2}\,.$$

The *Steinberg map*

$$\boldsymbol{\tau} : \mathbf{G}_2 \longrightarrow \mathbb{A}^2$$

is given [**12**, p. 44], [**25**], [**28**, 6.10] by

$$\boldsymbol{\tau}(g) = (\tau_1(g), \tau_2(g))\,, \quad g \in \mathbf{G}_2\,.$$

We denote by

$$\boldsymbol{\psi}(t) = (\psi_1(t), \psi_2(t))\,, \quad t \in \mathbf{T}\,,$$

the restriction to $\mathbf{T}$ of the Steinberg map. Then [**28**, 6.16], [**29**, Cor. 2, p. 89]:

THEOREM 5.1. *The algebra $\mathscr{C}[\mathbf{G}_2]^\circ$ of regular class functions is freely generated as a commutative algebra by τ_1, τ_2. Thus the Steinberg map induces an isomorphism*

$$\boldsymbol{\psi} : \mathbf{T}/W \longrightarrow \mathbb{A}^2. \qquad \square$$

Call $\mathsf{Char}(\mathbf{G}_2)$ the ring of *virtual characters* of $\mathbf{G}_2$, *i.e.* the linear combinations with integer coefficients of the characters of representations of $\mathbf{G}_2$. This is a polynomial ring:

$$\mathsf{Char}(\mathbf{G}_2) = \mathbb{Z}[\tau_1, \tau_2]\,.$$

EXAMPLE 5.2. Denote by

$$\mathrm{pc}_g(X) = \det(X.\mathbf{I} - g)$$

the characteristic polynomial of a matrix $g \in \mathbf{M}_n(\mathbb{C})$, and for $0 \le k \le n$, let $\sigma_k(g)$ be the elementary symmetric polynomial of degree k in the eigenvalues of g:

$$\mathrm{pc}_g(X) = \sum_{k=0}^{n} (-1)^k \sigma_k(g) X^{n-k}\,.$$

Assume $g \in \mathbf{G}_2$. Then its characteristic polynomial can be written

$$\mathrm{pc}_g(X) = (X^7 - 1) - \sigma_1(X^6 - X) + \sigma_2(X^5 - X^2) - \sigma_3(X^4 - X^3),$$

as it can be easily checked if g belongs to the maximal torus $\mathbf{T}$ of $\mathbf{G}_2$. Since

$$\sigma_k(g) = \mathrm{Tr}(\wedge^k g),$$

the polynomial σ_k belongs to $\mathsf{Char}(\mathbf{G}_2)$, and one finds

$$\sigma_1 = \tau_1, \quad \sigma_2 = \tau_1 + \tau_2, \quad \sigma_3 = \tau_1^2 - \tau_2.$$

EXAMPLE 5.3. Define [**7**, Ch.9, §7, no 4], [**12**, 4.23, p. 75]:

$$f(t) = \chi_0(t) \prod_{\alpha>0} (1-\chi_\alpha(t)^{-1}), \quad \text{where} \quad \chi_0(t) = \chi_{\varpi_1}(t) \dots \chi_{\varpi_r}(t) \quad (t \in \mathbf{T}). \tag{5.3}$$

This is a regular function on $\mathbf{T}$, and

$$f(t) = \sum_{w \in W} (\det w)\ w.\chi_0(t).$$

It is known that $f(t)$ is anti-invariant under the Weyl group. Then $f(t)^2$ is real, invariant under W, and

$$f(t)^2 = \prod_{\alpha>0} (1 - \chi_\alpha(t)^{-1}) \prod_{\alpha>0} (\chi_\alpha(t) - 1).$$

Hence,

$$f(t)^2 = \pm \prod_{\alpha \in \Phi} (\chi_\alpha(t) - 1). \tag{5.4}$$

According to Theorem 5.1, the function $f(t)^2$ can thus be written as a polynomial in the ψ_i; we call D this polynomial:

$$f(t)^2 = D(\psi_1(t), \psi_2(t)) = D \circ \boldsymbol{\psi}(t). \tag{5.5}$$

Precisely, if

$$c_1(x, y) = 4y - x^2 - 2x + 7,$$
$$c_2(x, y) = (y + 5(x + 1))^2 - 4(x + 2)^3,$$

we have

$$D(x, y) = c_1(x, y) c_2(x, y). \tag{5.6}$$

By a direct computation, one proves that the Jacobian determinant of the map

$$\boldsymbol{\psi} \circ \mathbf{t} : \mathbb{G}_m^2 \longrightarrow \mathbb{A}^2$$

is given by

$$J(a_1, a_2) = \mathrm{Jac}(\boldsymbol{\psi} \circ \mathbf{t})(a_1, a_2) = -\frac{1}{a_1 a_2} f \circ \mathbf{t}(a_1, a_2). \tag{5.7}$$

This formula is related to Steinberg Formula [**28**, §8, Lemma 8.2], [**29**, p. 125, Lemma]. From the decomposition

$$f(t) = f_S(t) f_L(t), \tag{5.8}$$
$$f_S(t) = \varpi_1(t) \prod_{\alpha \text{ short}} (1 - \alpha(t)^{-1}),$$
$$f_L(t) = \varpi_2(t) \prod_{\alpha \text{ long}} (1 - \alpha(t)^{-1}),$$

we deduce that if $a = (a_1, a_2) \in \mathbb{G}_m^2$, then

$$J(a) = J_S(a) J_L(a)\,, \tag{5.9}$$

$$J_S(a) = -\left(a_1 - \frac{1}{a_1} + a_2 - \frac{1}{a_2} - \left(a_1 a_2 - \frac{1}{a_1 a_2}\right)\right),$$

$$J_L(a) = -\left(\frac{a_1}{a_2} - \frac{a_2}{a_1} + a_1 a_2^2 - \frac{1}{a_1 a_2^2} - \left(a_1^2 a_2 - \frac{1}{a_1^2 a_2}\right)\right).$$

6. Maximal torus and alcove of $\mathbf{UG}_2$

We identify the Cartan algebra $\mathfrak{h}(\mathbb{R})$ to $\mathbb{R}^2$ by the isomorphism (3.6). The image of the Cartan subalgebra $\mathfrak{t} = i\mathfrak{h}(\mathbb{R})$ of $\mathfrak{ug}_2$ by the exponential map is a maximal torus T of $\mathbf{UG}_2$. The elements of $\mathfrak{t}$ are the matrices $H(i\theta_1, i\theta_2)$, where $H(x_1, x_2)$ is defined in (3.6) and $\theta = (\theta_1, \theta_2) \in \mathbb{R}^2$. We get a surjective map

$$\mathbf{u} : \mathbb{R}^2 \longrightarrow T$$

by putting

$$\mathbf{u}(\theta_1, \theta_2) = \exp H(i\theta_1, i\theta_2) = \mathbf{t}(e^{i\theta_1}, e^{i\theta_2})\,;$$

that is,

$$\mathbf{u}(\theta) = \begin{pmatrix} 1 &&&&&& \\ & e^{i\theta_1} &&&&& \\ && e^{i\theta_2} &&&& \\ &&& e^{-i(\theta_1+\theta_2)} &&& \\ &&&& e^{-i\theta_1} && \\ &&&&& e^{-i\theta_2} & \\ &&&&&& e^{i(\theta_1+\theta_2)} \end{pmatrix}. \tag{6.1}$$

The *nodal group* of T, denoted by $\Gamma(T)$, is the kernel of the exponential map $\mathfrak{t} \longrightarrow T$. It induces by passage to the quotient an isomorphism

$$\mathfrak{t}/\Gamma(T) \xrightarrow{\ \sim\ } T\,.$$

Let

$$\mathsf{K} = \frac{1}{2i\pi}\Gamma(T) = \{H \in \mathfrak{h} \mid \exp 2i\pi H = \mathbf{I}\}\,.$$

If $h_1 = H(1,0), h_2 = H(0,1)$ are the matrices defined in (3.5), then (h_1, h_2) is a basis of the discrete group K, hence, the coroot system $\Phi^\vee$ is included in K. We denote by

$$X_2 = [0, 1[^2$$

a fundamental domain of K.

COROLLARY 6.1. *The map $H \mapsto \exp 2i\pi H$ from $\mathfrak{h}$ to T is surjective and through the identification $(\mathbb{R}/\mathbb{Z})^2 \simeq \mathfrak{h}/\mathsf{K}$, it induces by passage to the quotient an isomorphism*

$$\mathbf{v} : (\mathbb{R}/\mathbb{Z})^2 \xrightarrow{\ \sim\ } T\,.$$

□

We have

$$\mathbf{v}(\theta_1, \theta_2) = \mathbf{u}(2\pi\theta_1, 2\pi\theta_2)\,. \tag{6.2}$$

If we put $d\theta = d\theta_1 d\theta_2$, the relation

$$\int_T \mathsf{F}(t)dt = \int_{[0,1]^2} \mathsf{F} \circ \mathbf{v}(\theta) d\theta, \quad \mathsf{F} \in \mathscr{C}(T)\,,$$

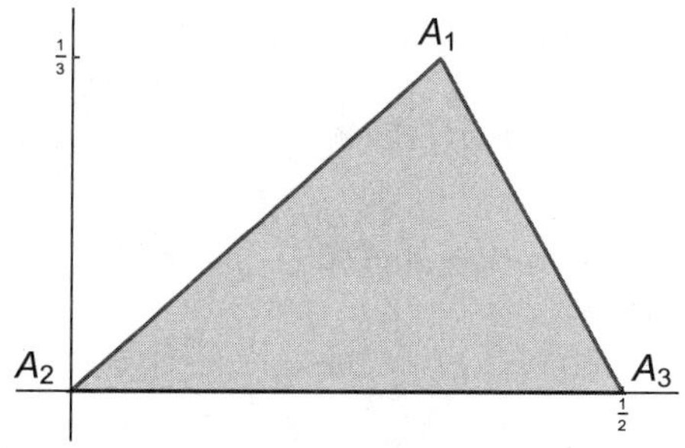

FIGURE 3. Fundamental alcove of $\Phi^\vee$

defines dt as the Haar mesure on T of total volume 1. The *fundamental alcove* [**6**, Ch. 6, §2] of the root system $\Phi(\mathfrak{g}_2, \mathfrak{h})$ is the standard fundamental domain A_0 of W, that is, the open simplex

$$A_0 = \{H \in \mathfrak{h} \mid 0 < \langle \alpha, H \rangle < 1 \quad \text{if } \alpha \in \Delta\}\,.$$

The map $\mathfrak{h} \longrightarrow T$ defined by $H \mapsto \exp iH$ induces a homeomorphism

$$\bar{A}_0 \xrightarrow{\ \sim\ } T/W\,;$$

see [**7**, Ch. 9, §5, no. 2, Cor. 1 of Prop. 2, p. 326]. One finds that A_0 is the intersection of three half-planes (see Figure 3):

$$P_1 : \theta_2 > 0\,, \quad P_2 : 1 - \theta_2 - 2\theta_1 > 0\,, \quad P_3 : \theta_1 - \theta_2 > 0\,.$$

This is a triangle with vertices

$$A_1 = \left(\frac{1}{3}, \frac{1}{3}\right), \quad A_2 = (0, 0)\,, \quad A_3 = \left(\frac{1}{2}, 0\right).$$

Notice that of course $A_0 \subset X_2$. If $\mathsf{F} \in \mathscr{C}(T)^W$, then

$$\frac{1}{|W|}\int_T \mathsf{F}(t)dt = \int_{A_0} \mathsf{F} \circ \mathbf{v}(\theta)d\theta\,. \tag{6.3}$$

7. The Steinberg map on $\mathbf{UG}_2$

We compute now the restriction to $\mathbf{UG}_2$ of the Steinberg map. Since $-1 \in W$, the characters τ_i are real valued functions. The map

$$\Psi(\theta_1, \theta_2) = \boldsymbol{\psi} \circ \mathbf{v}(\theta_1, \theta_2)$$

defines an application

$$\Psi : (\mathbb{R}/\mathbb{Z})^2 \longrightarrow \mathbb{R}^2.$$

Since

$$\boldsymbol{\psi} \circ \mathbf{v}(\theta_1, \theta_2) = \boldsymbol{\psi} \circ \mathbf{t}(e^{2i\pi\theta_1}, e^{2i\pi\theta_2})\,,$$

we have

$$\operatorname{Jac}\Psi(\theta_1, \theta_2) = -4\pi^2 e^{2i\pi\theta_1+\theta_2} \operatorname{Jac}(\boldsymbol{\psi} \circ \mathbf{t})(e^{2i\pi\theta_1}, e^{2i\pi\theta_2})\,.$$

By Equation (5.7), we have

$$\operatorname{Jac}(\boldsymbol{\psi} \circ \mathbf{t})(e^{2i\pi\theta_1}, e^{2i\pi\theta_2}) = -e^{-2i\pi\theta_1+\theta_2} f \circ \mathbf{v}(\theta_1, \theta_2)\,,$$

hence

$$\operatorname{Jac}\Psi(\theta_1, \theta_2) = 4\pi^2 \cdot f \circ \mathbf{v}(\theta_1, \theta_2)\,. \tag{7.1}$$

The following statement is a particular case of a result holding for any simply connected semi-simple compact group, see Steinberg [**28**] and Serre [**25**]. Recall that the set $\mathrm{Cl}(\mathbf{UG}_2)$ of conjugacy classes of $\mathbf{UG}_2$ is in natural correspondence with T/W.

THEOREM 7.1. *The restriction to* $\mathbf{UG}_2$ *of the Steinberg map* $\boldsymbol{\tau}$ *gives a map with the following properties*:

(i) *It gives a homeomorphism of the space* $\mathrm{Cl}(\mathbf{UG}_2)$ *onto a compact subset* $\Sigma \subset \mathbb{R}^r$.

(ii) *The restriction* $\boldsymbol{\psi}$ *to* T *of* $\boldsymbol{\tau}$ *induces a homeomorphism*

$$\Psi : A_0 \longrightarrow \Sigma$$

of the fundamental alcove $A_0 \subset \mathfrak{h}$ *onto its image* $\Sigma \subset \mathbb{R}^r$. *The boundary of* Σ *corresponds to the singular classes.*

(iii) *The restriction to* Σ *of the polynomial* $D(x_1, x_2)$ *of* (5.5) *is zero on the boundary and nowhere else.*

PROOF. Hint: we know the Jacobian matrix $\operatorname{Jac}\Psi$ by (7.1). Since $f > 0$ in A_0 by definition, we have also $\operatorname{Jac}\Psi > 0$ in A_0. Hence, Ψ is a local diffeomorphism at every point of A_0, and a diffeomorphism from the fundamental chamber A_0 to Σ, since it is injective on the open subset A_0. □

We deduce from (5.1) that we have, comp. [**15**, p. 260]:

$$\psi_1 \circ \mathbf{u}(\theta) = 1 + 2\cos\theta_1 + 2\cos\theta_2 + 2\cos(\theta_1 + \theta_2)\,.$$

One finds $\psi_1 \circ \mathbf{u}(0,0) = 7$ and $\psi_1 \circ \mathbf{u}(2\pi/3, 4\pi/3) = -2$. We deduce from (5.2):

$$\begin{aligned}\psi_2 \circ \mathbf{u}(\theta) &= 2(1 + 2\cos\theta_1\cos\theta_2 + 2(\cos\theta_1 + \cos\theta_2)\cos(\theta_1 + \theta_2)) \\ &= 2(1 + \cos\theta_1 + \cos\theta_2 + \cos(\theta_1 + \theta_2) \\ &\qquad + \cos(\theta_1 - \theta_2) + \cos(2\theta_1 - \theta_2) + \cos(\theta_1 - 2\theta_2))\,.\end{aligned}$$

Here $\psi_2 \circ \mathbf{u}(0,0) = 14$ and $\psi_2 \circ \mathbf{u}(0,\pi) = -2$, which is the minimum value predicted by Serre's theorem [**23**].

8. The Weyl integration formula revisited

Let G be a connected compact semisimple Lie group, and T a maximal torus of G, with Weyl group W. Denote by dg (resp. dt) the Haar measure on G (resp. T) with total mass 1. The restriction map defines an isomorphism

$$\mathscr{C}(G)^\circ \xrightarrow{\ \sim\ } \mathscr{C}(T)^W \tag{8.1}$$

from the vector space $\mathscr{C}(G)^\circ = \mathscr{C}(\mathrm{Cl}\,G)$ of *complex central continuous functions* on G to the space $\mathscr{C}(T)^W$ of complex continuous functions on T, invariant under the action

$$[w.f](t) = f({}^t w.t) \quad (w \in W)\,.$$

The *Weyl density* defined by the root system Φ is

$$\delta_G(t) = \prod_{\alpha\in\Phi} (\chi_\alpha(t) - 1)\,, \quad t \in T\,.$$

According to Equation (5.4), we have

$$\delta_G(t) = |f(t)|^2, \tag{8.2}$$

with f as in Equation (5.3). Let dg be the Haar measure of volume 1 on G. If $\mathsf{F} \in \mathscr{C}(G)^\circ$, then

$$\int_{\mathrm{Cl}\,G} \mathsf{F}(\dot{g})\, d\dot{g} = \int_G \mathsf{F}(g)\, dg\,,$$

where $d\dot{g}$ is the image measure on $\mathrm{Cl}\,G$ of the measure dg. The following result is classical [**7**, Ch. 9, § 6, Cor. 2, p. 337].

THEOREM 8.1 (Weyl integration formula). *If* $\mathsf{F} \in \mathscr{C}(G)^\circ$, *then*

$$\int_G \mathsf{F}(g)\, dg = \frac{1}{|W|} \int_T \mathsf{F}(t)\delta_G(t)dt$$

with the Weyl measure $\mu_G(t) = \delta_G(t)dt$. □

9. Image of the alcove

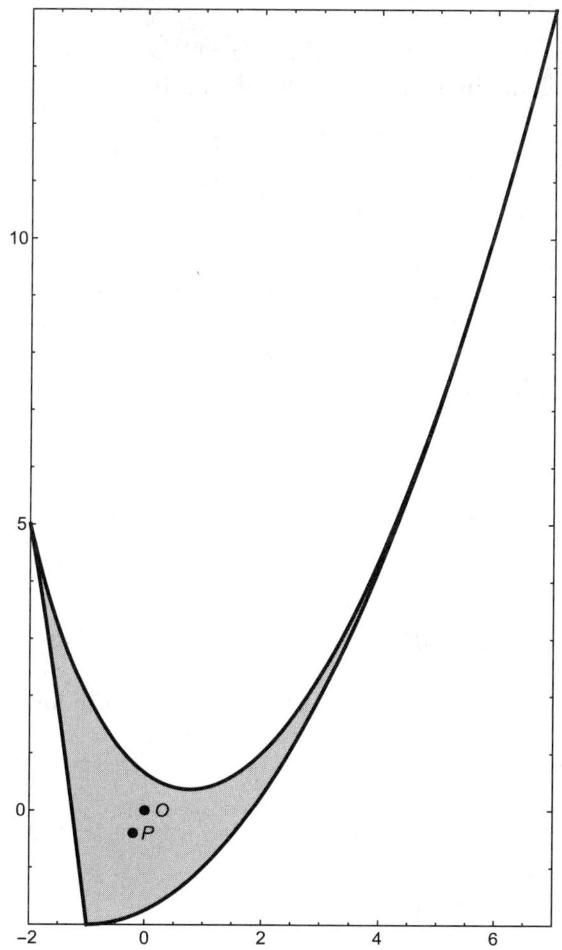

FIGURE 4. The compact set Σ

According to Theorem 7.1, the compact set Σ is defined by an inequation:

$$\Sigma\colon D(x, y) \geq 0\,.$$

By (5.6), a system of inequations for Σ is

$$c_1(x,y) > 0 \quad \text{and} \quad c_2(x,y) > 0\,,$$

where

$$c_1(x,y) = 4y - x^2 - 2x + 7, \quad c_2(x,y) = (y + 5(x+1))^2 - 4(x+2)^3.$$

Let

$$\begin{aligned} r_1(x) &= \frac{1}{4}(x^2 + 2x - 7)\,, \\ r_{21}(x) &= -5(x+1) + 2(x+2)^{3/2}, \\ r_{22}(x) &= -5(x+1) - 2(x+2)^{3/2}\,; \end{aligned}$$

then

$$c_1(x,y) = 4(y - r_1(x)), \quad c_2(x,y) = (r_{21}(x) - y)(y - r_{22}(x))\,,$$

and

$$D(x,y) = (y - r_1(x)(y - r_{21}(x))(y - r_{22}(x))\,.$$

In other words, Σ is the intersection of the open sets

$$P_1 : y > r_1(x)\,, \quad P_2 : y > r_{22}(x)\,, \quad P_3 : y < r_{21}(x)\,, \tag{9.1}$$

and the vertices are the points

$$A_1 = (-2,5)\,, \quad A_2 = (7,14)\,, \quad A_3 = (-1,-2)\,.$$

The domain Σ is drawn in Figure 4. The boundary of Σ is the locus $D(x,y) = 0$. The traces are mainly concentrated on the left: firstly, because the origin O is the center of gravity of Σ with the Weyl density $D^{1/2}$ as mass distribution (because the representations are real); secondly, because D reaches its maximum at the point $P = (-1/5, -2/5)$ which is also on the left. We have

$$\frac{D(P)^{1/2}}{4\pi^2} = \frac{2^2 \cdot 3^3}{5^{5/2} \cdot \pi^2} = 0.195749\ldots$$

10. Distribution of the trace

THEOREM 10.1 (Integration formula). *Let φ be any piecewise continuous function on Σ. Then*

$$\int_G \varphi \circ \boldsymbol{\tau}(g)\, dg = \frac{1}{4\pi^2} \int_\Sigma \varphi(x,y) D(x,y)^{1/2} dx\, dy\,.$$

PROOF. Call I the integral of the left hand side in the formula. Then

$$\begin{aligned} I = \frac{1}{|W|} &\int_T \varphi \circ \boldsymbol{\psi}(t)\, \delta_G(t)\, dt && \text{by Theorem 8.1,} \\ = &\int_{A_0} \varphi \circ \Psi(\theta)\; \delta_G \circ \mathbf{v}(\theta)\, d\theta && \text{by Equation (6.3),} \\ = &\int_A \varphi \circ \Psi(\theta)\, |f \circ \mathbf{v}(\theta)|^2\, d\theta && \text{by Equation (8.2),} \\ = \frac{1}{4\pi^2} &\int_A \varphi \circ \Psi(\theta)\, |f \circ \mathbf{v}(\theta)|\; |\mathrm{Jac}\, \Psi(\theta)|\, d\theta && \text{by Equation (7.1),} \\ = \frac{1}{4\pi^2} &\int_A \varphi \circ \Psi(\theta)\, |D \circ \Psi(\theta)|^{1/2}\; |\mathrm{Jac}\, \Psi(\theta)|\, d\theta && \text{by Equation (5.5),} \end{aligned}$$

and we conclude using the change of variables $(x,y) = \boldsymbol{\psi} \circ \mathbf{v}(\theta_1, \theta_2)$. □

Taking for φ the characteristic function of the set $\{x \leq t\}$, we get the function

$$F_\tau(t) = \frac{1}{4\pi^2} \int_{x \leq t} D(x,y)^{1/2} dx\, dy = \mathrm{vol}\,\{g \in G \mid \tau(g) \leq t\}$$

which is the cumulative distribution function of the fundamental character $\tau = \tau_1$. Hence, the density function of τ is given by the derivative

$$f_\tau(x) = F'_\tau(x) = \frac{1}{4\pi^2} \int_{(x,y)\in\Sigma} D(x,y)^{1/2} dy\,.$$

We find:

— If $-2 \leq x \leq -1$, then $f_\tau(x) = f_1(x)$, with

$$f_1(x) = \frac{1}{4\pi^2} \int_{r_{22}}^{r_{21}} \sqrt{(y-r_1)(y-r_{22})(r_{21}-y)}\; dy\,.$$

— If $-1 \leq x \leq 7$, then $f_\tau(x) = f_2(x)$, with

$$f_2(x) = \frac{1}{4\pi^2} \int_{r_1}^{r_{21}} \sqrt{(y-r_1)(y-r_{22})(r_{21}-y)}\; dy\,.$$

Recall that the *Gauss hypergeometric function* can be defined by its integral representation [**20**, p. 54]

$$ {}_2F_1(a,b;c;z) = \frac{\Gamma(c)}{\Gamma(b)\Gamma(c-b)} \int_0^1 t^{b-1}(1-t)^{c-b-1}(1-tz)^{-a}\; dt\,, \tag{10.1}$$

where $c > b > 0$. This is a holomorphic function of z in the complex plane cut along the positive real axis from 1 to ∞. The main result is the following. Let

$$H(z) = {}_2F_1\left(-\frac{1}{2}, \frac{3}{2}, 3; z\right).$$

Moreover put $y = y(x) = \sqrt{x+2}$ and

$$z(y) = \frac{16y^3}{(y+1)(3-y)^3}\,.$$

Then $z(y(x))$ is an increasing function from $[-2,7]$ to $[0,+\infty)$.

THEOREM 10.2. *Let*

$$f_1(x) = \frac{1}{2\pi} y^6 (3-y)^{3/2} (y+1)^{1/2} H(z(y(x)))\,,$$

$$f_2(x) = \frac{1}{128\pi} y^{3/2} (3-y)^6 (y+1)^2 H\left(\frac{1}{z(y(x))}\right).$$

The density function of the trace τ is given by

$$f_\tau(x) = \begin{cases} f_1(x) & \text{if } -2 \leq x \leq -1, \\ f_2(x) & \text{if } -1 \leq x \leq -7. \end{cases}$$

We postpone the proof of Theorem 10.2 to the end of this section, finding it worthwhile to firstly exploit this result. This statement can be simplified: define

$$K(z) = z^{-3/2} H\left(\frac{1}{z}\right).$$

The functions H and K are solutions of the *hypergeometric differential equation* [**20**, p. 42]:

$$z(1-z)\frac{d^2w}{dz^2} + (3-2z)\frac{dw}{dz} + \frac{3}{4}w = 0\,.$$

Let

$$Q(z) = \begin{cases} H(z) & \text{if } 0 \leq x < 1\,, \\ K(z) & \text{if } 1 \leq x < \infty\,. \end{cases}$$

Then

$$H(0) = 1\,, \quad H(1) = K(1) = \frac{32}{15}\,\pi\,, \quad \lim_{z\to\infty} K(z) = 0\,,$$

and $H'(1) = K'(1)$ but

$$\lim_{z\to 1, z<1} H''(z) = \lim_{z\to 1, z>1} K''(z) = -\infty\,.$$

COROLLARY 10.3. *If $-2 \leq x \leq 7$, then*

$$f_\tau(x) = \frac{2}{\pi}\left(\frac{y(x)^{15}}{z(y(x))}\right)^{1/2} Q(z(y(x)))\,.$$

The density function f_τ is of class C^1 in $[-2,7]$, and is a real analytic function in $(-2,7)$ outside $x=-1$. □

The graph of f_τ is drawn in Figure 6. The maximum of f_τ is reached if

$$x = x_{\max} = -0.736\ldots, \quad \text{and} \quad f(x_{\max}) = 0.481\ldots$$

If $\varepsilon > 0$, then

$$f_\tau(7-\varepsilon) = \frac{1}{2^9\cdot 3^4\cdot\sqrt{3}\cdot\pi}\,\varepsilon^6 + O(\varepsilon)^8,$$

and

$$f_\tau(-2+\varepsilon) = \frac{3\sqrt{3}}{2\pi}\,\varepsilon^3 + O(\varepsilon)^4.$$

The median is equal to $x_{\text{med}} = -0.203\,237$. A formula for the cumulative distribution function F_τ seems hard to obtain. Two tables of values are displayed in Figure 5, and the graph of F_τ, obtained by interpolation, is in Figure 7.

REMARK 10.4. The density distribution f_τ can be expressed by other special functions, with the help of the following relations:

(i) If $\mathfrak{P}_b^a(x)$ is the *Legendre function of the first kind*, then [**20**, p. 51]:

$${}_2F_1\left(-\frac{1}{2},\frac{3}{2},3;z\right) = \frac{4(1-z)^{3/4}}{z}\,\mathfrak{P}_{-5/2}^{-1}\left(\frac{1-(z/2)}{\sqrt{1-z}}\right).$$

(ii) If $E(z)$ and $K(z)$ are *Legendre's elliptic integrals*, then:

$${}_2F_1\left(-\frac{1}{2},\frac{3}{2},3;z\right) = \frac{16}{15\pi z^2}\left((z^2-z+1)E(z) + (z^2-3z+2)K(z)\right),$$

see [**2**, 07.23.03.2632.01].

(iii) If G is *Meijer G-function*, then [**8**, 9.34.1, p. 1035]:

$${}_2F_1\left(-\frac{1}{2},\frac{3}{2},3;z\right) = -\frac{2}{\pi}\,G_{2,2}^{1,2}\left(-z\,\middle|\,\begin{matrix}-\frac{1}{2}, & \frac{3}{2} \\ 0, & -2\end{matrix}\right).$$

x	$f_\tau(x)$	$F_\tau(x)$
−2.0	0	0
−1.5	0.083	0.011
−1.0	0.432	0.134
−0.5	0.462	0.370
0.0	0.366	0.578
0.5	0.260	0.735
1.0	0.172	0.842
1.5	0.107	0.911
2.0	0.062	0.953
2.5	0.034	0.977
3.0	0.017	0.989
⋮	⋮	⋮
7.0	0	1

$F_\tau(x)$	x
0%	−2
10%	−1.086
20%	−0.857
30%	−0.648
40%	−0.434
50%	−0.203
60%	0.058
70%	0.371
80%	0.776
90%	1.395
100%	7

FIGURE 5. Numerical values and quantiles

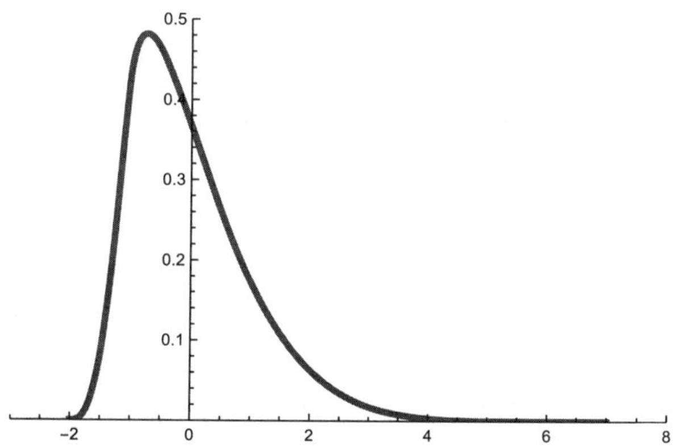

FIGURE 6. Probability density function $f_\tau(x)$

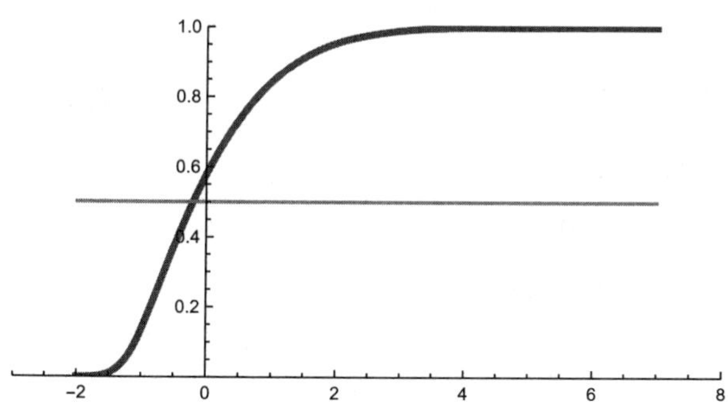

FIGURE 7. Cumulative distribution function $F_\tau(x)$

REMARK 10.5. A good approximation of the density distribution $f_\tau(x)$ is the *Gamma distribution*

$$\gamma(x) = \frac{8}{3}\,(x+2)^3 e^{-2(x+2)}, \quad x > -2\,.$$

Then

$$\max_x |f_\tau(x) - \gamma(x)| = 0.075\ldots, \quad \frac{1}{9}\int_{-2}^{7} |_\tau(x) - \gamma(x)|dx = 0.009\ldots$$

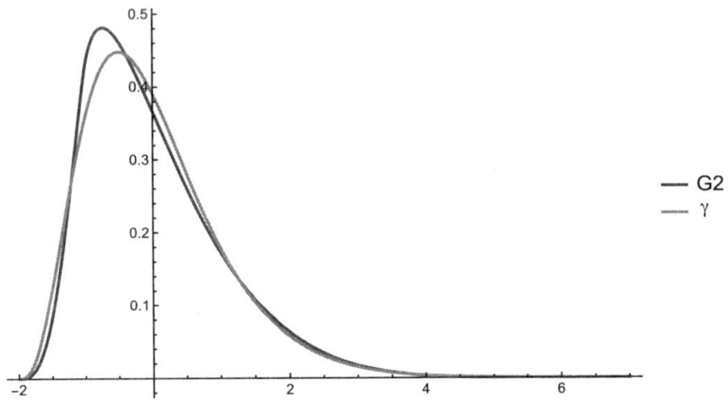

FIGURE 8. Comparison of f_τ and γ

Before proving Theorem 10.2, we first prove a lemma.

LEMMA 10.6. *If $c < a < b$ and if $b - a < a - c$, then*

$$\int_a^b \sqrt{(t-a)(t-c)(b-t)}\; dt = \frac{8}{\pi}\,(b-a)^2(a-c)^{1/2}\;{}_2F_1\left(-\frac{1}{2}, \frac{3}{2}, 3; \frac{b-a}{a-c}\right).$$

PROOF. Let I be the integral on the left. With $t = (b-a)u + a$, we get

$$I = (b-a)^2(a-c)^{1/2}\int_0^1 \sqrt{u(1-u)(1-uz)}\; du\,, \quad z = -\frac{b-a}{a-c}\,,$$

with $0 < z < 1$. Applying the integral representation (10.1), we obtain

$$\int_0^1 \sqrt{t(1-t)(1-tz)}\; dt = \frac{8}{\pi}\;{}_2F_1\left(-\frac{1}{2}, \frac{3}{2}, 3; z\right),$$

and we get the required result. □

PROOF OF THEOREM 10.2. We apply Lemma 10.6. If $-2 \le x \le -1$, we have

$$\begin{aligned} b - a &= r_{21} - r_{22} = 4(x+2)^{3/2}, \\ a - c &= r_{22} - r_1 = -\frac{1}{4}(\sqrt{x+2} - 1)(\sqrt{x+2} + 3)^3, \\ z_1(x) &= -\frac{r_{21} - r_{22}}{r_{22} - r_1} = -\frac{16(x+2)^{3/2}}{(\sqrt{x+2} - 1)(\sqrt{x+2} + 3)^3}\,. \end{aligned}$$

We find

$$f_1(x) = \frac{1}{2\pi}(x+2)^3(1 - \sqrt{x+2})^{1/2}(3 + \sqrt{x+2})^{3/2}{}_2F_1\left(-\frac{1}{2}, \frac{3}{2}, 3; z_1(x)\right).$$

In other words, $f_1(x) = g_1(\sqrt{x+2})$ with

$$g_1(y) = \frac{1}{2\pi} y^6 (1-y)^{1/2} (y+3)^{3/2} {}_2F_1\left(-\frac{1}{2}, \frac{3}{2}, 3; \frac{-16y^3}{(y-1)(y+3)^3}\right).$$

On the other hand, if $-1 \le x \le 7$, we have

$$\begin{aligned} b - a = r_{21} - r_1 &= -\frac{1}{4}(\sqrt{x+2}+1)(\sqrt{x+2}-3)^3, \\ a - c = r_1 - r_{22} &= \frac{1}{4}(\sqrt{x+2}-1)(\sqrt{x+2}+3)^3, \\ z_2(x) = -\frac{r_{21}-r_1}{r_1-r_{22}} &= -\frac{(\sqrt{x+2}+1)(\sqrt{x+2}-3)^3}{(\sqrt{x+2}-1)(\sqrt{x+2}+3)^3}. \end{aligned}$$

We find

$$\begin{aligned} f_2(x) = \frac{1}{512\pi} &(\sqrt{x+2}-3)^6(\sqrt{x+2}+1)^2 \\ &\times (\sqrt{x+2}-1)^{1/2}(\sqrt{x+2}+3)^{3/2} {}_2F_1\left(-\frac{1}{2}, \frac{3}{2}, 3; z_2(x)\right). \end{aligned}$$

By performing a linear transformation [**20**, p. 47], we obtain Theorem 10.2. □

11. Moments

If $n \ge 0$, the n-th moment of τ_1 is

$$M_n = \int_G \boldsymbol{\tau}(g)^n \, dg\,.$$

If n is given, these moments can be computed: with the definition of the density function f_τ given in Theorem 10.2, we have

$$M_n = \int_{-2}^{7} x^n f_\tau(x)\, dx\,.$$

The sequence begins as

$$1, 0, 1, 1, 4, 10, 35, 120, 455, 1792, 7413, \ldots$$

The random variable τ_1 is *standardized, i.e.* the *mean* $\mu = M_1$ is equal to zero and the *variance* $\sigma^2 = M_2$ is equal to one (σ is the *standard deviation*). More generally, M_n is the dimension of the space of invariants of n-th tensor power of the 7-dimensional irreducible representation of $\mathbf{G}_2$. The corresponding sequence is labeled A059710 in the OEIS [**1**]. According to Mihailovs, this sequence can be defined by the third-order recurrence formula $M_0 = 1, M_1 = 0, M_2 = 1$ and if $n \ge 3$:

$$\begin{aligned} (n+5)(n+6)M_n = 2&(n-1)(2n+5)M_{n-1} \\ &+ (n-1)(19n+18)M_{n-2} + 14(n-1)(n-2)M_{n-3}\,. \end{aligned}$$

REMARK 11.1. Define the two Laurent polynomials

$$\begin{aligned} X &= \frac{x^2}{y} + \frac{y}{x^2} + \frac{x}{y} + \frac{y}{x} + x + \frac{1}{x} + 1\,, \\ Y &= x - \frac{y^2}{x^3} + \frac{y^3}{x^6} - \frac{y^3}{x^8} + \frac{y^2}{x^8} - \frac{1}{x^6} + \frac{1}{x^3y^2} - \frac{x}{y^4} + \frac{x^4}{y^5} - \frac{x^6}{y^5} + \frac{x^6}{y^4} - \frac{x^4}{y^2}\,. \end{aligned}$$

If $n \ge 2$, then M_n is the constant term in YX^{n-1}, see Westbury [**30**].

The *skewness* of τ is

$$\gamma_1 = \int_G \left(\frac{\tau(g) - \mu}{\sigma}\right)^3 dg = M_3 = 1\,.$$

Since the skewness is > 0, the right tail is longer; the mass of the distribution is concentrated on the left of the figure, and f is said to be skewed to the right.

The *(Pearson) kurtosis* of τ is

$$\beta_2 = \int_G \left(\frac{\tau(g) - \mu}{\sigma}\right)^4 dg = M_4 = 4\,.$$

The *kurtosis excess* of τ is

$$\gamma_2 = \beta_2 - 3 = 1\,.$$

Since $\gamma_2 > 0$, the density distribution of τ is *leptokurtic* (a high peak).

References

[1] *The On-Line Encyclopedia of Integer Sequences*, published at https://oeis.org, 2015.

[2] *The Wolfram functions site*, published at http://functions.wolfram.com, Wolfram Research, 2015.

[3] Ilka Agricola, *Old and new on the exceptional group G_2*, Notices Amer. Math. Soc. **55** (2008), no. 8, 922–929. MR2441524

[4] Armand Borel, *Linear algebraic groups*, 2nd ed., Graduate Texts in Mathematics, vol. 126, Springer-Verlag, New York, 1991. MR1102012

[5] Nicolas Bourbaki, *Commutative algebra. Chapters 1–7*, Elements of Mathematics (Berlin), Springer-Verlag, Berlin, 1989. Translated from the French; Reprint of the 1972 edition. MR979760

[6] Nicolas Bourbaki, *Lie groups and Lie algebras. Chapters 4–6*, Elements of Mathematics (Berlin), Springer-Verlag, Berlin, 2002. Translated from the 1968 French original by Andrew Pressley. MR1890629

[7] Nicolas Bourbaki, *Lie groups and Lie algebras. Chapters 7–9*, Elements of Mathematics (Berlin), Springer-Verlag, Berlin, 2005. Translated from the 1975 and 1982 French originals by Andrew Pressley. MR2109105

[8] I. S. Gradshteyn and I. M. Ryzhik, *Table of integrals, series, and products*, 7th ed., Elsevier/Academic Press, Amsterdam, 2007. Translated from the Russian; Translation edited and with a preface by Alan Jeffrey and Daniel Zwillinger; With one CD-ROM (Windows, Macintosh and UNIX). MR2360010

[9] William Fulton and Joe Harris, *Representation theory*, Graduate Texts in Mathematics, vol. 129, Springer-Verlag, New York, 1991. A first course; Readings in Mathematics. MR1153249

[10] Sigurdur Helgason, *Differential geometry, Lie groups, and symmetric spaces*, Pure and Applied Mathematics, vol. 80, Academic Press, Inc. [Harcourt Brace Jovanovich, Publishers], New York-London, 1978. MR514561

[11] James E. Humphreys, *Introduction to Lie algebras and representation theory*, Graduate Texts in Mathematics, vol. 9, Springer-Verlag, New York-Berlin, 1978. Second printing, revised. MR499562

[12] James E. Humphreys, *Conjugacy classes in semisimple algebraic groups*, Mathematical Surveys and Monographs, vol. 43, American Mathematical Society, Providence, RI, 1995. MR1343976

[13] Nicholas M. Katz and Peter Sarnak, *Random matrices, Frobenius eigenvalues, and monodromy*, American Mathematical Society Colloquium Publications, vol. 45, American Mathematical Society, Providence, RI, 1999. MR1659828

[14] Nicholas M. Katz, *Letter to J.-P. Serre, June 16, 2003.*

[15] Nicholas M. Katz, *Notes on G_2, determinants, and equidistribution*, Finite Fields Appl. **10** (2004), no. 2, 221–269, DOI 10.1016/j.ffa.2003.11.002. MR2045016

[16] Nicholas M. Katz, *G_2 and some exceptional Witt vector identities*, Finite Fields Appl. **47** (2017), 125–144, DOI 10.1016/j.ffa.2017.06.008. MR3681084

[17] J. P. Keating, N. Linden, and Z. Rudnick, *Random matrix theory, the exceptional Lie groups and L-functions*, J. Phys. A **36** (2003), no. 12, 2933–2944, DOI 10.1088/0305-4470/36/12/305. Random matrix theory. MR1986400

[18] Tatsuo Kimura, *Introduction to prehomogeneous vector spaces*, Translations of Mathematical Monographs, vol. 215, American Mathematical Society, Providence, RI, 2003. Translated from the 1998 Japanese original by Makoto Nagura and Tsuyoshi Niitani and revised by the author. MR1944442

[19] Gilles Lachaud, *On the distribution of the trace in the unitary symplectic group and the distribution of Frobenius*, Frobenius distributions: Lang-Trotter and Sato-Tate conjectures, Contemp. Math., vol. 663, Amer. Math. Soc., Providence, RI, 2016, pp. 185–221, DOI 10.1090/conm/663/13355. MR3502944

[20] Wilhelm Magnus, Fritz Oberhettinger, and Raj Pal Soni, *Formulas and theorems for the special functions of mathematical physics*, Third enlarged edition. Die Grundlehren der mathematischen Wissenschaften, Band 52, Springer-Verlag New York, Inc., New York, 1966. MR0232968

[21] M. Sato and T. Kimura, *A classification of irreducible prehomogeneous vector spaces and their relative invariants*, Nagoya Math. J. **65** (1977), 1–155. MR0430336

[22] Jean-Pierre Serre, *Letter to N. Katz.* June 14, 2003.

[23] Jean-Pierre Serre, *On the values of the characters of compact Lie groups.* Oberwolfach Reports 1 (2004), 666–667.

[24] Jean-Pierre Serre, *Lectures on $N_X(p)$*, Chapman & Hall/CRC Research Notes in Mathematics, vol. 11, CRC Press, Boca Raton, FL, 2012. MR2920749

[25] Jean-Pierre Serre, *Letter to N. Katz.* April 12, 2015.

[26] D. Speiser, *Fundamental representations of Lie groups*, Helv. Phys. Acta **38** (1965), 73–97. MR0177708

[27] Tonny A. Springer and Ferdinand D. Veldkamp, *Octonions, Jordan algebras and exceptional groups*, Springer Monographs in Mathematics, Springer-Verlag, Berlin, 2000. MR1763974

[28] Robert Steinberg, *Regular elements of semisimple algebraic groups*, Inst. Hautes Études Sci. Publ. Math. **25** (1965), 49–80. MR0180554

[29] Robert Steinberg, *Conjugacy classes in algebraic groups*, Lecture Notes in Mathematics, Vol. 366, Springer-Verlag, Berlin-New York, 1974. Notes by Vinay V. Deodhar. MR0352279

[30] Bruce W. Westbury, *Enumeration of non-positive planar trivalent graphs*, J. Algebraic Combin. **25** (2007), no. 4, 357–373, DOI 10.1007/s10801-006-0041-4. MR2320368

Aix Marseille Université, CNRS, Centrale Marseille, I2M, UMR 7373, 13453 Marseille, France

Contemporary Mathematics
Volume **722**, 2019
https://doi.org/10.1090/conm/722/14537

The de Rham cohomology of the Suzuki curves

Beth Malmskog, Rachel Pries, and Colin Weir

ABSTRACT. For a natural number m, let $\mathcal{S}_m/\mathbf{F}_2$ be the mth Suzuki curve. We study the mod 2 Dieudonné module of $\mathcal{S}_m$, which gives the equivalent information as the Ekedahl-Oort type or the structure of the 2-torsion group scheme of its Jacobian. We accomplish this by studying the de Rham cohomology of $\mathcal{S}_m$. For all m, we determine the structure of the de Rham cohomology as a 2-modular representation of the mth Suzuki group and the structure of a submodule of the mod 2 Dieudonné module. For $m = 1$ and 2, we determine the complete structure of the mod 2 Dieudonné module.

1. Introduction

The structure of the de Rham cohomology of the Hermitian curves as a representation of $\mathrm{PGU}(3, q)$ was studied in [**3**, **4**, **12**]. The mod p Dieudonné module and the Ekedahl-Oort type of the Hermitian curves were determined in [**22**]. In this paper, we study the analogous structures for the Suzuki curves.

For $m \in \mathbb{N}$, let $q_0 = 2^m$, and let $q = 2^{2m+1}$. The Suzuki curve $\mathcal{S}_m$ is the smooth projective connected curve over $\mathbb{F}_2$ given by the affine equation:

$$z^q + z = y^{q_0}(y^q + y).$$

It has genus $g_m = q_0(q - 1)$.

The number of points of $\mathcal{S}_m$ over $\mathbb{F}_q$ is $\#\mathcal{S}_m(\mathbb{F}_q) = q^2 + 1$; which is *optimal* in that it reaches Serre's improvement to the Hasse-Weil bound [**14**, Proposition 2.1]. In fact, $\mathcal{S}_m$ is the unique $\mathbb{F}_q$-optimal curve of genus g_m [**8**]. Because of the large number of rational points relative to their genus, the Suzuki curves provide good examples of Goppa codes [**9**],[**10**], [**14**].

The automorphism group of $\mathcal{S}_m$ is the Suzuki group $\mathrm{Sz}(q)$. The order of $\mathrm{Sz}(q)$ is $q^2(q-1)(q^2+1)$ which is very large compared with g_m. In fact, $\mathcal{S}_m$ is the Deligne-Lusztig curve associated with the group $\mathrm{Sz}(q) = {}^2B_2(q)$ [**13**, Proposition 4.3].

The L-polynomial of $\mathcal{S}_m/\mathbb{F}_q$ is $(1 + \sqrt{2q}t + qt^2)^{g_m}$ and so $\mathcal{S}_m$ is supersingular for each $m \in \mathbb{N}$ [**13**, Proposition 4.3]. This implies that the Jacobian $\mathrm{Jac}(\mathcal{S}_m)$ is isogenous over $\bar{\mathbb{F}}_2$ to a product of supersingular elliptic curves. In particular, $\mathrm{Jac}(\mathcal{S}_m)$ has 2-rank 0; it has no points of order 2 over $\overline{\mathbb{F}}_2$.

2010 *Mathematics Subject Classification.* Primary 11G10, 11G20, 14F40, 14H40, 20C20; Secondary 14L15, 20C33.

Key words and phrases. Suzuki curve, Suzuki group, Ekedahl-Oort type, de Rham cohomology, Dieudonné module, modular representation.

The 2-torsion group scheme $\mathrm{Jac}(\mathcal{S}_m)[2]$ is a BT_1-group scheme of rank 2^{2g_m}. By [**7**], the a-number of $\mathrm{Jac}(\mathcal{S}_m)[2]$ is $a_m = q_0(q_0+1)(2q_0+1)/6$; in particular, $\lim_{m\to\infty} a_m/g_m = 1/6$. However, the Ekedahl-Oort type of $\mathrm{Jac}(\mathcal{S}_m)[2]$ is not known. Understanding the Ekedahl-Oort type is equivalent to understanding the structure of the de Rham cohomology or the mod 2 reduction of the Dieudonné module as a module under the actions of the operators Frobenius F and Verschiebung V.

In this paper, we study the de Rham cohomology group $H^1_{\mathrm{dR}}(\mathcal{S}_m)$ of the Suzuki curves. The 2-modular representations of the Suzuki group are understood from [**2, 16, 18, 23**]. Using results about the cohomology of Deligne-Lusztig varieties from [**17**] and [**11**], we determine the multiplicity of each irreducible 2-modular representation of $\mathrm{Sz}(q)$ in $H^1_{\mathrm{dR}}(\mathcal{S}_m)$ in Corollary 2.2.

Let D_m denote the mod 2 reduction of the Dieudonné module of (the Jacobian of) $\mathcal{S}_m$. It is an $\mathbb{E}$-module where $\mathbb{E}$ is the non-commutative ring generated over $\bar{\mathbb{F}}_2$ by F and V with the relations $FV = VF = 0$. As explained in Section 3.1, there is an $\mathbb{E}$-module decomposition $D_m = D_{m,0} \oplus D_{m,\neq 0}$, where the $\mathbb{E}$-submodule $D_{m,0}$ is the trivial eigenspace for the action of an automorphism τ of order $q-1$.

In Proposition 3.1, we determine the structure of $D_{m,0}$ completely by finding that its Ekedahl-Oort type is $[0, 1, 1, 2, 2, \ldots, q_0 - 1, q_0]$. This yields the following corollary.

COROLLARY 1.1. *(Corollary 3.10) If $2^m \equiv 2^e \bmod 2^{e+1}+1$, then the $\mathbb{E}$-module $\mathbb{E}/\mathbb{E}(V^{e+1}+F^{e+1})$ occurs as an $\mathbb{E}$-submodule of the mod 2 Dieudonné module D_m of $\mathcal{S}_m$. In particular,*

(1) *$\mathbb{E}/\mathbb{E}(V^{m+1}+F^{m+1})$ occurs as an $\mathbb{E}$-submodule of D_m for all m;*
(2) *$\mathbb{E}/\mathbb{E}(V+F)$ occurs as an $\mathbb{E}$-submodule of D_m if m is even; and*
(3) *$\mathbb{E}/\mathbb{E}(V^2+F^2)$ occurs as an $\mathbb{E}$-submodule of D_m if $m \equiv 1 \bmod 4$.*

We have less information about $D_{m,\neq 0}$, the sum of the non-trivial eigenspaces for τ. In Section 3.3, we explain a connection between the Ekedahl-Oort type and irreducible subrepresentations of $H^1_{\mathrm{dR}}(\mathcal{S}_m)$. This motivates Conjecture 3.2, in which we conjecture that the $\mathbb{E}$-module $\mathbb{E}/\mathbb{E}(V^{2m+1}+F^{2m+1})$ occurs with multiplicity 4^m in D_m.

We determine the complete structure of the mod 2 Dieudonné module D_m for $m = 1$ and $m = 2$ in Propositions 3.3-3.4. To do this, we explicitly compute a basis for $H^1_{\mathrm{dR}}(\mathcal{S}_m)$ for all $m \in \mathbb{N}$ in Section 4 and, for $m = 1, 2$, we compute the actions of F and V on this basis.

There is a similar result in [**5**] for the first Ree curve, which is defined over $\mathbb{F}_3$, namely the authors determine its mod 3 Dieudonné module.

Malmskog was partially supported by NSA grant H98230-16-1-0300. Pries was partially supported by NSF grant DMS-15-02227. We would like to thank Jeff Achter for helpful comments.

1.1. Notation. We begin by establishing some notation regarding p-torsion group schemes, mod p Dieudonné modules, and Ekedahl-Oort types, taken directly from [**22**, Section 2].

Let k be an algebraically closed field of characteristic $p > 0$. Suppose A is a principally polarized abelian variety of dimension g defined over k. Consider the multiplication-by-p morphism $[p] : A \to A$ which is a finite flat morphism of degree p^{2g}. It factors as $[p] = V \circ F$. Here $F : A \to A^{(p)}$ is the relative Frobenius morphism coming from the p-power map on the structure sheaf; it is purely inseparable of degree p^g. The Verschiebung morphism $V : A^{(p)} \to A$ is the dual of $F_{A^{\mathrm{dual}}}$.

The *p-torsion group scheme* of A, denoted $A[p]$, is the kernel of $[p]$. It is a finite commutative group scheme annihilated by p, again having morphisms F and V, with $\mathrm{Ker}(F) = \mathrm{Im}(V)$ and $\mathrm{Ker}(V) = \mathrm{Im}(F)$. The principal polarization of A induces a symmetry on $A[p]$ as defined in [**20**, 5.1]; when $p = 2$, there are complications with the polarization which are resolved in [**20**, 9.2, 9.5, 12.2].

There are two important invariants of (the p-torsion of) A: the p-rank and a-number. The *p-rank* of A is $f = \dim_{\mathbb{F}_p} \mathrm{Hom}(\mu_p, A[p])$ where μ_p is the kernel of Frobenius on $\mathbb{G}_m$. Then p^f is the cardinality of $A[p](k)$. The *a-number* of A is $a = \dim_k \mathrm{Hom}(\alpha_p, A[p])$ where α_p is the kernel of Frobenius on $\mathbb{G}_a$.

One can describe the group scheme $A[p]$ using the *mod p Dieudonné module*, i.e., the modulo p reduction of the covariant Dieudonné module, see e.g., [**20**, 15.3]. More precisely, there is an equivalence of categories between finite commutative group schemes over k annihilated by p and left $\mathbb{E}$-modules of finite dimension. Here $\mathbb{E} = k[F, V]$ denotes the non-commutative ring generated by semi-linear operators F and V with the relations $FV = VF = 0$ and $F\lambda = \lambda^p F$ and $\lambda V = V\lambda^p$ for all $\lambda \in k$. Let $\mathbb{E}(A_1, \dots)$ denote the left ideal of $\mathbb{E}$ generated by $A_1, \dots$.

Furthermore, there is a bijection between isomorphism classes of $2g$ dimensional left $\mathbb{E}$-modules and *Ekedahl-Oort types*. To find the Ekedahl-Oort type, let N be the mod p Dieudonné module of $A[p]$. The canonical filtration of N is the smallest filtration of N stabilized by the action of F^{-1} and V; denote it by

$$0 = N_0 \subset N_1 \subset \cdots N_z = N.$$

The canonical filtration can be extended to a final filtration; the Ekedahl-Oort type is the tuple $[\nu_1, \dots, \nu_g]$, where the ν_i are the dimensions of the images of V on the subspaces in the final filtration.

For example, let $I_{t,1}$ denote the p-torsion group scheme of rank p^{2t} having p-rank 0 and a-number 1. Then $I_{t,1}$ has Dieudonné module $\mathbb{E}/\mathbb{E}(F^t + V^t)$ and Ekedahl-Oort type $[0, 1, \dots, t-1]$ [**21**, Lemma 3.1].

For a smooth projective curve X, by [**19**, Section 5], there is an isomorphism of $\mathbb{E}$-modules between the contravariant mod p Dieudonné module of the p-torsion group scheme $\mathrm{Jac}(X)[p]$ and the de Rham cohomology $H^1_{\mathrm{dR}}(X)$.[1]

In the rest of the paper, $p = 2$.

2. The de Rham cohomology as a representation for the Suzuki group

In this section, we analyze the de Rham cohomology $H^1_{\mathrm{dR}}(\mathcal{S}_m)$ of the Suzuki curve as a 2-modular representation of the Suzuki group.

2.1. Some ordinary representations. Suzuki determined the irreducible ordinary characters and representations of $\mathrm{Sz}(q)$ [**24**]. Consider the following four unipotent representations of $\mathrm{Sz}(q)$. Let W_S denote the Steinberg representation of dimension q^2. Let W_0 be the trivial representation of dimension 1. Let W_+ and W_- be the two unipotent cuspidal representations of $\mathrm{Sz}(q)$, associated to the two ordinary characters of $\mathrm{Sz}(q)$ of degree $q_0(q-1)$ [**24**]. Then W_+ and W_- each have dimension $q_0(q-1)$.

[1] Differences between the covariant and contravariant theory do not cause a problem in this paper since all objects we consider are symmetric.

In [**17**, Theorem 6.1], Lusztig studied the compactly supported ℓ-adic cohomology of the affine Deligne-Lusztig curves. For the Suzuki curves, he proved that the ordinary representations W_S, W_+, W_-, W_0 are the eigenspaces under Frobenius and that each appears with multiplicity 1.

2.2. Modular representations of the Suzuki group. The absolutely irreducible 2-modular representations of $\mathrm{Sz}(q)$ are well-understood [**2, 16, 18, 23**].

Let $q = 2^{2m+1}$. We recall some results about the 2-modular representations of the Suzuki group $\mathrm{Sz}(q)$ from [**18**]. Fix a generator ζ of $\mathbb{F}_q^*$. Let $\theta \in \mathrm{Aut}(\mathbb{F}_q)$ be such that $\theta^2(\alpha) = \alpha^2$ for all $\alpha \in \mathbb{F}_q$, i.e., θ is the square root of Frobenius.

The Suzuki group acts on $\mathcal{S}_m$. Let $\tau \in \mathrm{Sz}(q)$ be an element of order $q-1$; without loss of generality, we suppose that τ acts on $\mathcal{S}_m$ by

$$\tau : y \mapsto \zeta y,\ z \mapsto \zeta^{2^m+1} z.$$

Then $\mathrm{Sz}(q)$ has an irreducible 4-dimensional 2-modular representation V_0 in which $\tau \mapsto M$, where $M \in \mathrm{GL}_4(\mathbb{F}_q)$ is the matrix

$$M = \begin{pmatrix} \zeta^{\theta+1} & 0 & 0 & 0 \\ 0 & \zeta & 0 & 0 \\ 0 & 0 & \zeta^{-1} & 0 \\ 0 & 0 & 0 & \zeta^{-(\theta+1)} \end{pmatrix}.$$

For $0 \le i \le 2m$, consider the automorphism α_i of $\mathrm{Sz}(q)$ induced by the automorphism $x \mapsto x^{2^i}$ of $\mathbb{F}_q$. Let V_i be the 4-dimensional $\mathbb{F}_q\mathrm{Sz}(q)$-module where $g \in \mathrm{Sz}(q)$ acts as g^{α_i} on V_0.

Let I be a subset of $N = \mathbb{Z}/(2m+1)\mathbb{Z}$. Define $V_I = \otimes_{j\in I} V_j$, with $V_\emptyset$ being the trivial module. Then V_I is an absolutely irreducible 2-modular representation of $\mathrm{Sz}(q)$. By [**18**, Lemma 1], if $I \neq J$ then V_I and V_J are geometrically non-isomorphic and $\{V_I \mid I \subset N\}$ is the complete set of simple $\overline{\mathbb{F}}_2\mathrm{Sz}(q)$-modules. Note that V_I has dimension $4^{|I|}$ and that V_N is the Steinberg module.

By [**23**, Theorem, page 1], for $I, J \subset N$, there are no non-trivial extensions of V_I by V_J, namely $\mathrm{Ext}^1_{\overline{\mathbb{F}}_2\mathrm{Sz}(q)}(V_I, V_J) = 0$.

The Frobenius $x \mapsto x^2$ on $\mathbb{F}_q$ acts on $\{V_i\}$ taking $V_i \mapsto V_{i+1 \bmod 2m+1}$. Note that $\oplus_{I\in\mathcal{I}} V_I$ is an $\mathbb{F}_2\mathrm{Sz}(q)$-module if and only if $\mathcal{I}$ is invariant under Frobenius or, equivalently, if and only if $\{I \mid I \in \mathcal{I}\}$ is invariant under the translation $i \mapsto i+1 \bmod 2m+1$.

For $i \in N$, let ϕ_i denote the Brauer character associated to the 4-dimensional module V_i. For $I \subseteq N$, let $\phi_I = \prod_{i\in I} \phi_i$, so ϕ_I is the character associated to the module V_I. Then $\{\phi_I : I \subseteq N\}$ is a complete set of Brauer characters for $\mathrm{Sz}(q)$.

By [**2**, Theorem 3.4], $\phi_i^2 = 4 + 2\phi_{i+m+1} + \phi_{i+1}$. Using this relation, Liu constructs a graph with vertex set N and edge set $\{(i, i+1), (i, i+1+m) : i \in N\}$. Edges of the form $(i, i+1)$ are called short edges and edges of the form $(i, i+1+m)$ are called long edges. Two vertices i, j are called adjacent if they are connected by a long edge, i.e., if $i - j \equiv \pm m \bmod 2m+1$. A set $I' \subseteq N$ is called circular if no vertices of $I = N \setminus I'$ are adjacent. A set $I \subseteq S$ is called good if $I' = N \setminus I$ is circular.

The decompositions of W_+ and W_- into irreducible 2-modular representations are known.

THEOREM 2.1. *Liu* [**16**, Theorem 3.4] *The irreducible* 2*-modular representation* V_I *occurs in* $W_\pm$ *if and only if* I *is good, i.e., if and only if there do not exist* $i, j \in I$

such that $j - i \equiv \pm m \bmod 2m + 1$. *In this case, the multiplicity of* V_I *in* $W_\pm$ *is* $2^{m-|I|}$.

2.3. Modular representation of the de Rham cohomology. The de Rham cohomology $H^1_{\mathrm{dR}}(\mathcal{S}_m)$ is an $\mathbb{F}_2[\mathrm{Sz}(q)]$-module of dimension $2g_m = 2q_0(q-1)$. We consider the decomposition of $H^1_{\mathrm{dR}}(\mathcal{S}_m)$ into irreducible 2-modular representations of the Suzuki group $\mathrm{Sz}(q)$.

COROLLARY 2.2. *The irreducible* 2*-modular representation* V_I *occurs in* $H^1_{\mathrm{dR}}(\mathcal{S}_m)$ *if and only if there do not exist* $i, j \in I$ *such that* $j - i \equiv \pm m \bmod 2m+1$. *If* V_I *occurs in* $H^1_{\mathrm{dR}}(\mathcal{S}_m)$ *then its multiplicity is* $2^{m+1-|I|}$. *Thus the* 2*-modular* $\mathrm{Sz}(q)$*-representation of* $H^1_{\mathrm{dR}}(\mathcal{S}_m)$ *is:*

$$H^1_{\mathrm{dR}}(\mathcal{S}_m) \simeq \bigoplus_{I \text{ good}} V_I^{2^{m+1-|I|}}. \tag{2.1}$$

PROOF. In [**11**, page 2535], Gross uses [**17**, Theorem 6.1] to prove that, as a $\mathrm{Sz}(q)$-representation, the ℓ-adic cohomology of the smooth projective curve $\mathcal{S}_m$ is:

$$H^1(\mathcal{S}_{m,\bar{\mathbb{F}}_2}, \bar{\mathbb{Q}}_\ell) \simeq W_+ \oplus W_-.$$

By [**15**, Theorem 2], the characters of $H^1(\mathcal{S}_{m,\bar{\mathbb{F}}_2}, \bar{\mathbb{Q}}_\ell)$ and $H^1_{\mathrm{crys}}(\mathcal{S}_m, \mathrm{Frac}(W(\bar{\mathbb{F}}_2)))$ as representations of $\mathrm{Sz}(q)$ are the same, and thus the representations are isomorphic. The de Rham cohomology is the reduction modulo 2 of the crystalline cohomology. Thus the result follows from Theorem 2.1. □

EXAMPLE 2.3. When $m = 1$, then $H^1_{\mathrm{dR}}(\mathcal{S}_m) \simeq (V_0 \oplus V_1 \oplus V_2)^2 \oplus V_\emptyset^4$.

EXAMPLE 2.4. When $m = 2$, then

$$H^1_{\mathrm{dR}}(\mathcal{S}_m) \simeq \left(V_{\{0,1\}} \oplus V_{\{1,2\}} \oplus V_{\{2,3\}} \oplus V_{\{3,4\}} \oplus V_{\{4,0\}}\right)^2 \oplus (V_0 \oplus V_1 \oplus V_2 \oplus V_3 \oplus V_4)^4 \oplus V_\emptyset^8.$$

REMARK 2.5. For $m \le 10$, we verified Corollary 2.2 using the multiplicity of the eigenvalues for τ on $H^1_{\mathrm{dR}}(\mathcal{S}_m)$.

3. The Dieudonné module and de Rham cohomology

In this section, we study the structure of the mod 2 Dieudonné module D_m of the Suzuki curve $\mathcal{S}_m$ or, equivalently, the structure of $H^1_{\mathrm{dR}}(\mathcal{S}_m)$ as an $\mathbb{E}$-module.

3.1. Results and conjectures. The chosen element $\tau \in \mathrm{Sz}(q)$ of order $q-1$ acts on the mod 2 Dieudonné module D_m. Let $D_{m,0}$ denote the trivial eigenspace and $D_{m,\neq 0}$ denote the direct sum of the non-trivial eigenspaces. Since F and V commute with τ, they stabilize $D_{m,0}$ and $D_{m,\neq 0}$; thus there is an $\mathbb{E}$-module decomposition $D_m = D_{m,0} \oplus D_{m,\neq 0}$.

In Section 3.2, we prove the next proposition; it determines the $\mathbb{E}$-module structure of $D_{m,0}$.

PROPOSITION 3.1. *Let* $m \in \mathbb{N}$ *and let* $q_0 = 2^m$. *The trivial eigenspace* $D_{m,0}$ *of the mod* 2 *Dieudonné module of* $\mathcal{S}_m$ *has Ekedahl-Oort type* $[0, 1, 1, 2, 2, \ldots, q_0-1, q_0]$*; in particular, it has rank* $2q_0$, 2*-rank* 0, *and a-number* 2^{m-1}.

We have less information about the $\mathbb{E}$-module structure of $D_{m,\neq 0}$. In Section 3.3, we explain how the non-trivial representations V_I in $H^1_{\mathrm{dR}}(\mathcal{S}_m)$ lead to $\mathbb{E}$-submodules D_I of the mod 2 Dieudonné module of $H^1_{\mathrm{dR}}(\mathcal{S}_m)$. We would like to understand how to determine the $\mathbb{E}$-module structure of D_I from the representation

V_I for the subset $I \subset N = \mathbb{Z}/(2m+1)\mathbb{Z}$. In Section 3.3, we consider a particular representation W_m, and make the following conjecture.

CONJECTURE 3.2. The multiplicity of $\mathbb{E}/\mathbb{E}(F^{2m+1} + V^{2m+1})$ in the mod 2 Dieudonné module D_m of $\mathcal{S}_m$ is 4^m.

We verify Conjecture 3.2 for $m = 1$ and $m = 2$ in Propositions 3.3 and 3.4. In fact, for $m = 1$ and $m = 2$, we determine the mod 2 Dieudonné module D_m completely. To do this, we find a basis for $H^1_{\mathrm{dR}}(\mathcal{S}_m)$ for all m in Section 4. For $m = 1$, we explicitly compute the action of F and V on this basis, proving that:

PROPOSITION 3.3. *When $m = 1$, then the mod 2 Dieudonné module of $\mathcal{S}_1$ is*

$$D_1 = (\mathbb{E}/\mathbb{E}(F^3 + V^3))^4 \oplus \mathbb{E}/\mathbb{E}(F^2 + V^2).$$

For $m = 2$, we determine the action of F and V on $H^1_{\mathrm{dR}}(\mathcal{S}_m)$ using Magma [**1**]. Consider the $\mathbb{E}$-module $\mathbb{E}(Z)$ generated by X_1, X_2, X_3 with the following relations: $V^3X_1 - F^3X_2 = 0$; $V^4X_2 - F^3X_3 = 0$; and $V^3X_3 - F^4X_1 = 0$. Then $\mathbb{E}(Z)$ is symmetric and has rank 20, p-rank 0, and a-number 3.

PROPOSITION 3.4. *When $m = 2$, then the mod 2 Dieudonné module of $\mathcal{S}_2$ is*

$$D_2 = \left(\mathbb{E}/\mathbb{E}(F^5 + V^5)\right)^{16} \oplus (\mathbb{E}(Z))^4 \oplus (\mathbb{E}/\mathbb{E}(F^3 + V^3) \oplus \mathbb{E}/\mathbb{E}(F + V)).$$

3.2. The trivial eigenspace. The eigenspace $D_{m,0}$ is the subspace of $H^1_{\mathrm{dR}}(\mathcal{S}_m)$ of elements fixed by τ. Since τ acts fixed point freely on the 4-dimensional module V_i for each i [**18**, proof of Lemma 3], the generators of $H^1_{\mathrm{dR}}(\mathcal{S}_m)$ which are fixed by τ are exactly those in V_I for $I = \emptyset$. In other words, the representation for $D_{m,0}$ consists of the $2^{m+1} = 2q_0$ copies of the trivial representation in (2.1).

PROOF. (Proof of Proposition 3.1) Let $\mathcal{C}_{m,0}$ be the quotient curve of $\mathcal{S}_m$ by the subgroup $\langle\tau\rangle$. Then $\mathcal{C}_{m,0}$ is a hyperelliptic curve of genus q_0 [**10**, Theorem 6.9].

The de Rham cohomology $H^1_{\mathrm{dR}}(\mathcal{C}_{m,0})$ of $\mathcal{C}_{m,0}$ is isomorphic as an $\mathbb{E}$-module to $D_{m,0}$. Thus the trivial eigenspace $D_{m,0}$ for the mod 2 Dieudonné module of $\mathcal{S}_m$ is isomorphic to the mod 2 Dieudonné module of $\mathcal{C}_{m,0}$; in particular, it has rank $2q_0$.

Since $\mathcal{S}_m$ has 2-rank 0, so does $\mathcal{C}_{m,0}$. Thus $\mathcal{C}_{m,0}$ is a hyperelliptic curve of 2-rank 0. By [**6**, Corollary 5.3], the Ekedahl-Oort type of $\mathcal{C}_{m,0}$ is $[0, 1, 1, 2, 2, \ldots, q_0 - 1, q_0]$; this implies that the a-number is 2^{m-1}. □

We determine the $\mathbb{E}$-module structure of $D_{m,0}$ by applying results from [**6**, Section 5].

PROPOSITION 3.5. [**6**, Proposition 5.8] *The mod 2 Dieudonné module $D_{m,0}$ is the $\mathbb{E}$-module generated as a k-vector space by $\{X_1, \ldots, X_{q_0}, Y_1, \ldots, Y_{q_0}\}$ with the actions of F and V given by:*

(1) $F(Y_j) = 0$.

(2) $V(Y_j) = \begin{cases} Y_{2j} & \textit{if } j \le q_0/2, \\ 0 & \textit{if } j > q_0/2. \end{cases}$

(3) $F(X_i) = \begin{cases} X_{j/2} & \textit{if } j \textit{ is even}, \\ Y_{q_0-(j-1)/2} & \textit{if } j \textit{ is odd}. \end{cases}$

(4) $V(X_j) = \begin{cases} 0 & \textit{if } j \le (q_0-1)/2, \\ -Y_{2q_0-2j+1} & \textit{if } j > (q_0-1)/2. \end{cases}$

We have an explicit description of the generators and relations of $D_{m,0}$ as follows.

NOTATION 3.6. [**6**, Notation 5.9] Fix $c = q_0 \in \mathbb{N}$. Consider the set

$$I = \{j \in \mathbb{N} \mid \lceil (c+1)/2 \rceil \leq j \leq c\},$$

which has cardinality $\lfloor (c+1)/2 \rfloor$. For $j \in I$, let $\ell(j)$ be the odd part of j and let $e(j) \in \mathbb{Z}^{\geq 0}$ be such that $j = 2^{e(j)}\ell(j)$. Let $s(j) = c - (\ell(j) - 1)/2$. Then $\{s(j) \mid j \in I\} = I$. Also, let $m(j) = 2c - 2j + 1$ and let $\epsilon(j) \in \mathbb{Z}^{\geq 0}$ be such that $t(j) := 2^{\epsilon(j)} m(j) \in I$. Then $\{t(j) \mid j \in I\} = I$. Thus, there is a unique bijection $\iota : I \to I$ such that $t(\iota(j)) = s(j)$ for each $j \in I$.

PROPOSITION 3.7. [**6**, Proposition 5.10] *The set* $\{X_j \mid j \in I\}$ *generates the mod 2 Dieudonné module* $D_{m,0}$ *as an* $\mathbb{E}$*-module subject to the following relations, for* $j \in I$*:* $F^{e(j)+1}(X_j) + V^{\epsilon(\iota(j))+1}(X_{\iota(j)})$.

EXAMPLE 3.8.
(1) When $m = 1$ and the Ekedahl-Oort type is $[0, 1]$, then $D_{1,0} \simeq \mathbb{E}/\mathbb{E}(F^2 + V^2)$ (group scheme $I_{2,1}$).
(2) When $m = 2$ and the Ekedahl-Oort type is $[0, 1, 1, 2]$, then one checks that $D_{2,0} \simeq \mathbb{E}/\mathbb{E}(F + V) \oplus \mathbb{E}/\mathbb{E}(F^3 + V^3)$ (group scheme $I_{1,1} \oplus I_{3,1}$).

In the next result, we determine some $\mathbb{E}$-submodules of $D_{m,0}$ for general m.

PROPOSITION 3.9. *The* $\mathbb{E}$*-module* $\mathbb{E}/\mathbb{E}(V^{e+1} + F^{e+1})$ *occurs as an* $\mathbb{E}$*-submodule of* $D_{m,0}$ *if and only if* $2^m \equiv 2^e \bmod 2^{e+1} + 1$. *In particular:*

(1) $\mathbb{E}/\mathbb{E}(V^{m+1} + F^{m+1})$ *occurs for all* m*;*
(2) $\mathbb{E}/\mathbb{E}(V + F)$ *occurs if and only if* m *is even; and*
(3) $\mathbb{E}/\mathbb{E}(V^2 + F^2) = 0$ *occurs if and only if* $m \equiv 1 \bmod 4$.

PROOF. Let $e \in \mathbb{Z}^{\geq 0}$. By Proposition 3.7, the relation $(V^{e+1} + F^{e+1})X_j = 0$ is only possible if $j = 2^e \ell$ where ℓ is odd. Write $s(j) = c - (\ell - 1)/2$. Then $F^{e+1}(X_j) = F(X_\ell) = Y_{s(j)}$. Now $V(X_j) = -Y_{m(j)}$ where $m(j) = 2c - 2j + 1$. Also $V^{e+1}(X_j) = 2^e m(j)$. Thus we need $s(j) = 2^e m(j)$. This is equivalent to $2^{e+1}c - (j - 2^e) = 2^{2e+1}(2c - 2j + 1)$, which is equivalent to

$$j = \frac{c2^{e+1}(2^{e+1} - 1) + 2^e(2^{2e+1} - 1)}{2^{2e+2} - 1} = \frac{c2^{e+1} + 2^e}{2^{e+1} + 1}.$$

This value of j is integral if and only if $c \equiv 2^e \bmod 2^{e+1} + 1$. Thus, the relation $(V^{e+1} + F^{e+1})X_j = 0$ occurs if and only if $2^m \equiv 2^e \bmod 2^{e+1} + 1$ and also $j = (2^{e+1}q_0 + 2^e)/(2^{e+1} + 1)$. In particular, one checks that:

(1) $(V^{m+1} + F^{m+1})X_{2^m} = 0$;
(2) the relation $(V + F)X_j = 0$ occurs if and only if m is even and $j = (2 \cdot 2^m + 1)/3$;
(3) the relation $(V^2 + F^2)X_j = 0$ occurs if and only if $m \equiv 1 \bmod 4$ and $j = (4 \cdot 2^m + 2)/5$.

□

As a corollary, we determine cases when the $\mathbb{E}$-module $\mathbb{E}/\mathbb{E}(V^{e+1} + F^{e+1})$ occurs in D_m.

COROLLARY 3.10. *If* $2^m \equiv 2^e \bmod 2^{e+1}+1$, *then* $\mathbb{E}/\mathbb{E}(V^{e+1}+F^{e+1})$ *occurs as an* $\mathbb{E}$*-submodule of the mod 2 Dieudonné module* D_m *of* $\mathcal{S}_m$. *In particular,*

(1) $\mathbb{E}/\mathbb{E}(V^{m+1}+F^{m+1})$ *occurs as an* $\mathbb{E}$*-submodule of* D_m *for all* m*;*
(2) $\mathbb{E}/\mathbb{E}(V+F)$ *occurs as an* $\mathbb{E}$*-submodule of* D_m *if* m *is even; and*
(3) $\mathbb{E}/\mathbb{E}(V^2+F^2)$ *occurs as an* $\mathbb{E}$*-submodule of* D_m *if* $m \equiv 1 \bmod 4$.

PROOF. By Proposition 3.9, $\mathbb{E}/\mathbb{E}(V^{e+1}+F^{e+1})$ occurs as an $\mathbb{E}$-submodule of the mod 2 Dieudonné module $D_{m,0}$. The result follows since $D_{m,0}$ is an $\mathbb{E}$-submodule of D_m. □

3.3. The nontrivial eigenspaces. Recall that $D_{m,\neq 0}$ is the direct sum of the non-trivial eigenspaces for τ. Consider the canonical filtration of $D_{m,\neq 0}$, which is the smallest filtration stabilized under the action of F^{-1} and V; denote it by

$$0 = N_0 \subset N_1 \subset \cdots N_t = N.$$

By [**20**, Chapter 2] (see also [**5**, Section 2.2]), the blocks $B_i = N_{i+1}/N_i$ in the canonical filtration are representations for $H^1_{\mathrm{dR}}(\mathcal{S}_m)$. On each block B_i, either (i) $V|_{B_i} = 0$ in which case $B_i \subset \mathrm{Im}(F)$ and $F^{-1} : B_i^{(p)} \to B_j$ is an isomorphism to another block with index $j > i$; or (ii) $V : B_i^{(p)} \to B_j$ is an isomorphism to another block with index $j < i$. This action of V and F^{-1} yields a permutation π of the set of blocks B_i. Cycles in the permutation are in bijection with orbits $\mathcal{O}$ of the blocks under the action of V and F^{-1}.

Fix an orbit $\mathcal{O}$ of a block B_i under the action of F^{-1} and V. As in [**22**, Section 5.2], this yields a word w in F^{-1} and V. From this, we produce a symmetric $\mathbb{E}$-module $\mathbb{E}(w)$ whose dimension over k is the length of w. Then $\mathbb{E}(w)$ is an isotypic component of $D_{m,0}$. The multiplicity of $\mathbb{E}(w)$ in $D_{m,\neq 0}$ is the dimension of the block B_i in $\mathcal{O}$.

By Corollary 2.2, the representations occurring in $H^1_{\mathrm{dR}}(\mathcal{S}_m)$ are the representations in $W_\pm$, namely the representations V_I for I a good subset of $N = \mathbb{Z}/(2m+1)\mathbb{Z}$.

We now explain the motivation for Conjecture 3.2. Let $I_m = \{0, \dots, m-1\}$. The smallest power of F that stabilizes I_m is $2m+1$. Consider the 2-modular representation of Sz(q) given by $W_m = \oplus_{i=0}^{2m} F^i(V_{I_m})$. For example, when $m = 1$ then $W_1 = V_0 \oplus V_1 \oplus V_2$ and when $m = 2$ then

$$W_2 = (V_0 \otimes V_1) \oplus (V_1 \otimes V_2) \oplus (V_2 \otimes V_3) \oplus (V_3 \otimes V_4) \oplus (V_4 \otimes V_0).$$

By definition, W_m is an $\mathbb{F}_2\mathrm{Sz}(q)$-module of dimension $(2m+1)4^m$. By Corollary 2.2, the 2-modular representation W_m appears with multiplicity 2 in $H^1_{\mathrm{dR}}(\mathcal{S}_m)$. Consider the $\mathbb{E}$-module $\mathbb{E}/\mathbb{E}(F^{2m+1}+V^{2m+1})$; it has dimension $2(2m+1)$ over k.

The idea behind Conjecture 3.2 is that the subrepresentation W_m^2 of $H^1_{\mathrm{dR}}(\mathcal{S}_m)$ should correspond to a submodule of D_m with structure $(\mathbb{E}/\mathbb{E}(F^{2m+1}+V^{2m+1}))^{4^m}$. More precisely, Conjecture 3.2 would follow from the claims that there is a unique i such that V_{I_m} is a subrepresentation of B_i, that B_i is irreducible and thus equal to V_{I_m}, and that the word w on the orbit of B_i is $(F^{-1})^{2m+1}V^{2m+1}$.

4. An explicit basis for the de Rham cohomology

In this section, we compute an explicit basis for $H^1_{\mathrm{dR}}(\mathcal{S}_m)$ for all m. This material is needed to determine the mod 2 Dieudonné module of $\mathcal{S}_m$ when $m = 1$ and $m = 2$ in Propositions 3.3 and 3.4. We determine the action of F and V on the basis elements explicitly here when $m = 1$ and using Magma [**1**] when $m = 2$.

4.1. Preliminaries. Consider the affine equation $z^q + z = y^{q_0}(y^q + y)$ for $\mathcal{S}_m$. Let P_∞ be the point at infinity on $\mathcal{S}_m$. Let $P_{(y,z)}$ denote the point (y, z) on $\mathcal{S}_m$.

Define the functions $h_1, h_2 \in \mathbb{F}_2(\mathcal{S}_m)$ by:

$$h_1 := z^{2q_0} + y^{2q_0+1}, \; h_2 := z^{2q_0}y + h_1^{2q_0}.$$

LEMMA 4.1. (1) *The function y has divisor*

$$\operatorname{div}(y) = \sum_{z \in \mathbb{F}_q} P_{(0,z)} - qP_\infty.$$

(2) *The function z has divisor*

$$\operatorname{div}(z) = \sum_{y \in \mathbb{F}_q^\times} P_{(y,0)} + (q_0 + 1)P_{(0,0)} - (q + q_0)P_\infty.$$

(3) *Let $S = \{(y, z) \in \mathbb{F}_q^2 : y^{2q_0+1} = z^{2q_0}, (y, z) \neq (0, 0)\}$. The function h_1 has divisor*

$$\operatorname{div}(h_1) = \sum_{(y,z) \in S} P_{(y,z)} + (2q_0 + 1)P_{(0,0)} - (q + 2q_0)P_\infty.$$

(4) *The function h_2 has divisor*

$$\operatorname{div}(h_2) = (q + 2q_0 + 1)(P_{(0,0)} - P_\infty).$$

PROOF. The pole orders of these functions are determined in [**14**, Proposition 1.3]. The orders of the zeros can be determined using the equation for the curve and the definitions of h_1 and h_2. □

Let $\mathcal{E}_m$ be the set of $(a, b, c, d) \subset \mathbb{Z}^4$ satisfying

$$\begin{aligned} &0 \le a, \quad 0 \le b \le 1, \quad 0 \le c \le q_0 - 1, \quad 0 \le d \le q_0 - 1, \\ &aq + b(q + q_0) + c(q + 2q_0) + d(q + 2q_0 + 1) \le 2g - 2. \end{aligned}$$

LEMMA 4.2. *The following set is a basis of $H^0(\mathcal{S}_m, \Omega^1)$:*

$$\mathcal{B}_m := \left\{ g_{a,b,c,d} := y^a z^b h_1^c h_2^d \, dy \mid (a, b, c, d) \in \mathcal{E}_m \right\}.$$

PROOF. See [**7**, Proposition 3.7]. □

A basis for $H^1(\mathcal{S}_m, \mathcal{O})$ can be built similarly. Define the map

$$\pi : \mathcal{S}_m \to \mathbb{P}_y^1, \; (y, z) \mapsto y, \quad P_\infty \mapsto \infty_y.$$

Let 0_y be the point on $\mathbb{P}_y^1$ with $y = 0$. Then $\pi^{-1}(0_y) = \{(0, z) : z \in \mathbb{F}_q\}$ has cardinality q.

LEMMA 4.3. *The following set represents a basis of $H^1(\mathcal{S}_m, \mathcal{O})$:*

$$\mathcal{A}_m := \left\{ f_{a,b,c,d} := \frac{1}{y^a z^b h_1^c h_2^d} \frac{z h_1^{q_0-1} h_2^{q_0-1}}{y} \mid (a, b, c, d) \in \mathcal{E}_m \right\}.$$

PROOF. Let $U_\infty = \mathcal{S}_m \setminus \pi^{-1}(\infty_y) = \mathcal{S}_m \setminus P_\infty$ and $U_0 = \mathcal{S}_m \setminus \pi^{-1}(0_y)$. The elements of $H^1(\mathcal{S}_m, \mathcal{O})$ can be represented by classes of functions that are regular on $U_\infty \cap U_0$, but are not regular on U_∞ or regular on U_0. In other words, these functions have a pole at P_∞ and at some point in $\pi^{-1}(0_y)$.

Let $f = f_{a,b,c,d}$ for some $(a,b,c,d) \in \mathcal{E}_m$. Then f has poles only in $\{P_\infty, \pi^{-1}(0_y)\}$ by Lemma 4.1. Let $Q = (0,\alpha)$ for some $\alpha \in \mathbb{F}_q^\times$. Then $v_Q(f) = -(a+1) \leq -1$. Also, let $t = q + 2q_0 + 1$, then

$$\begin{aligned} v_{P_\infty}(f) &= (a+1)(q) - (1-b)(q+q_0) - (q_0 - 1 - c)(q + 2q_0) - (q_0 - 1 - d)t \\ &= aq + b(q+q_0) + c(q+2q_0) + d(q+2q_0+1) + (2q_0 - 2q_0q + 1) \\ &\leq 2g_m - 2 + (2q_0 - 2q_0q + 1) \\ &= (2q_0q - 2q_0 - 2) + (2q_0 - 2q_0q + 1) \\ &= -1. \end{aligned}$$

So f is regular on $U_\infty \cap U_0$ but not on U_∞ or U_0. By a calculation similar to [**7**, Proposition 3.7], the elements of $\mathcal{A}_m$ are independent because each element has a different pole order at P_∞. The cardinality of $\mathcal{A}_m$ is $g_m = \dim(H^1(\mathcal{S}_m, \mathcal{O}))$. Thus $\mathcal{A}$ is a basis for $H^1(\mathcal{S}_m, \mathcal{O})$.

□

4.2. Constructing the de Rham cohomology. Let $\mathcal{U}$ be the open cover of $\mathcal{S}_m$ given by U_∞ and U_0 from the proof of Lemma 4.3. For a sheaf $\mathcal{F}$ on $\mathcal{S}_m$, let

$$\begin{aligned} C^0(\mathcal{U}, \mathcal{F}) &:= \{g = (g_\infty, g_0) \mid g_i \in \Gamma(U_i, \mathcal{F})\}, \\ C^1(\mathcal{U}, \mathcal{F}) &:= \{\phi \in \Gamma(U_\infty \cap U_0, \mathcal{F})\}. \end{aligned}$$

Define the coboundary operator $\delta : C^0(\mathcal{U}, \mathcal{F}) \to C^1(\mathcal{U}, \mathcal{F})$ by $\delta g = g_\infty - g_0$. The closed de Rham cocycles are the set

$$Z^1_{\mathrm{dR}}(\mathcal{U}) := \{(f,g) \in C^1(\mathcal{U}, \mathcal{O}) \times C^0(\mathcal{U}, \Omega^1) : df = \delta g\}.$$

The de Rham coboundaries are the set

$$B^1_{\mathrm{dR}}(\mathcal{S}_m) := \{(\delta\kappa, d\kappa) \in Z^1_{\mathrm{dR}}(\mathcal{U}) : \kappa \in C^0(\mathcal{U}, \mathcal{O})\},$$

where $d\kappa = (d(\kappa_0), d(\kappa_\infty))$. The de Rham cohomology $H^1_{\mathrm{dR}}(\mathcal{S}_m)$ is

$$H^1_{\mathrm{dR}}(\mathcal{S}_m) \cong H^1_{\mathrm{dR}}(\mathcal{S}_m)(\mathcal{U}) := Z^1_{\mathrm{dR}}(\mathcal{U}) / B^1_{\mathrm{dR}}(\mathcal{U}).$$

There is an injective homomorphism $\lambda : H^0(\mathcal{S}_m, \Omega^1) \to H^1_{\mathrm{dR}}(\mathcal{S}_m)$ denoted informally by $g \mapsto (0, \mathbf{g})$, where the second coordinate is a tuple $\mathbf{g} = (g_\infty, g_0)$ defined by $g_i = g|_{U_i}$. Define another homomorphism $\gamma : H^1_{\mathrm{dR}}(\mathcal{S}_m) \to H^1(\mathcal{S}_m, \mathcal{O})$ with $(f, \mathbf{g}) \mapsto f$. These create a short exact sequence

$$0 \longrightarrow H^0(\mathcal{S}_m, \Omega^1) \xrightarrow{\lambda} H^1_{\mathrm{dR}}(\mathcal{S}_m) \xrightarrow{\gamma} H^1(\mathcal{S}_m, \mathcal{O}) \longrightarrow 0. \tag{4.1}$$

Let A be a basis for $H^1(\mathcal{S}_m, \mathcal{O})$ and B a basis for $H^0(\mathcal{S}_m, \Omega^1)$. A basis for $H^1_{\mathrm{dR}}(\mathcal{S}_m)$ is then given by $\psi(A) \cup \lambda(B)$, where ψ is defined as follows. Given $f \in H^1(\mathcal{S}_m, \mathcal{O})$, one can write $df = df_\infty + df_0$, where $df_i \in \Gamma(U_i, \Omega^1)$ for $i \in \{0, \infty\}$. For convenience, define $\mathbf{df} = (df_\infty, df_0)$. Define a section of (4.1) by:

$$\psi : H^1(\mathcal{S}_m, \mathcal{O}) \to H^1_{\mathrm{dR}}(\mathcal{S}_m), \ \psi(f) = (f, \mathbf{df}).$$

The image of ψ is a complement in $H^1_{\mathrm{dR}}(\mathcal{S}_m)$ to $\lambda(H^0(\mathcal{S}_m, \Omega^1))$.

4.2.1. *The Frobenius and Verschiebung operators.* The Frobenius F and Verschiebung V act on $H^1_{\mathrm{dR}}(\mathcal{S}_m)$ by

$$F(f,\mathbf{g}) := (f^p,(0,0)) \text{ and } V(f,\mathbf{g}) := (0,\mathscr{C}(\mathbf{g}))$$

where $\mathscr{C}$ is the Cartier operator, which acts componentwise on $\mathbf{g}$. The Cartier operator is defined by the properties that it annihilates exact differentials, preserves logarithmic differentials, and is p^{-1}-linear. It follows from the definitions that

$$\ker(F) = \lambda(H^0(\mathcal{S}_m,\Omega^1)) = \mathrm{im}(V).$$

4.3. The case $m = 1$. When $m = 1$, then $q_0 = 2$, $q = 8$, and $g = 14$. The Suzuki curve $\mathcal{S}_1$ has affine equation

$$z^8 + z = y^2(y^8 + y).$$

The set $\mathcal{E}_1$ consists of the 14 tuples

$$\begin{aligned}\mathcal{E}_1 \;=\; \{&(0,0,0,0),(0,0,0,1),(0,0,1,0),(0,0,1,1),(0,1,0,0),(0,1,0,1),(0,1,1,0),\\ &(1,0,0,0),(1,0,0,1),(1,0,1,0),(1,1,0,0),(2,0,0,0),(2,1,0,0),(3,0,0,0)\}.\end{aligned}$$

By Lemmas 4.2 and 4.3, $\mathcal{B}_1$ is a basis for $H^0(\mathcal{S}_1,\Omega^1)$ and $\mathcal{A}_1$ is a basis for $H^1(\mathcal{S}_1,\mathcal{O})$. Based on the action of Frobenius and Verschiebung, the following sets make more convenient bases:

LEMMA 4.4. (1) *A basis for $H^1(\mathcal{S}_1,\mathcal{O})$ is given by the set*

$$\begin{aligned}A \;=\; \{&f_{(0,0,0,0)}, f_{(2,0,0,0)}, f_{(0,1,0,0)} + f_{(3,0,0,0)}, f_{(2,1,0,0)} + f_{(0,0,1,0)},\\ &f_{(0,0,0,1)} + f_{(1,0,1,0)}, f_{(1,0,0,0)}, f_{(2,1,0,0)}, f_{(1,0,0,1)}, f_{(0,0,1,1)},\\ &f_{(1,0,1,0)}, f_{(3,0,0,0)}, f_{(1,1,0,0)}, f_{(0,1,1,0)}, f_{(0,1,0,1)}\}.\end{aligned}$$

(2) *A basis for $H^0(\mathcal{S}_1,\Omega^1)$ is given by the set*

$$\begin{aligned}B \;=\; \{&g_{(0,0,0,0)}, g_{(2,0,0,0)}, g_{(0,1,0,0)} + g_{(3,0,0,0)}, g_{(2,1,0,0)} + g_{(0,0,1,0)},\\ &g_{(0,0,0,1)} + g_{(1,0,1,0)}, g_{(1,0,0,0)}, g_{(2,1,0,0)}, g_{(1,0,0,1)}, g_{(0,0,1,1)}, g_{(1,0,1,0)},\\ &g_{(3,0,0,0)}, g_{(1,1,0,0)}, g_{(0,1,1,0)}, g_{(0,1,0,1)}\}.\end{aligned}$$

PROOF. By Lemma 4.2 (resp. 4.3), these 1-forms (resp. functions) have distinct pole orders at P_∞, are therefore linearly independent, and thus form a basis of $H^1(\mathcal{S}_1,\mathcal{O})$ (resp. $H^0(\mathcal{S}_1,\Omega^1)$). □

It is now possible to calculate the action of F and V on $\psi(A)\cup\lambda(B)$, a basis for $H^1_{\mathrm{dR}}(\mathcal{S}_1)$.

4.3.1. *The action of Frobenius when $m = 1$.* The action of F is summarized in the right column of Table 2. Note that $F(g) = 0$ for $g \in B$ since $\ker(F) = \mathrm{im}(V) \cong H^0(\mathcal{S}_1,\Omega^1)$. For the action of F on $\psi(f)$ for $f \in A$, note that $F(\psi(f)) = (f^2,(0,0))$. Then

$$\begin{aligned}f^2 = (f_{(a,b,c,d)})^2 &= (y^{-1-a}z^{1-b}h_1^{1-c}h_2^{1-d})^2\\ &= (y^{-2})^{1+a}(yh_1+h_2)^{1-b}(z+y^3)^{1-c}(h_1+zy^2)^{1-d}.\end{aligned}$$

To do these calculations, we simplify f^2 and write it as a sum of quotients of monomials in $\{y,z,h_1,h_2\}$. These monomials can then be classified as belonging to $\Gamma(U_0)$ or $\Gamma(U_\infty)$, or can otherwise be rewritten in terms of the basis for $H^1(\mathcal{S}_1,\mathcal{O})$. It is then possible to use coboundaries to write $(f^2,(0,0))$ in terms of the given basis for $H^1_{\mathrm{dR}}(\mathcal{S}_1)$.

EXAMPLE 4.5. To compute that $F(\psi(f_{(0,1,0,1)})) = \lambda(g_{(0,0,0,0)})$, note first that
$$\left(f_{(0,1,0,1)}\right)^2 = y^{-2}(z+y^3) = \frac{z}{y^2} + y.$$
Also,
$$d\left(\frac{z}{y^2}\right) = \frac{1}{y^2}dz - 2\frac{z}{y^3}dy = dy \text{ and } d(y) = dy.$$
Since $y \in \Gamma(U_\infty, \mathcal{O})$ and $\frac{z}{y^2} \in \Gamma(U_0, \mathcal{O})$, the pair $\left(\frac{z}{y^2}, y\right)$ is in $C^0(\mathcal{U}, \mathcal{O})$ and one sees that $(\frac{z}{y^2} + y, (dy, dy))$ is a coboundary. Thus
$$\begin{aligned} F\left(\psi\left(f_{(0,1,0,1)}\right)\right) &= \left(\frac{z}{y^2} + y, (0,0)\right) + \left(\frac{z}{y^2} + y, (dy, dy)\right) \\ &= (0, (dy, dy)) = \lambda(dy) = (0, \mathbf{g}_{(\mathbf{0},\mathbf{0},\mathbf{0},\mathbf{0})}). \end{aligned}$$

EXAMPLE 4.6. We compute that $F\left(\psi(f_{(0,0,1,1)})\right) = \psi\left(f_{(0,1,0,1)}\right)$. This is true because
$$\left(f_{(0,0,1,1)}\right)^2 = y^{-2}(yh_1 + h_2) = \frac{h_1}{y} + \frac{h_2}{y^2}.$$
Note that $\frac{h_2}{y^2} \in \Gamma(U_0, \mathcal{O})$, so $(\frac{h_2}{y^2}, 0) \in C^0(\mathcal{U}, \mathcal{O})$, and $d\left(\frac{h_2}{y^2}\right) = \frac{z^4}{y^2}dy$. So one sees that $\left(\frac{h_2}{y^2}, (\frac{z^4}{y^2}dy, 0)\right)$ is a coboundary. Also, $d\left(\frac{h_1}{y}\right) = \frac{z^4}{y^2}dy$. Thus
$$\begin{aligned} F\left(\psi\left(f_{(0,0,1,1)}\right)\right) &= \left(\frac{h_1}{y} + \frac{h_2}{y^2}, (0,0)\right) + \left(\frac{h_2}{y^2}, \left(\frac{z^4}{y^2}dy, 0\right)\right) \\ &= \left(\frac{h_1}{y}, \left(\frac{z^4}{y^2}dy, 0\right)\right) = \psi\left(f_{(0,1,0,1)}\right). \end{aligned}$$

4.3.2. *The action of Verschiebung when $m = 1$.* The action of V is summarized in the middle column of Table 2. In [**7**], the authors calculate the action of the Cartier operator $\mathscr{C}$ (see Table 1). This determines the action of V on $\lambda(g)$ for $g \in B$. It also helps determine the action of V on $\psi(f)$ for $f \in A$.

EXAMPLE 4.7. We compute that $V\left(\psi(f_{(0,1,0,1)})\right) = (0, \mathbf{0})$. Writing
$$f = f_{(0,1,0,1)} = \frac{h_1}{y} = \frac{z^4}{y} + y^4,$$
then
$$df = \frac{\partial f}{\partial y}dy + \frac{\partial f}{\partial z}dz = \left(-\frac{z^4}{y^2} + 4y^3\right)dy + 4\frac{z^3}{y}dz = \frac{z^4}{y^2}dy.$$
Considering the pole orders of y, z, and dy, define $df = df_0 \in \Omega_0$ and $df_\infty = 0$, so $\mathbf{df} = (0, df)$. Thus $\mathscr{C}(df) = \frac{z^2}{y}\mathscr{C}(dy) = 0$. Thus $\mathscr{C}(\mathbf{df}) = (0,0) = \mathbf{0}$ and $V(\psi(f_{(0,1,0,1)})) = (0, \mathbf{0})$.

EXAMPLE 4.8. We compute that $V\left(\psi(f_{(2,1,0,0)})\right) = (0, \mathbf{g}_{(0,1,0,0)})$. This is because
$$f_{(2,1,0,0)} = \frac{h_1 h_2}{y^3},$$
so
$$df = y^{-3}d(h_1h_2) + y^{-4}h_1h_2dy = y^{-3}h_1d(h_2) + y^{-3}h_2d(h_1) + y^{-4}h_1h_2dy.$$
Then
$$d(h_1) = d(z^4 + y^5) = y^4dy \text{ and } d(h_2) = d(z^4y + h_1^4) = z^4dy,$$

so

$$\begin{aligned} df &= y^{-3}z^4 h_1 dy + yh_2 dy + y^{-4}h_1 h_2 \\ &= y^{-3}h_1(h_1 + y^5)dy + yh_2 dy + y^{-4}h_1 h_2 \\ &= \frac{h_1^2}{y^3}dy + \frac{h_1 h_2}{y^4}dy + y^2 h_1 dy + yh_2 dy, \end{aligned}$$

using the fact that $z^4 = h_1 + y^5$. Considering the orders of the poles, define $df_0 = \frac{h_1^2}{y^3}dy + \frac{h_1 h_2}{y^4}dy \in \Omega_0$ and $df_\infty = y^2 h_1 dy + yh_2 dy \in \Omega_\infty$. Using Table 1 and the fact that $h_1^2 = z + y^3$, then

$$\begin{aligned} \mathscr{C}(df_\infty) &= y\mathscr{C}(h_1 dy) + \mathscr{C}(yh_2 dy) \\ &= y^3 dy + h_1^2 dy = (y^3 + z + y^3)dy = zdy. \end{aligned}$$

Thus $V\left(\psi(f_{(2,1,0,0)})\right) = (0, \mathbf{g}_{(0,1,0,0)})$.

The actions of F and V are summarized in Table 2.

TABLE 1. Cartier Operator on $H^0(\mathcal{S}_1, \Omega^1)$

f	$\mathscr{C}(fdy)$
1	0
y	dy
z	$y^{q_0/2}\,dy$
h_1	$y^{q_0}\,dy$
h_2	$\left((yh_1)^{q_0/2} + h_2\right)\,dy$
yz	$h_1^{q_0/2}\,dy$
yh_1	$\left((yh_1)^{q_0/2} + h_2\right)\,dy$
zh_1	$(yh_2)^{q_0/2}\,dy$
zh_2	$(h_1 h_2)^{q_0/2}\,dy$
$h_1 h_2$	$(h_1 + zy^{q_0})\,dy$
yzh_1	$\left(y^{q_0/2}z + (h_1 h_2)^{q_0/2}\right)\,dy$
yzh_2	$\left(zh_1^{q_0/2} + y^{q_0/2+1}h_2^{q_0/2}\right)\,dy$
$zh_1 h_2$	$\left(zy^{q_0/2}h_2^{q_0/2} + h_1^{q_0/2+1}\right)\,dy$
$yh_1 h_2$	$\left((yh_1)^{q_0/2}z + h_2^{q_0/2}z\right)\,dy$
$yzh_1 h_2$	$\left(y^{q_0/2}h_2 + zh_1^{q_0/2}h_2^{q_0 2}\right)\,dy$

To conclude, we use the tables to give an explicit proof of Proposition 3.3.

PROPOSITION 4.9. *When $m = 1$, then the mod 2 Dieudonné module of $\mathcal{S}_1$ is*

$$D_1 \simeq \mathbb{E}/\mathbb{E}(F^2 + V^2) \oplus (\mathbb{E}/\mathbb{E}(F^3 + V^3))^4.$$

PROOF. As an $\mathbb{E}$-module, D_1 is isomorphic to $H^1_{\mathrm{dR}}(\mathcal{S}_1)$. From Table 2, $H^1_{\mathrm{dR}}(\mathcal{S}_1)$ has a summand of rank 4 generated by $X_1 = \psi(f_{(1,0,1,0)})$ with the relation given by $(F^2 + V^2)X_1 = 0$. There are 4 summands of rank 6 generated by $X_2 = \psi(f_{(2,1,0,0)})$, $X_3 = \psi(f_{(2,0,0,0)})$, $X_4 = \psi(f_{(3,0,0,0)})$, and $X_5 = \psi(f_{(0,0,0,0)})$ with the relations given by $(F^3 + V^3)X_i = 0$. This yields the $\mathbb{E}$-module $\mathbb{E}/\mathbb{E}(F^2 + V^2) \oplus \left(\mathbb{E}/\mathbb{E}(F^3 + V^3)\right)^4$. $\square$

TABLE 2. Action of Verschiebung and Frobenius on $H^1_{\mathrm{dR}}(\mathcal{S}_1)$

$(f, \mathbf{g})$	$V(f, \mathbf{g})$	$F(f, \mathbf{g})$
$(0, \mathbf{g}_{(0,0,0,0)})$	$(0, \mathbf{0})$	$(0, \mathbf{0})$
$(0, \mathbf{g}_{(2,0,0,0)})$	$(0, \mathbf{0})$	$(0, \mathbf{0})$
$(0, \mathbf{g}_{(0,1,0,0)} + \mathbf{g}_{(3,0,0,0)})$	$(0, \mathbf{0})$	$(0, \mathbf{0})$
$(0, \mathbf{g}_{(2,1,0,0)} + \mathbf{g}_{(0,0,1,0)})$	$(0, \mathbf{0})$	$(0, \mathbf{0})$
$(0, \mathbf{g}_{(0,0,0,1)} + \mathbf{g}_{(1,0,1,0)})$	$(0, \mathbf{0})$	$(0, \mathbf{0})$
$(0, \mathbf{g}_{(1,0,0,0)})$	$(0, \mathbf{g}_{(0,0,0,0)})$	$(0, \mathbf{0})$
$(0, \mathbf{g}_{(0,0,1,0)})$	$(0, \mathbf{g}_{(2,0,0,0)})$	$(0, \mathbf{0})$
$(0, \mathbf{g}_{(1,0,0,1)})$	$(0, \mathbf{g}_{(0,1,0,0)} + \mathbf{g}_{(3,0,0,0)})$	$(0, \mathbf{0})$
$(0, \mathbf{g}_{(0,0,1,1)})$	$(0, \mathbf{g}_{(2,1,0,0)} + \mathbf{g}_{(0,0,1,0)})$	$(0, \mathbf{0})$
$(0, \mathbf{g}_{(1,0,1,0)})$	$(0, \mathbf{g}_{(0,0,0,1)} + \mathbf{g}_{(1,0,1,0)})$	$(0, \mathbf{0})$
$(0, \mathbf{g}_{(0,1,0,0)})$	$(0, \mathbf{g}_{(1,0,0,0)})$	$(0, \mathbf{0})$
$(0, \mathbf{g}_{(1,1,0,0)})$	$(0, \mathbf{g}_{(0,0,1,0)})$	$(0, \mathbf{0})$
$(0, \mathbf{g}_{(0,1,1,0)})$	$(0, \mathbf{g}_{(1,0,0,1)})$	$(0, \mathbf{0})$
$(0, \mathbf{g}_{(0,1,0,1)})$	$(0, \mathbf{g}_{(0,0,1,1)})$	$(0, \mathbf{0})$
$\psi(f_{(0,1,0,1)})$	$(0, \mathbf{0})$	$(0, \mathbf{g}_{(0,0,0,0)})$
$\psi(f_{(0,1,1,0)})$	$(0, \mathbf{0})$	$(0, \mathbf{g}_{(2,0,0,0)})$
$\psi(f_{(1,1,0,0)})$	$(0, \mathbf{0})$	$(0, \mathbf{g}_{(0,1,0,0)} + \mathbf{g}_{(3,0,0,0)})$
$\psi(f_{(0,1,0,0)} + f_{(3,0,0,0)})$	$(0, \mathbf{0})$	$(0, \mathbf{g}_{(2,1,0,0)} + \mathbf{g}_{(0,0,1,0)})$
$\psi(f_{(0,0,0,1)} + f_{(1,0,1,0)})$	$(0, \mathbf{0})$	$(0, \mathbf{g}_{(0,0,0,1)} + \mathbf{g}_{(1,0,1,0)})$
$\psi(f_{(0,0,1,1)})$	$(0, \mathbf{0})$	$\psi(f_{(0,1,0,1)})$
$\psi(f_{(1,0,0,1)})$	$(0, \mathbf{0})$	$\psi(f_{(0,1,1,0)})$
$\psi(f_{(2,1,0,0)} + f_{(0,0,1,0)})$	$(0, \mathbf{0})$	$\psi(f_{(1,1,0,0)})$
$\psi(f_{(1,0,0,0)})$	$(0, \mathbf{0})$	$\psi(f_{(0,1,1,0)})$
$\psi(f_{(1,0,1,0)})$	$(0, \mathbf{g}_{(1,0,1,0)})$	$\psi(f_{(0,0,0,1)} + f_{(1,0,1,0)})$
$\psi(f_{(2,1,0,0)})$	$(0, \mathbf{g}_{(0,1,0,0)})$	$\psi(f_{(0,0,1,1)})$
$\psi(f_{(3,0,0,0)})$	$(0, \mathbf{g}_{(1,1,0,0)})$	$\psi(f_{(1,0,0,1)})$
$\psi(f_{(2,0,0,0)})$	$(0, \mathbf{g}_{(0,1,1,0)})$	$\psi(f_{(2,1,0,0)} + f_{(0,0,1,0)})$
$\psi(f_{(0,0,0,0)})$	$(0, \mathbf{g}_{(0,1,0,1)})$	$\psi(f_{(1,0,0,0)})$

Note that the trivial eigenspace $D_{1,0}$ appears as the summand $\mathbb{E}/(F^2 + V^2)$. It is spanned by

$$\{\psi(f_{(1,0,1,0)}), \psi(f_{(0,0,0,1)} + f_{(1,0,1,0)}), (0, \mathbf{g}_{(1,0,1,0)}), (0, \mathbf{g}_{(0,0,0,1)} + \mathbf{g}_{(1,0,1,0)})\}.$$

References

[1] Wieb Bosma, John Cannon, and Catherine Playoust, *The Magma algebra system. I. The user language*, J. Symbolic Comput. **24** (1997), no. 3-4, 235–265, DOI 10.1006/jsco.1996.0125. Computational algebra and number theory (London, 1993). MR1484478

[2] Leonard Chastkofsky and Walter Feit, *On the projective characters in characteristic 2 of the groups* Suz(2^m) *and* Sp$_4$(2^n), Inst. Hautes Études Sci. Publ. Math. **51** (1980), 9–35. MR573820

[3] Neil Dummigan, *The determinants of certain Mordell-Weil lattices*, Amer. J. Math. **117** (1995), no. 6, 1409–1429, DOI 10.2307/2375024. MR1363073

[4] Neil Dummigan, *Complete p-descent for Jacobians of Hermitian curves*, Compositio Math. **119** (1999), no. 2, 111–132, DOI 10.1023/A:1001721808335. MR1723124

[5] Iwan Duursma and Dane Skabelund, *The de Rham cohomology of the Ree curve*, preprint.

[6] Arsen Elkin and Rachel Pries, *Ekedahl-Oort strata of hyperelliptic curves in characteristic 2*, Algebra Number Theory **7** (2013), no. 3, 507–532, DOI 10.2140/ant.2013.7.507. MR3095219

[7] Holley Friedlander, Derek Garton, Beth Malmskog, Rachel Pries, and Colin Weir, *The a-numbers of Jacobians of Suzuki curves*, Proc. Amer. Math. Soc. **141** (2013), no. 9, 3019–3028, DOI 10.1090/S0002-9939-2013-11581-9. MR3068955

[8] Rainer Fuhrmann and Fernando Torres, *On Weierstrass points and optimal curves*, Rend. Circ. Mat. Palermo (2) Suppl. **51** (1998), 25–46. MR1631013

[9] Massimo Giulietti and Gábor Korchmáros, *On automorphism groups of certain Goppa codes*, Des. Codes Cryptogr. **47** (2008), no. 1-3, 177–190, DOI 10.1007/s10623-007-9110-5. MR2375466

[10] Massimo Giulietti, Gábor Korchmáros, and Fernando Torres, *Quotient curves of the Suzuki curve*, Acta Arith. **122** (2006), no. 3, 245–274, DOI 10.4064/aa122-3-3. MR2239917

[11] Benedict H. Gross, *Rigid local systems on $\mathbb{G}_m$ with finite monodromy*, Adv. Math. **224** (2010), no. 6, 2531–2543, DOI 10.1016/j.aim.2010.02.008. MR2652215

[12] Burkhard Haastert and Jens Carsten Jantzen, *Filtrations of the discrete series of* $\mathrm{SL}_2(q)$ *via crystalline cohomology*, J. Algebra **132** (1990), no. 1, 77–103, DOI 10.1016/0021-8693(90)90253-K. MR1060833

[13] Johan P. Hansen, *Deligne-Lusztig varieties and group codes*, Coding theory and algebraic geometry (Luminy, 1991), Lecture Notes in Math., vol. 1518, Springer, Berlin, 1992, pp. 63–81, DOI 10.1007/BFb0087993. MR1186416

[14] Johan P. Hansen and Henning Stichtenoth, *Group codes on certain algebraic curves with many rational points*, Appl. Algebra Engrg. Comm. Comput. **1** (1990), no. 1, 67–77, DOI 10.1007/BF01810849. MR1325513

[15] Nicholas M. Katz and William Messing, *Some consequences of the Riemann hypothesis for varieties over finite fields*, Invent. Math. **23** (1974), 73–77, DOI 10.1007/BF01405203. MR0332791

[16] Li-Qian Liu, *The decomposition numbers of* $\mathrm{Suz}(q)$, J. Algebra **172** (1995), no. 1, 1–31, DOI 10.1006/jabr.1995.1045. MR1320616

[17] G. Lusztig, *Coxeter orbits and eigenspaces of Frobenius*, Invent. Math. **38** (1976/77), no. 2, 101–159, DOI 10.1007/BF01408569. MR0453885

[18] R. P. Martineau, *On 2-modular representations of the Suzuki groups*, Amer. J. Math. **94** (1972), 55–72, DOI 10.2307/2373593. MR0360777

[19] Tadao Oda, *The first de Rham cohomology group and Dieudonné modules*, Ann. Sci. École Norm. Sup. (4) **2** (1969), 63–135. MR0241435

[20] Frans Oort, *A stratification of a moduli space of abelian varieties*, Moduli of abelian varieties (Texel Island, 1999), Progr. Math., vol. 195, Birkhäuser, Basel, 2001, pp. 345–416, DOI 10.1007/978-3-0348-8303-0_13. MR1827027

[21] Rachel Pries, *A short guide to p-torsion of abelian varieties in characteristic p*, Computational arithmetic geometry, Contemp. Math., vol. 463, Amer. Math. Soc., Providence, RI, 2008, pp. 121–129, DOI 10.1090/conm/463/09051. MR2459994

[22] Rachel Pries and Colin Weir, *The Ekedahl-Oort type of Jacobians of Hermitian curves*, Asian J. Math. **19** (2015), no. 5, 845–869, DOI 10.4310/AJM.2015.v19.n5.a3. MR3431681

[23] Peter Sin, *Extensions of simple modules for* $\mathrm{Sp}_4(2^n)$ *and* $\mathrm{Suz}(2^m)$, Bull. London Math. Soc. **24** (1992), no. 2, 159–164, DOI 10.1112/blms/24.2.159. MR1148676

[24] Michio Suzuki, *A new type of simple groups of finite order*, Proc. Nat. Acad. Sci. U.S.A. **46** (1960), 868–870. MR0120283

Department of Mathematics and Computer Science, Colorado College, Colorado Springs, Colorado 80903

Email address: beth.malmskog@gmail.com

Department of Mathematics, Colorado State University, Fort Collins, Colorado 80523

Email address: pries@math.colostate.edu

The Tutte Institute for Mathematics and Computing, Ottawa, Ontario, Canada

Email address: colinoftheweirs@gmail.com

Contemporary Mathematics
Volume **722**, 2019
https://doi.org/10.1090/conm/722/14529

Décompositions en hauteurs locales

Fabien Pazuki

Résumé. Soit A la jacobienne d'une courbe hyperelliptique définie sur un corps de nombres k. On donne une formule de décomposition de la hauteur de Faltings de A et de la hauteur de Néron-Tate des points k-rationnels de A. On propose de plus en 3.4 une question de type Bogomolov sur l'espace de modules $\mathcal{A}_g$ des variétés abéliennes principalement polarisées de dimension g.

Abstract. Let A be the jacobian variety of a hyperelliptic curve defined over a number field k. We provide a decomposition formula for the Faltings height of A and for the Néron-Tate height of k-rational points on A. We formulate in 3.4 a question of Bogomolov type on the space $\mathcal{A}_g$ of principally polarized abelian varieties of dimension g.

1. Introduction

Soit E une courbe elliptique définie sur $\overline{\mathbb{Q}}$ et donnée dans un modèle de Weierstrass. L'article **[CoSi86]** est dévolu à montrer une formule de décomposition en composantes locales de la hauteur de Faltings stable de E en fonction des invariants classiques du modèle de Weierstrass choisi, voir l'énoncé du théorème 2.1 ci-après.

Le présent texte propose une généralisation en dimension supérieure de cette formule. On traite des jacobiennes de courbes hyperelliptiques, un cadre où la définition d'un discriminant est aisée et où ce discriminant joue le même rôle que dans le cas elliptique, reliant notamment des propriétés de bonne réduction de la variété et des formules closes aux places archimédiennes en termes de fonctions thêta. A ce titre, les jacobiennes de courbes hyperelliptiques sont une généralisation naturelle.

Plus généralement, soit A une variété abélienne définie sur $\overline{\mathbb{Q}}$, principalement polarisée, semi-stable de dimension $g \geq 1$, munie d'un fibré L ample et symétrique. On s'intéresse à trois questions étroitement liées.

Question 1 : peut-on décomposer en composantes locales la hauteur différentielle de A/k ? On s'accordera sur le fait qu'une composante locale en la place v

2010 *Mathematics Subject Classification.* 11G50, 14G40, 14G05, 11G30, 11G10.
Key words and phrases. Heights, abelian varieties, torsion points, rational points, hyperelliptic curves.

d'un corps k de définition de A est un nombre réel calculable à partir des seules données locales de la variété A en v.

Question 2 : peut-on donner des formules explicites pour les hauteurs locales canoniques de Néron d'un point $\overline{\mathbb{Q}}$-rationnel sur A ? L'existence de cette décomposition remonte à Néron [**Nér65**].

Question 3 : étant donné un point $P \in A(\overline{\mathbb{Q}}) \subset \mathbb{P}^N(\overline{\mathbb{Q}})$, comment estimer la différence $\hat{h}(P) - h(P)$ pour un choix de hauteur projective h ? Citons par exemple les travaux [**CrPrSi06**] et [**Bru13**] en dimension 1, [**FlySm97**] et [**Sto02**] en dimension 2. L'objectif étant d'obtenir en dimension quelconque de meilleures bornes que celles existantes dans [**ZaMa72**].

On cherche ici à apporter une réponse possible à ces trois questions, réponse provenant de la construction des modèles de Moret-Bailly des variétés abéliennes. On trouvera le nécessaire les concernant dans la section 4 du présent texte. Ces modèles de Moret-Bailly (ou MB-modèles) jouaient déjà un rôle important dans le travail [**Paz12, BoDa99**] où leur définition est aussi rappelée en détails.

L'article se veut accessible et comporte ainsi plusieurs paragraphes de rappels. Il est organisé comme suit. On présente les formules en détails dans la section 2. La section 3 décrit les énoncés de la théorie des hauteurs de points algébriques utiles à la suite. La section 4 présente une partie de la théorie des modèles de Moret-Bailly, permettant de calculer la hauteur d'un point algébrique par la formule clef dans la section 5. Dans la section 6 on donne une formule explicite valable dans le cas des jacobiennes de courbes hyperelliptiques. Finalement la section 7 montre que toutes les formules proposées pour la hauteur d'une courbe elliptique fournissent le même résultat.

2. Présentation des décompositions

Dans tout le texte on note M_k l'ensemble des places du corps de nombres k et M_k^∞ l'ensemble de ses places archimédiennes, M_k^0 désignant l'ensemble de ses places finies. On note $d = [K : \mathbb{Q}]$ son degré. Pour toute place v de k on note k_v le complété de k pour la valuation $|.|_v$ associée où on fixe $|p|_v = p^{-1}$ pour toute place finie v au-dessus d'un nombre premier p. On note de plus $d_v = [k_v : \mathbb{Q}_v]$ pour le degré local. Pour tout vecteur $x = (x_1, ..., x_n)$ de k_v^n on pose

$$\|x\|_v = \begin{cases} \left(\displaystyle\sum_{i=1}^n |x_i|_v^2\right)^{\frac{1}{2}} & \text{si } v \text{ est archimédienne,} \\ \displaystyle\max_{1\leq i\leq n} \{|x_i|_v\} & \text{sinon.} \end{cases}$$

On travaillera avec la *hauteur différentielle positive* ou *hauteur de Faltings positive* définie par $h_{\mathrm{F}^+}(A) = \frac{g}{2}\log(2\pi^2) + h_{\mathrm{F}}(A)$, où $h_{\mathrm{F}}(A)$ est la hauteur introduite par Faltings dans [**Fa84**]. Cette version de la hauteur différentielle possède l'agréable propriété d'être positive, voir à ce sujet la remarque 3.3 ci-dessous.

Le point de départ de cette étude et l'origine de la question 1 est le cas de la dimension 1 où on dispose de la formule suivante exprimant la hauteur différentielle positive d'une courbe elliptique, montrée par Silverman dans l'ouvrage [**CoSi86**] page 254 (on a corrigé une puissance de 2π dans la définition du discriminant, voir par exemple la proposition 8.2 de [**deJo05**], et changé la normalisation des métriques ici) :

THÉORÈME 2.1. *(Silverman) Soit E une courbe elliptique sur un corps de nombres k de degré d. Alors on a*

$$h_{\mathrm{F}^+}(E) = \frac{1}{12d}\left[\log N_{k/\mathbb{Q}}(\Delta_E) - \sum_{v\in M_k^\infty} d_v \log\Big(|\Delta(\tau_v)|(2\operatorname{Im}\tau_v)^6\Big)\right],$$

où Δ_E est le discriminant minimal de E ; aux places archimédiennes τ_v est une matrice de périodes associée à $E(\overline{k}_v)$ et $\Delta(\tau_v) = q\prod_{n=1}^{+\infty}(1-q^n)^{24}$ est le discriminant modulaire, en ayant posé $q = e^{2\pi i\tau_v}$.

On cherche à généraliser cette formule en toute dimension. Dans le texte [**Aut06**], Autissier prouve une formule valable en dimension générale dans le cas où la variété a potentiellement bonne réduction partout :

THÉORÈME 2.2. *(Autissier) Soit A une variété abélienne de dimension g sur un corps de nombres k. Supposons que A a potentiellement bonne réduction partout. Soit Θ un diviseur symétrique et ample sur A, définissant une polarisation principale λ. On pose $L = \mathcal{O}_A(\Theta)$ et on note μ la mesure de Haar de $A(\mathbb{C})$ de masse 1. Alors*

$$h_{\mathrm{F}^+}(A) = 2g\hat{h}_L(\Theta) + \frac{2}{d}\sum_{v\in M_k^\infty} d_v I(A_v,\lambda_v),$$

où $I(A_v,\lambda_v) = -\displaystyle\int_{A(\overline{k}_v)}\log\|s\|_v\mu + \frac{1}{2}\log\int_{A(\overline{k}_v)}\|s\|_v^2\mu$ est positif et indépendant du choix de section s non nulle du fibré L_v.

L'article [**Aut06**] contient de plus un résultat de décomposition de $h_{F^+}(A) - 2g\hat{h}_L(\Theta)$ inconditionnel lorsque A est un produit de courbes elliptiques et de surfaces abéliennes. On voudrait arriver à une formule explicite tout en se passant de l'hypothèse de bonne réduction.

On traite partiellement la première question dans la partie 6 avec une décomposition explicite de la hauteur de Faltings pour les jacobiennes de courbes hyperelliptiques de genre g. Cette formule n'apparaît pas dans la littérature mais est probablement connue des experts. Comme en dimension 1, le calcul est facilité par un choix de section très agréable et basé sur l'existence d'un discriminant de la courbe caractérisant la mauvaise réduction aux places finies et décrit comme une forme modulaire aux places infinies. Il faut cependant tenir compte du comportement de cette section le long du bord de l'espace de modules de courbes. C'est la généralisation de la formule de Ueno de l'article [**Uen88**], établie ici grâce à l'utilisation des articles de Lockhart [**Loc94**], Kausz [**Kau99**] et de Jong [**deJo07**]. Quelques notations tout d'abord : pour $m \in \frac{1}{2}\mathbb{Z}^{2g}$ on pose $\varphi_m(\tau) = \theta_m(0,\tau)^8$, où $\theta_m(z,\tau)$ est la fonction thêta de caractéristique m associée au réseau de dimension g dont la définition est rappelée en (3.1). Si S est un sous-ensemble de $\{1,2,...,2g+1\}$ on définit alors $m_S = \displaystyle\sum_{i\in S} m_i \in \frac{1}{2}\mathbb{Z}^{2g}$ avec :

$$m_{2i-1} \;=\; \begin{bmatrix} {}^t(0 & \dots & 0 & \frac{1}{2} & 0 & \dots & 0) \\ {}^t(\frac{1}{2} & \dots & \frac{1}{2} & 0 & 0 & \dots & 0) \end{bmatrix}, \quad 1\leq i\leq g+1,$$

$$m_{2i} = \begin{bmatrix} {}^t(0 & \ldots & 0 & \frac{1}{2} & 0 & \ldots & 0) \\ {}^t(\frac{1}{2} & \ldots & \frac{1}{2} & \frac{1}{2} & 0 & \ldots & 0) \end{bmatrix}, \quad 1 \leq i \leq g,$$

où le coefficient non nul de la première ligne est en i-ème position. Soit alors $\mathcal{T}$ la collection des sous-ensembles de $\{1, ..., 2g+1\}$ de cardinal $g+1$. Soit $U = \{1, 3, ..., 2g+1\}$ et notons $\circ$ l'opérateur de différence symétrique. On définit alors :

$$\varphi(\tau) = \prod_{T \in \mathcal{T}} \varphi_{m_{T \circ U}}(\tau). \tag{2.1}$$

THÉORÈME 2.3. *Soient k un corps de nombres et C une courbe hyperelliptique définie sur k, semi-stable et de genre $g \geq 1$. On note J_C sa jacobienne. On pose $l = \binom{2g+1}{g+1}$. Pour chaque place archimédienne v on note τ_v une matrice de l'espace de Siegel[1] $\mathfrak{S}_g$ telle que $\mathcal{A}_v(\mathbb{C}) \simeq \mathbb{C}^g/(\mathbb{Z}^g + \tau_v \mathbb{Z}^g)$ comme variétés abéliennes principalement polarisées et $\Delta_{\min}$ le discriminant minimal de C/k. Il existe des entiers $e_v \geq 0$ et des réels $f_v \geq 0$ tels que $(8g+4) \cdot f_v = g \cdot \mathrm{ord}_v(\Delta_{\min}) - (8g+4)e_v$ et tels que la hauteur de Faltings de J_C soit donnée par la formule :*

$$d \cdot h_{\mathrm{F}^+}(J_C) = \sum_{v | \Delta_{\min}} d_v f_v \log \mathrm{N}_{k/\mathbb{Q}}(v) - \sum_{v \in M_k^\infty} d_v \log \left(2^{-\frac{2g}{8g+4}} |\varphi(\tau_v)|^{\frac{1}{4l}} \det(\mathrm{Im}\, \tau_v)^{\frac{1}{2}} \right),$$

où φ est un produit explicite de constantes thêta donné par la formule (2.1). Dans le cas particulier des surfaces abéliennes, on obtient

$$d \cdot h_{\mathrm{F}^+}(J_C) = \sum_{v | \Delta_{\min}} d_v f_v \log \mathrm{N}_{k/\mathbb{Q}}(v) - \sum_{v \in M_k^\infty} d_v \log \left(2^{-\frac{1}{5}} |J_{10}(\tau_v)|^{\frac{1}{10}} \det(\mathrm{Im}\, \tau_v)^{\frac{1}{2}} \right),$$

avec $J_{10}(\tau) = \prod_{m \in Z_2} \theta_m(0, \tau_v)^2$, le produit portant sur l'ensemble Z_2 des 10 caractéristiques thêta paires en dimension 2.

Pour une courbe elliptique, toutes ces décompositions coïncident avec le théorème 2.1, ce fait est vérifié dans la section 7. Pour les surfaces abéliennes, cela donne une version démontrée de la formule qu'on peut trouver dans l'article [**Uen88**] page 765, en précisant les contributions aux places finies et les conventions pour J_{10}. La comparaison des termes locaux entre la formule d'Autissier et les formules de Ueno et du théorème 2.3 est moins aisée car le terme $\hat{h}_L(\Theta)$ n'est pas trivial en dimension $g \geq 2$.

On se donne à présent un MB-modèle de niveau $r = 4$ sur k, qui est un bon cadre pour obtenir l'énoncé suivant. Un tel modèle existe toujours modulo une éventuelle extension finie de corps de nombres. On suppose en particulier que les points de 16-torsion de A, dont l'ensemble est noté $A[16]$, sont rationnels sur k. On pourra consulter la section 4 (voir aussi l'article [**Paz12**] et ses références) pour une présentation détaillée de ces modèles et de leurs propriétés. On montre alors comme corollaire direct du théorème 2.3 et de la proposition 5.4 :

COROLLAIRE 2.4. *Soit A une jacobienne de courbe hyperelliptique de dimension g définie sur $\overline{\mathbb{Q}}$, semi-stable et munie d'un fibré L ample et symétrique portant une*

1. La définition de cet espace est rappelée dans la section suivante.

polarisation principale. Soit $R \in A(\overline{\mathbb{Q}})$, *soit* k *un corps de définition de* A, $A[16]$, R. *Alors on a la formule*

$$\widehat{h}_{A,L^{\otimes 16}}(R) - h_{A,L^{\otimes 16}}(R) = \frac{1}{d} \sum_{v \in M_k} d_v \alpha_{A,L,v}(R)$$

où $\alpha_{A,L,v}(R)$ *est donné par la différence des termes locaux dans la définition* 5.3 *et dans le théorème* 2.3, *en tenant compte du facteur* $1/2$.

Ce corollaire est donc une réponse possible pour les questions 2 et 3 proposées plus haut. Obtenir de telles décompositions permet de mener des calculs explicites place par place. Cela a des applications, par exemple dans le procédé de saturation du groupe de Mordell-Weil (*i.e.* la recherche de générateurs explicites du groupe des points rationnels), où il est important de savoir estimer la différence entre hauteur canonique et hauteur naïve d'un point rationnel. On sait depuis Manin-Zarhin **[ZaMa72]** pour les points et David-Philippon **[DaPh02]** pour les sous-variétés que la valeur absolue de cette différence est *majorée* par la hauteur de la variété abélienne ambiante, à une constante explicite près. On trouve dans la proposition 5.4 (ou corollaire 2.4 ci-dessus) une *égalité* permettant d'affirmer, avec le théorème 2.3, qu'on peut estimer cette différence en menant des calculs locaux.

Remerciements. L'auteur remercie chaleureusement Pascal Autissier et Gaël Rémond pour leurs conseils précieux. Merci à l'arbitre de publication pour son travail. L'auteur est soutenu par le programme ANR-14-CE25-0015 Gardio, par ANR-17-CE40-0012 Flair et par la chaire Niels Bohr DNRF de Lars Hesselholt.

3. Hauteurs globales, hauteurs locales

On rappelle dans cette section un théorème de Néron, la définition de l'espace de Siegel et la définition de la hauteur de Faltings.

Théorème de Néron. Une hauteur de Weil $h_{A,\mathcal{D}}$ associée à un diviseur $\mathcal{D}$ sur une variété abélienne A/k est par définition une somme indexée par les places de k de fonctions $\lambda_{\mathcal{D},v}$ à valeurs réelles (définies hors du diviseur $\mathcal{D}$). C'est une fonction vérifiant la relation suivante (issue du théorème du cube) : il existe une constante c telle que pour tous points $P, Q, R \in A(k)$, et en notant temporairement $h = h_{A,\mathcal{D}}$:

$$\Big| h(P+Q+R) - h(P+Q) - h(Q+R) - h(R+P) + h(P) + h(Q) + h(R) \Big| \leq c.$$

Si on suppose de plus que le diviseur $\mathcal{D}$ est symétrique on obtient (en prenant $R = -Q$) une relation de quasi-parallélogramme :

$$\Big| h(P+Q) + h(P-Q) - 2h(P) - 2h(Q) \Big| \leq c.$$

Le passage à la limite effectué pour définir la hauteur de Néron-Tate permet d'obtenir $c = 0$. Cette construction offre donc l'avantage suivant : la hauteur de Néron-Tate devient une forme quadratique, dont le cône isotrope est le sous-groupe de torsion de la variété abélienne. Le théorème suivant de Néron offre la possibilité de décomposer cette hauteur canonique aussi (voir **[HiSi00]** page 242) :

THÉORÈME 3.1. *(Néron) Soit* A/k *une variété abélienne définie sur un corps de nombres* k. *Soit* M_k *l'ensemble des places de* k. *Pour tout diviseur* $\mathcal{D}$ *sur* A

on note $A_{\mathcal{D}} = A\backslash \operatorname{supp}(\mathcal{D})$. Alors pour toute place $v \in M_k$ il existe une fonction hauteur locale, unique à une fonction constante près :

$$\widehat{\lambda}_{\mathcal{D},v} : A_{\mathcal{D}}(k_v) \longrightarrow \mathbb{R},$$

appelée hauteur locale canonique, dépendant du choix de $\mathcal{D}$ et vérifiant les propriétés suivantes, avec $\gamma_{i,v}$ des constantes dépendant de v :

(1) $\widehat{\lambda}_{\mathcal{D},v} - \lambda_{\mathcal{D},v}$ est une fonction bornée.

(2) Pour tous diviseurs $\mathcal{D}_1$ et $\mathcal{D}_2$ sur A, on a $\widehat{\lambda}_{\mathcal{D}_1+\mathcal{D}_2,v} = \widehat{\lambda}_{\mathcal{D}_1,v} + \widehat{\lambda}_{\mathcal{D}_2,v} + \gamma_{1,v}$.

(3) Si $\mathcal{D} = \operatorname{div}(f)$, alors $\widehat{\lambda}_{\mathcal{D},v} = v \circ f + \gamma_{2,v}$, où $v(.) = -\log |.|_v$.

(4) Si $\Phi : B \to A$ est un morphisme entre deux variétés abéliennes alors on a la relation $\widehat{\lambda}_{\Phi^*\mathcal{D},v} = \widehat{\lambda}_{\mathcal{D},v} \circ \Phi + \gamma_{3,v}$.

(5) Soit $Q \in A(k)$ et soit $t_Q : A \to A$ la translation par Q. Alors on a la relation : $\widehat{\lambda}_{t_Q^*\mathcal{D},v} = \widehat{\lambda}_{\mathcal{D},v} \circ t_Q + \gamma_{4,v}$.

(6) Soit $\widehat{h}_{A,\mathcal{D}}$ la hauteur globale canonique de A associée à $\mathcal{D}$. Il existe une constante $\hat{c}$ telle que, pour tout $P \in A_{\mathcal{D}}(k)$:

$$\widehat{h}_{A,\mathcal{D}}(P) = \frac{1}{d} \sum_{v\in M_k} d_v \widehat{\lambda}_{\mathcal{D},v}(P) + \hat{c}.$$

(7) Si $\mathcal{D}$ vérifie $[2]^*\mathcal{D} = 4\mathcal{D} + \operatorname{div}(f)$ pour f une fonction rationnelle sur A et si l'on fixe les constantes de telle sorte que, pour tout $P \in A_{\mathcal{D}}$ avec $[2]P \in A_{\mathcal{D}}$, on ait la relation $(*)$ $\widehat{\lambda}_{\mathcal{D},v}([2]P) = 4\widehat{\lambda}_{\mathcal{D},v}(P) + v(f(P))$, alors pour tout $P \in A_{\mathcal{D}}$:

$$\widehat{h}_{A,\mathcal{D}}(P) = \frac{1}{d} \sum_{v\in M_k} d_v \widehat{\lambda}_{\mathcal{D},v}(P).$$

(Notons que f est unique à multiplication par une constante $a \in k^*$ près. Notons aussi que la relation $(*)$ permet de fixer la constante $\hat{c} = 0$ dans l'item précédent.)

Espace de Siegel et fonctions thêta. Soit v une place archimédienne. On notera $\mathfrak{S}_g$ l'espace de Siegel associé aux variétés abéliennes sur $\overline{k}_v$ principalement polarisées de dimension g et munies d'une base symplectique (on pourra consulter **[BiLa04]** page 213). C'est l'ensemble des matrices $\tau = \tau_v$ de taille $g \times g$ symétriques à coefficients complexes et vérifiant la condition $\operatorname{Im} \tau > 0$ (*i.e.* définies positives). Cet espace est muni d'une action transitive du groupe symplectique $\Gamma = \operatorname{Sp}(2g, \mathbb{R})$ donnée par :

$$\begin{bmatrix} A & B \\ C & D \end{bmatrix} \cdot \tau = (A\tau + B)(C\tau + D)^{-1}.$$

On considère alors F_g un domaine fondamental pour l'action du sous-groupe $\operatorname{Sp}_{2g}(\mathbb{Z})$. On peut choisir F_g de telle sorte qu'une matrice τ de ce domaine vérifie en particulier les conditions suivantes (voir **[Fre83]** page 34) :

- S1 : Pour tout $\sigma \in \operatorname{Sp}_{2g}(\mathbb{Z})$ on a : $\det(\operatorname{Im} \sigma \cdot \tau) \leq \det(\operatorname{Im} \tau)$. On dira que $\operatorname{Im} \tau$ est *maximale* pour l'action de $\operatorname{Sp}_{2g}(\mathbb{Z})$.
- S2 : Si $\operatorname{Re} \tau = (a_{i,j})_{1\leq i,j\leq g}$ alors $|a_{i,j}| \leq \frac{1}{2}$ pour tous $1 \leq i, j \leq g$.
- S3 : Si $\operatorname{Im} \tau = (b_{i,j})$ alors pour tout $l \in \{1, ..., g\}$ et tout $\zeta = (\zeta_1, ..., \zeta_g) \in \mathbb{Z}^g$ tel que $\operatorname{pgcd}(\zeta_l, ..., \zeta_g) = 1$ on a ${}^t\zeta(\operatorname{Im} \tau)\zeta \geq b_{l,l}$. De plus pour tout $i \in \{1, ..., g\}$ on a $b_{i,i+1} \geq 0$.

Ces conditions impliquent $b_{g,g} \geq ... \geq b_{1,1} \geq \sqrt{3}/2$ et $b_{i,i}/2 \geq |b_{i,j}|$. On définit alors pour $z \in \mathbb{C}^g$ et $\tau \in \mathfrak{S}_g$ les séries thêta avec caractéristiques $a, b \in \mathbb{R}^g$:

$$\theta_{a,b}(z,\tau) = \sum_{n \in \mathbb{Z}^g} e^{i\pi\,(n+a)'\tau(n+a)+2i\pi\,(n+a)'(z+b)}, \tag{3.1}$$

où x' est le vecteur ligne transposé du vecteur colonne $x \in \mathbb{R}^g$. On dira que les matrices de F_g sont réduites au sens de Siegel.

Hauteur de Faltings positive. Soient k un corps de nombres de degré d et $S = \mathrm{Spec}(\mathcal{O}_k)$ le spectre de son anneau d'entiers. Un *fibré vectoriel métrisé* de rang r sur S est un $\mathcal{O}_k$-module projectif $\mathcal{L}$ de rang r muni d'une collection $\{||\cdot||_v\}_{v \in M_k^\infty}$ telle que $||.||_v$ soit une norme hermitienne sur le k_v-espace vectoriel $\mathcal{L} \otimes_{\mathcal{O}_k} \overline{k}_v$, vérifiant $||x||_v = ||\overline{x}||_{\overline{v}}$ pour tout plongement $v : k \hookrightarrow \mathbb{C}$.

Le *degré d'Arakelov* d'un fibré en droites métrisé $(\mathcal{L}, ||.||_v)$ sur S est défini, en prenant un élément non nul $s \in \mathcal{L}$:

$$\widehat{\deg}(\mathcal{L}) = \log \mathrm{Card}\left(\mathcal{L}/s\mathcal{O}_k\right) - \sum_{v \in M_k^\infty} d_v \log ||s||_v.$$

La formule du produit nous assure que ce degré ne dépend pas globalement du choix de section s non nulle (mais les formules locales dépendent de la section bien entendu).

Soit alors A/k une variété abélienne de dimension $g \geq 1$. Soient $\mathcal{A} \to S$ son modèle de Néron, $\varepsilon : S \to \mathcal{A}$ sa section neutre et $\Omega^g_{\mathcal{A}/S}$ le faisceau des g-formes différentielles, qui est localement libre de rang 1. On pose $\omega_{\mathcal{A}/S} = \varepsilon^*(\Omega^g_{\mathcal{A}/S})$; c'est un fibré en droites sur $S = \mathrm{Spec}(\mathcal{O}_k)$ qu'on peut identifier au module de ses sections globales. On munit ce fibré des métriques suivantes :

$$\forall \alpha \in \omega_{\mathcal{A}/S} \otimes_v \mathbb{C}, \quad ||\alpha||_v^2 = \frac{i^{g^2}}{(2\pi)^{2g}} \int_{A_v(\mathbb{C})} \alpha \wedge \overline{\alpha}, \tag{3.2}$$

où on a identifié α à une section globale de $\Omega^g_{\mathcal{A}/S}$. On notera qu'on a choisi ici d'élever 2π à la puissance $2g$, voir la discussion plus bas. On définit alors :

DÉFINITION 3.2. *Soit A/k une variété abélienne semi-stable définie sur un corps de nombres k. On appelle* hauteur de Faltings positive *la quantité :*

$$h_{\mathrm{F}^+}(A) = \frac{1}{d}\,\widehat{\deg}(\omega_{\mathcal{A}/S}).$$

On a donc la relation

$$h_{\mathrm{F}^+}(A) = \frac{g}{2}\log(2\pi^2) + h_{\mathrm{F}}(A),$$

où $h_{\mathrm{F}}(A)$ est la hauteur de Faltings de **[Fa84]**.

REMARQUE 3.3. Un point sur la normalisation des métriques hermitiennes.

La puissance i^{g^2} au numérateur est en fait $i^g(-1)^{\frac{g(g-1)}{2}}$, le terme -1 provient du caractère alterné du produit extérieur. Il y a ensuite (au moins) cinq possibilités intéressantes pour le dénominateur.

A. 2^g *: cela permet de simplifier le passage des coordonnées complexes aux coordonnées réelles dans les calculs de volumes, voir par exemple l'article de Chai dans* **[CoSi86]** *page 250. C'est le choix de* **[Fa84]**.

B. $(2\pi)^g$ *: permet d'obtenir une formule de hauteur dans le cas CM qui ne fait pas intervenir de puissance de* π*, d'après la formule de Chowla-Selberg. Voir par exemple* [**Del85**] *page 29.*

C. $(2\pi)^{2g}$ *: assure que la hauteur de Faltings est toujours positive, d'après une inégalité de Bost, voir* [**Bo96a**] *ou* [**Aut06**]*, ainsi que des détails de preuve dans* [**GaRé14**].

D. $(4\pi)^{\frac{2g}{3}}$ *: supprime le terme constant dans la formule de Noether donnée dans* [**MB89**]*, si on garde la normalisation traditionnelle du* δ *de Faltings.*

E. 1 *: évite les constantes dans la définition initiale. C'est le choix fait dans* [**Col98**].

Dans ce travail, nous avons opté pour le choix C. Un choix qui conduit plus naturellement à la question : les variétés de hauteur minimale ont-elle une structure particulière ? Ce que l'on pourrait formuler de la manière suivante :

Question 3.4. *(Bogomolov sur* $\mathcal{A}_g$*) Peut-on trouver un réel explicite* ***optimal*** $\varepsilon_g > 0$ *qui ne dépende que de* g *et tel que pour toute variété abélienne* A *sur* $\overline{\mathbb{Q}}$ *de dimension* g *on ait* $h_{F^+}(A) \geq \varepsilon_g$ *? Pour quelle dépendance en* g *? Est-ce que* ε_g *est toujours atteint pour une variété admettant des multiplications complexes ?*

Nous savons d'ores et déjà que l'additivité de la hauteur va fournir des contraintes sur la famille $(\varepsilon_g)_{g\geq 1}$. Par exemple si E est une courbe elliptique sur $\overline{\mathbb{Q}}$ on aura $h_{F^+}(E^g) \geq \varepsilon_g$, ce qui implique $g\varepsilon_1 \geq \varepsilon_g$. Une formule du type $\varepsilon_g = c_0 g$, avec c_0 une constante universelle, serait fonctorielle. La littérature nous renseigne dans la remarque suivante sur l'existence de "petits points".

Remarque 3.5. *On donne ici des exemples de calculs de la hauteur de Faltings avec le choix de métrique fait ici :*

(1) (D'après Bost, Mestre, Moret-Bailly [**BoMeMB90**] *page 93) On observe la courbe* C *de genre 2 donnée par l'équation affine* $y^2 + y = x^5$ *sur un corps de nombres sur laquelle elle est semi-stable, alors*

$$h_{\mathrm{F}^+}(J_C) = 3\log 2\pi - \frac{1}{2}\log\left(\Gamma\left(\frac{1}{5}\right)^5\Gamma\left(\frac{2}{5}\right)^3\Gamma\left(\frac{3}{5}\right)\Gamma\left(\frac{4}{5}\right)^{-1}\right),$$

donc de valeur approchée $h_{\mathrm{F}^+}(J_C) = 0,38537... \geq \varepsilon_2$.

(2) (D'après Chowla-Selberg, voir Deligne [**Del85**] *page 29) Si* E *est une courbe elliptique semi-stable sur un corps de nombres et à multiplication complexe par l'anneau des entiers de* $\mathbb{Q}(\sqrt{-D})$ *où* $-D$ *est le discriminant, on note* ϵ *le caractère quadratique de Dirichlet,* w *le nombre d'unités et* h *le nombre de classes, alors*

$$h_{\mathrm{F}^+}(E) = \frac{1}{2}\log 2\pi - \frac{1}{2}\log\left(\frac{1}{\sqrt{D}}\left[\prod_{0<a<D}\Gamma\left(\frac{a}{D}\right)^{\epsilon(a)}\right]^{\frac{w}{2h}}\right),$$

en particulier si on choisit $D = 3$ *on trouve la valeur approchée* $h_{\mathrm{F}^+}(E) = 0,16993... \geq \varepsilon_1$.

4. Modèles de Moret-Bailly des variétés abéliennes

On rappelle dans ce paragraphe une partie de la théorie des modèles de Moret-Bailly, voir [**MB85**] et [**MB90**]. Cela reprend des travaux de [**Bo96b**], § 4. 2, dans la même veine que [**Paz12**, **BoDa99**].

4.1. Définitions. Soit k un corps de nombres, $\mathcal{O}_k$ son anneau d'entiers et $\pi\colon \mathcal{A} \longrightarrow \operatorname{Spec}(\mathcal{O}_k)$ un schéma en groupes semi-stable, *i. e.* un schéma en groupes lisse de type fini et séparé sur $\operatorname{Spec}(\mathcal{O}_k)$, tel que les composantes de ses fibres sont des extensions de variétés abéliennes par des tores.

Soit $\mathcal{L}$ un fibré en droite sur $\mathcal{A}$. Notons

$$p_i\colon \mathcal{L}^3 := \mathcal{L} \times_{\mathcal{O}_k} \mathcal{L} \times_{\mathcal{O}_K} \mathcal{L} \longrightarrow \mathcal{L}\,, \qquad i = 1,2,3$$

les projections sur les trois facteurs, notons

$$p_I\colon \mathcal{A}^3 \longrightarrow \mathcal{A}$$

le morphisme envoyant un point géométrique (x_1, x_2, x_3) sur $\sum_{i\in I} x_i$, pour tout sous-ensemble non vide I de $\{1,2,3\}$, et notons par $\overline{\mathcal{O}_{\mathcal{A}^3}}$ le fibré hermitien trivial $(\mathcal{O}_{\mathcal{A}^3}, \|.\|)$ donné par $\|1\| = 1$. Alors, un fibré en droites hermitien $\overline{\mathcal{L}}$ sur $\mathcal{A}$ est cubiste si et seulement s'il existe un isomorphisme isométrique

$$\mathcal{D}_3(\overline{\mathcal{L}}) := \bigotimes_{I\subset\{1,2,3\}, I\neq\emptyset} \left(p_I^\star\overline{\mathcal{L}}\right)^{\otimes(-1)^{\# I}} \xrightarrow{\sim} \overline{\mathcal{O}_{\mathcal{A}^3}} \tag{4.1}$$

de fibrés en droites hermitiens sur $\mathcal{A}^3$ satisfaisant de bonnes conditions de symétrie et de cocycle (*confer* **[MB85]**, I. 2. 4. 5., (i) et (iii)). La relation (4.1) implique que si $\varepsilon\colon \operatorname{Spec}(\mathcal{O}_k) \longrightarrow \mathcal{A}$ est la section neutre,

$$\varepsilon^\star\overline{\mathcal{L}} \simeq \overline{\mathcal{O}}_{\operatorname{Spec}(\mathcal{O}_k)}\ ,$$

et si $\mathcal{A}_k$ est une variété abélienne, la $(1,1)$ forme $c_1(\overline{\mathcal{L}})$ est invariante par translation sur tous les tores complexes $\mathcal{A}_\sigma(\mathbb{C})$, pour $\sigma\colon k \hookrightarrow \mathbb{C}$.

Réciproquement, si $\mathcal{A}$ est un schéma abélien sur $\mathcal{O}_k$, ces deux dernières propriétés caractérisent les fibrés en droites hermitiens sur $\mathcal{A}$.

Soit $\pi\colon \mathcal{A} \longrightarrow \operatorname{Spec}(\mathcal{O}_k)$ un schéma en groupes semi-stable dont la fibre générique $\mathcal{A}_k$ est une variété abélienne. Pour tout fibré en droites $\mathcal{M}$ sur $\mathcal{A}$, l'image directe $\pi_\star\mathcal{M}$ est cohérente (voir **[MB85]**, lemme VI, I. 4. 2) et sans torsion, donc localement libre. Si $\overline{\mathcal{L}_k}$ est un fibré en droites cubiste hermitien sur $\mathcal{A}$ et si $\mathcal{L}_k$ est ample sur $\mathcal{A}_k$, alors $\mathcal{L}$ est ample on $\mathcal{A}$ (voir **[Ra85]**, théorème VIII. 2, et **[MB85]**, proposition VI. 2. 1) et $c_1(\mathcal{L})$ est strictement positif sur $\mathcal{A}(\mathbb{C})$. En effet, elle est invariante par translation sur chaque composante de $\mathcal{A}(\mathbb{C})$ et cohomologue à une forme $(1,1)$ strictement positive. On peut ainsi définir $\pi_\star(\overline{\mathcal{L}})$ comme le fibré vectoriel hermitien de rang

$$\rho(\mathcal{L}_{\overline{\mathbb{Q}}}) := \frac{1}{g!} c_1(\mathcal{L}_{\overline{\mathbb{Q}}})^g$$

sur $\operatorname{Spec}(\mathcal{O}_k)$ tel que $\pi_\star(\mathcal{L})$ muni de la structure hermitienne héritée de la métrique L^2 notée $\|.\|$ sur $\overline{\mathcal{L}}$ et la mesure de Haar normalisée sur le tore complexe $\mathcal{A}_\sigma(\mathbb{C})$. En d'autres termes, pour toute section $s \in \pi_\star\mathcal{L} \otimes_\sigma \mathbb{C} \simeq \mathrm{H}^2(\mathcal{A}_\sigma, \mathcal{L}_\sigma)$, on pose

$$\|s\|_\sigma^2 = \int_{\mathcal{A}_\sigma(\mathbb{C})} \|s(x)\|_{\overline{\mathcal{L}}}^2 d\mu(x)\ ,$$

où $d\mu$ désigne la mesure de Haar normalisée sur $\mathcal{A}_\sigma(\mathbb{C})$.

Définition 4.1. *Soit A une variété abélienne sur $\overline{\mathbb{Q}}$, soit L un fibré en droites ample et symétrique sur A et F un sous-ensemble fini de $A(\overline{\mathbb{Q}})$. On appelle MB-*modèle *de (A, L, F) sur un corps de nombres $k \subset \overline{\mathbb{Q}}$ la donnée suivante :*

- *un schéma en groupes semi-stable* $\pi\colon \mathcal{A} \longrightarrow \operatorname{Spec}(\mathcal{O}_k)$,
- *un isomorphisme $i\colon A \xrightarrow{\sim} \mathcal{A}_{\overline{\mathbb{Q}}}$ de variétés abeliennes sur $\overline{\mathbb{Q}}$,*

- *un fibré hermitien cubiste* $\overline{\mathcal{L}}$ *sur* $\mathcal{A}$,
- *un isomorphisme* φ *comme précisé en* **[Paz12]** *dans la proposition 2.5.*
- *pour tout* $P \in F$, *une section* $\varepsilon_P\colon \mathrm{Spec}(\mathcal{O}_k) \longrightarrow \mathcal{A}$ *de l'application* π *telle que le point géométrique* $\varepsilon_{P,\overline{\mathbb{Q}}} \in \mathcal{A}(\overline{\mathbb{Q}})$ *égale le point* $i(P)$,

satisfaisant de plus les conditions : il existe un sous-schéma $\mathcal{K}$ *de* $\mathcal{A}$, *plat et fini sur* $\mathrm{Spec}(\mathcal{O}_k)$, *tel que* $i^{-1}(\mathcal{K}_{\overline{\mathbb{Q}}})$ *égale le groupe de Mumford* $K(L^{\otimes 2})$, *i.e. le sous-groupe algébrique fini de* A *dont les points rationels* x *sur* $\overline{\mathbb{Q}}$ *sont caractérisés par l'existence d'un isomorphisme de fibrés en droites sur* A

$$t_x^\star L^{\otimes 2} \simeq L^{\otimes 2} .$$

DÉFINITION 4.2. *Pour tout triplet* (A, L, r) *avec* A *une variété abélienne sur* $\overline{\mathbb{Q}}$ *et* L *un fibré en droites symétrique et ample,* $r > 0$ *un entier pair, on dit qu'un corps de nombres* k *est* MB *s'il existe un* MB*–modèle du type* $(\pi\colon \mathcal{A} \longrightarrow \mathrm{Spec}(\mathcal{O}_k), i, \overline{\mathcal{L}}, \varphi, (\varepsilon_P)_{P\in A_{r^2}})$ *rationnel sur* k.

REMARQUE 4.3. *On peut construire des corps de nombres* MB *en utilisant le théorème de réduction semi-stable (confer* **[MB85]** *Theorem 3.5 page 58).*

4.2. Propriétés des MB**–modèles.** Les propriétés principales des MB–modèles sont montrées par Moret-Bailly en **[MB85]** et **[MB90]**. Voir de plus Breen **[Br83]** et Mumford **[Mum83]**. En voici une présentation concise :

THÉORÈME 4.4. *Soit* A *une variété abélienne de dimension* g *sur* $\overline{\mathbb{Q}}$, *soit* L *un fibré en droites symétrique et ample sur* A *et* F *un sous-ensemble fini de* $A(\overline{\mathbb{Q}})$. *On a*

(i) Existence. *Pour tout corps de nombres* k_0, *il existe un corps de nombres* k *contenant* k_0 *et un* MB*–modèle* $(\pi\colon \mathcal{A} \longrightarrow \mathrm{Spec}(\mathcal{O}_k), i, \overline{\mathcal{L}}, \varphi, (\varepsilon_P)_{P\in F})$ *pour les données* (A, L, F).

(ii) Hauteur de Néron-Tate. *Pour tout* MB*–modèle comme en* (i) *et pour tout point* $P \in F$, *la hauteur normalisée* $[k : \mathbb{Q}]^{-1}\widehat{\deg}(\varepsilon_P^\star\overline{\mathcal{L}})$ *égale la valeur en* P *de la hauteur de Néron-Tate associée à* L *et notée* $\widehat{h}_L(P)$.

(iii) Indépendance des MB–modèles. *Pour toute paire de* MB*–modèles*

$$(\pi\colon \mathcal{A} \longrightarrow \mathrm{Spec}(\mathcal{O}_k), i, \overline{\mathcal{L}}, \varphi, (\varepsilon_P)_{P\in F})$$

et

$$(\pi'\colon \mathcal{A} \longrightarrow \mathrm{Spec}(\mathcal{O}_k), i', \overline{\mathcal{L}'}, \varphi', (\varepsilon'_P)_{P\in F})$$

de (A, L, F) *sur un corps de nombres* k, *les isomorphismes canoniques* i, φ, i' *et* φ' :

$$(\pi_\star\mathcal{L})_{\overline{\mathbb{Q}}} \simeq \mathrm{H}^0(A, L) \simeq (\pi'_\star\mathcal{L})'_{\overline{\mathbb{Q}}}$$

et

$$(\varepsilon_P^\star\mathcal{L})_{\overline{\mathbb{Q}}} \simeq L_{|P} \simeq ({\varepsilon'_P}^\star\mathcal{L}')_{\overline{\mathbb{Q}}} \qquad (\forall P \in F)$$

s'étendent en isomorphismes isométriques de fibrés en droites hermitiens sur la base $\mathrm{Spec}(\mathcal{O}_k)$:

$$\pi_\star(\overline{\mathcal{L}}) \simeq \pi'_\star(\overline{\mathcal{L}'})$$

et

$${\varepsilon_P}^\star(\overline{\mathcal{L}}) \simeq {\varepsilon'_P}^\star(\overline{\mathcal{L}'}) .$$

(iv) Extensions des scalaires. *Soit*

$$(\pi\colon \mathcal{A} \longrightarrow \operatorname{Spec}(\mathcal{O}_k), i, \overline{\mathcal{L}}, \varphi, (\varepsilon_P)_{P\in F})$$

un MB-modèle sur un corps de nombres k et soit k' un corps de nombres tel que $k \subset k' \subset \overline{\mathbb{Q}}$. On obtient par extension des scalaires de $\mathcal{O}_k$ à $\mathcal{O}_{k'}$ un schéma en groupes semi-stable

$$\tilde{\pi}\colon \tilde{\mathcal{A}} := \mathcal{A} \times_{\mathcal{O}_k} \mathcal{O}_{k'} \longrightarrow \operatorname{Spec}(\mathcal{O}_{k'}) \ ,$$

un fibré en droites hermitien $\overline{\tilde{\mathcal{L}}}$ sur $\tilde{\mathcal{A}}$ (en tirant en arrière $\overline{\mathcal{L}}$ par la première projection $\mathcal{A} \times_{\mathcal{O}_k} \mathcal{O}_{k'} \longrightarrow \mathcal{A}$), des sections

$$\tilde{\varepsilon}_P := \varepsilon_P \otimes_{\mathcal{O}_k} \mathcal{O}_{k'}\colon \operatorname{Spec}(\mathcal{O}_{k'}) \longrightarrow \tilde{\mathcal{A}} \ ,$$

et les isomorphismes i et φ déterminent des isomorphismes :

$$\tilde{i}\colon A \xrightarrow{\sim} \tilde{\mathcal{A}}_{\overline{\mathbb{Q}}} \quad \text{et} \quad \tilde{\varphi}\colon L \xrightarrow{\sim} \tilde{i}^\star \mathcal{L}_{\overline{\mathbb{Q}}} \ .$$

Le 5-uplet $(\tilde{\pi}\colon \tilde{\mathcal{A}} \longrightarrow \operatorname{Spec}(\mathcal{O}_{k'}), \tilde{i}, \overline{\tilde{\mathcal{L}}}, \tilde{\varphi}, (\tilde{\varepsilon}_P)_{P\in F})$ est un MB-modèle de (A, L, F) sur k'. De plus, si $j\colon \operatorname{Spec}(\mathcal{O}_{k'}) \longrightarrow \operatorname{Spec}(\mathcal{O}_k)$ est l'application inclusion $\mathcal{O}_k \hookrightarrow \mathcal{O}_{k'}$, alors l'isomorphisme canonique :

$$j^\star \pi_\star \mathcal{L} \longrightarrow \tilde{\pi}_\star \tilde{\mathcal{L}}$$

définit un isomorphisme isométrique de fibrés vectoriels hermitiens sur la base $\operatorname{Spec}(\mathcal{O}_{k'})$ *:*

$$j^\star \pi_\star \overline{\mathcal{L}} \longrightarrow \tilde{\pi}_\star \overline{\tilde{\mathcal{L}}} \ .$$

(v) Pente de $\pi_\star\overline{\mathcal{L}}$. *Pour tout MB-modèle comme en* (i) *on a $\pi_\star\overline{\mathcal{L}}$ semi-stable et :*

$$\frac{\widehat{\deg} \pi_\star \overline{\mathcal{L}}}{[K:\mathbb{Q}]\rho(L)} = -\frac{1}{2} h_F(A) + \frac{1}{4} \log\left(\frac{\rho(L)}{(2\pi)^g}\right) .$$

(vi) Points base. *Pour tout MB-modèle comme en* (i)*, et tout $n \in \mathbb{N}^\star$, soit $\mathcal{A}^{[n]}$ le plus petit schéma en groupes ouvert de $\mathcal{A}$ contenant $K(\mathcal{L}_{\overline{\mathbb{Q}}}^{\otimes n})$. Si n est pair et si la clôture de $K(\mathcal{L}_{\overline{\mathbb{Q}}}^{\otimes n})$ dans $\mathcal{A}$ est finie sur* $\operatorname{Spec}(\mathcal{O}_k)$*, alors les sections globales $\mathrm{H}^0(\mathcal{A}, \mathcal{L}^{\otimes n})$ engendrent $\mathcal{L}^{\otimes n}$ sur $\mathcal{A}^{[n]}$.*

DÉMONSTRATION. Les preuves de (i)-(v) sont données dans [**Bo96b**], § 4. 3. 2. Le (vi) vient de [**MB85**], VI. 3.4 et VI. 2. 2. □

REMARQUE 4.5. *On remarque que si $(\pi\colon \mathcal{A} \longrightarrow \operatorname{Spec}(\mathcal{O}_k), i, \overline{\mathcal{L}}, \varphi, (\varepsilon_P)_{P\in F})$ est un MB-modèle, alors $(\pi\colon \mathcal{A} \longrightarrow \operatorname{Spec}(\mathcal{O}_{k'}), i, \overline{\mathcal{L}}^{\otimes r^2}, \varphi, (\varepsilon_P)_{P\in F})$ est aussi un MB-modèle pour une extension k' de k.*

5. Hauteur d'un point par la formule clef

Soit (A, L) une variété abélienne principalement polarisée et définie sur un corps de nombres k et soit $F = A[4]$ le sous-groupe de 4-torsion de A. On se donne un MB-modèle associé à (A, L, F) comme dans la partie précédente $(\pi\colon \mathcal{A} \longrightarrow \operatorname{Spec}(\mathcal{O}_K), i, \overline{\mathcal{L}}, \varphi, (\varepsilon_P)_{P\in F})$. On rappelle ici quelques notations de l'article [**Paz12**].

Soit Γ un système de représentants de A_{r^2}/A_r, le quotient des points de r^2-torsion par le groupe des points de r-torsion. Notons $\mathcal{F} = (\pi_\star\mathcal{L})^\Gamma$ la somme directe de r^{2g} copies de $\pi_\star\mathcal{L}$, indexée par Γ. Le sous-schéma $\mathcal{B}_\mathcal{F}$ des points bases du système linéaire $\mathcal{F}$ de sections de $\mathcal{L}^{\otimes r^2}$ est défini comme le sous-schéma fermé de $\mathcal{A}$ dont le faisceau d'idéaux $I_{\mathcal{B}_\mathcal{F}}$ est tel que l'image de l'application canonique $\pi^\star\mathcal{F} \longrightarrow \mathcal{L}^{\otimes r^2}$ soit $I_{\mathcal{B}_\mathcal{F}}.\mathcal{L}^{\otimes r^2}$.

Comme $\mathcal{B}_\mathcal{F}$ ne coupe pas la fibre générique $\mathcal{A}_k$, pour toute section P de π, le sous-schéma $P^\star\mathcal{B}_\mathcal{F}$ de $\mathrm{Spec}(\mathcal{O}_k)$ est un diviseur. On notera :

DÉFINITION 5.1.

$$P^\star\mathcal{B}_\mathcal{F} = \sum_{\substack{\mathfrak{p}\text{ premier}\\ \text{de } \mathcal{O}_k,\\ \mathfrak{p}\nmid\infty}} \beta_\mathfrak{p}(\mathcal{L}^{\otimes r^2}, \mathcal{F}, P)\mathfrak{p}\,. \tag{5.1}$$

Dans cette définition, les $\beta_\mathfrak{p}$ sont des entiers positifs presque tous nuls. Le fibré $\mathcal{F}$ varie avec le fibré $\mathcal{L}$.

DÉFINITION 5.2. *Soit (A, L) une variété abélienne principalement polarisée sur un corps de nombres k, de dimension g, avec L symétrique et ample. Soit P un point de $\mathcal{A}(\mathcal{O}_k)$ et $\sigma\colon k \hookrightarrow \mathbb{C}$ un plongement complexe. Soit τ_σ un point de l'espace de Siegel $\mathfrak{S}_g$ tel que :*

$$\mathcal{A}_\sigma(\mathbb{C}) \simeq \mathbb{C}^g/(\mathbb{Z}^g + \tau_\sigma\mathbb{Z}^g) \tag{5.2}$$

comme variétés abéliennes principalement polarisées, et soit $z \in \mathbb{C}^g$ tel que $[z] \in \mathbb{C}^g/(\mathbb{Z}^g + \tau_\sigma\mathbb{Z}^g)$ soit l'image de P_σ par l'application (5.2). Alors on pose :

$$\beta_\sigma(\overline{\mathcal{L}}^{\otimes r^2}, \overline{\mathcal{F}}, P) = -\frac{1}{2}\log\left(2^{\frac{g}{2}} \sum_{e\in\mathcal{Z}_r(\tau_\sigma)} \|\theta\|^2(rz + e, \tau_\sigma)\right), \tag{5.3}$$

où on a noté $\mathcal{Z}_r(\tau_\sigma)$ l'ensemble $\frac{1}{r}(\mathbb{Z}^g + \tau_\sigma\mathbb{Z}^g)/(\mathbb{Z}^g + \tau_\sigma\mathbb{Z}^g)$ et où on utilise

$$\|\theta\|(z,\tau) = \det(\mathrm{Im}\,\tau)^{\frac{1}{4}}\,\exp(-\pi\,{}^t\,\mathrm{Im}\,z(\mathrm{Im}\,\tau)^{-1}\,\mathrm{Im}\,z)\;|\theta(z,\tau)|.$$

Dans cette dernière définition, la dépendance en $\mathcal{F}$ est cachée dans le choix des fonctions thêta, on trouvera plus de détails dans le paragraphe 4 de **[Paz12]**. La définition (5.3) est la définition des β_σ de **[Bo96b]**.

DÉFINITION 5.3. *Pour v une place finie associée à un premier $\mathfrak{p}$ de $\mathcal{O}_k$, on notera*

$$\beta_{L^{\otimes r^2},v}(P) = \beta_\mathfrak{p}(\mathcal{L}^{\otimes r^2}, \mathcal{F}, P)\log \mathrm{N}_{k/\mathbb{Q}}(\mathfrak{p}).$$

Pour v une place archimédienne associée à un plongement σ, on notera

$$\beta_{L^{\otimes r^2},v}(P) = \beta_\sigma(\overline{\mathcal{L}}^{\otimes r^2}, \overline{\mathcal{F}}, P).$$

PROPOSITION 5.4. *Soit (A, L) une variété abélienne principalement polarisée sur un corps de nombres k de degré d, de dimension g, semi-stable, avec L symétrique et ample. Pour toute section P de $\pi\colon \mathcal{A} \longrightarrow \mathrm{Spec}(\mathcal{O}_k)$, on a :*

$$\widehat{h}_{A,L^{\otimes r^2}}(P) = h_{A,L^{\otimes r^2}}(P) - \frac{1}{2}h_{\mathrm{F}^+}(A) + \frac{1}{d}\sum_{v\in M_k} d_v\beta_{L^{\otimes r^2},v}(P)$$

où $\beta_{L^{\otimes r^2},v}$ est défini par (5.3).

DÉMONSTRATION. La preuve de cette proposition est donnée dans **[Paz12]** au lemme 5.2, modulo le passage à la hauteur de Faltings positive qui simplifie ici la formule en vertu de $h_{\mathrm{F}^+}(A) = h_\mathrm{F}(A) + \frac{g}{2}\log(2\pi^2)$. □

6. Décomposition de la hauteur de Faltings d'une jacobienne hyperelliptique

On s'intéresse dans cette partie aux jacobiennes de courbes hyperelliptiques. On donne une formule explicite pour la hauteur de Faltings d'une telle jacobienne A, généralisant ainsi la formule proposée par Ueno dans [**Uen88**] pour la dimension 2. Dans toute la suite k désigne un corps de nombres, $\mathcal{O}_k$ son anneau d'entiers et on note $S = \mathrm{Spec}(\mathcal{O}_k)$. Soient C/k une courbe hyperelliptique et $A = J_C$ sa jacobienne.

Equations de Weierstrass et discriminants. Soient $g \geq 2$ et k un corps de nombres. Soit $v \in M_k$. On note k_v le complété de k en la place v, vu comme corps de fractions de l'anneau de valuation discrète $\mathcal{O}_v$. Nous allons travailler avec les équations de Weierstrass des courbes hyperelliptiques et leur discriminant associé. Pour obtenir une meilleure formule close, nous allons utiliser les travaux de Lockhart [**Loc94**] aux places archimédiennes, qui fournissent via la théorie de Mumford une contribution explicitée en termes de fonctions thêta. Pour les places finies nous utiliserons les travaux de Kausz [**Kau99**], Maugeais [**Mau03**] et Liu [**Liu96**, **Liu02**] qui impliqueront le discriminant minimal des modèles de Weierstrass.

DÉFINITION 6.1. *(Lockhart) Un modèle de Weierstrass (E) d'une courbe hyperelliptique C sur k_v est une équation du type*

$$(E)\ :\ y^2 + Q(x)\,y = P(x),$$

où P et Q sont des polynômes à coefficients dans k_v tels que $\deg Q \leq g$ *et* $\deg P = 2g+1$. *On choisit P unitaire. Une telle équation est unique modulo les changements de variables*

$$(*)\begin{cases} x = u^g x' + s \\ \\ y = u^{2g+1} y' + t(x') \end{cases}$$

où $u \in k_v^*$, $s \in k_v$ et t est un polynôme à coefficients dans k_v de degré inférieur ou égal à g (voir proposition 1.2 page 730 de [**Loc94**]*).*

Le discriminant de (E) est défini par $\Delta_{Lock}(E) = 2^{4g}\,\mathrm{disc}\left(P(x) + \frac{1}{4}\,Q(x)^2\right)$. *Le discriminant minimal de Lockhart $\Delta_{Lock}(C)$ de C/k_v sera le discriminant de valuation minimale parmi tous les modèles (E) relativement aux changements de variables $(*)$.*

DÉFINITION 6.2. *(Liu) Soient $P, Q \in \mathcal{O}_v[x]$ des polynômes avec* $\deg P \leq 2g+2$ *and* $\deg Q \leq g+1$ *tels que*

$$(\mathcal{E}) : y^2 + Q(x)y = P(x) \tag{6.1}$$

est une équation entière *pour C où C/k_v est hyperelliptique de genre g. On note* $\Delta_{Liu}(\mathcal{E}) = 2^{-(4g+4)}\,\mathrm{disc}_{2g+2}(4P+Q^2)$ *le discriminant de (*6.1*). Si on écrit de plus*

$$P = p_{2g+2}x^{2g+2} + \cdots + p_0 \quad \text{et} \quad Q = q_{g+1}x^{g+1} + \cdots + q_0$$

où p_{2g+2} et q_{g+1} sont des éléments (éventuellement nuls) de $\mathcal{O}_v$. On a unicité de *ce modèle modulo les changements de variables (voir* **[Liu02]**, *corollaire 4.33 page 296)*

$$(**)\begin{cases} x = \dfrac{ax'+b}{cx'+d} \\ \\ y = \dfrac{H(x')+ey'}{(cx'+d)^{g+1}} \end{cases}$$

où $\begin{pmatrix} a & b \\ c & d \end{pmatrix} \in GL_2(K)$, $e \in K^*$, $H \in K[x']$ *avec* $\deg H \leq g+1$.

Le discriminant minimal de Liu $\Delta_{Liu}(C)$ *de* C *sera le discriminant de valuation minimale parmi tous les modèles* $(\mathcal{E})$ *relativement aux changements de variables* $(**)$, *voir* **[Liu02]** *1.26 page 464 and 1.9 page 480.*

Lorsqu'on passe d'une équation E à une équation E' par le changement de coordonnées $(*)$ on obtient la relation $\Delta_E = u^{4g(2g+1)}\Delta_{E'}$.

Formes différentielles. Soit C une courbe hyperelliptique donnée par un modèle de Weierstrass E. On peut alors exhiber une base de $H^0(C, \Omega^1_{C/k})$, donnée par les formes :

$$\omega_i = \frac{x^{i-1}dx}{2y+Q(x)},\ 1 \leq i \leq g.$$

Considérons alors la g-forme $\alpha = \omega_1 \wedge ... \wedge \omega_g$. On vérifie que lorsqu'on change de coordonnées en utilisant $(*)$ dans le modèle de Weierstrass pour passer de E à E' on obtient la relation $\alpha = u^{-g^2}\alpha'$. Ceci va nous permettre de trouver une g-forme différentielle η ne dépendant pas du modèle de Weierstrass. On va chercher η de la forme :

$$\eta = \Delta_E^a(\omega_1 \wedge ... \wedge \omega_g)^b, \text{ avec } a, b \in \mathbb{Z}.$$

Un changement de modèle de E vers E' de la forme $(*)$ conduit alors à :

$$\eta = u^{4g(2g+1)a - g^2 b}\eta'.$$

On fait le choix $a = g$ et $b = 4(2g+1)$ et on obtient :

PROPOSITION 6.3. *La g-forme* $\eta = \Delta_E^g\,(\omega_1 \wedge ... \wedge \omega_g)^{\otimes 4(2g+1)}$ *est indépendante du modèle de Weierstrass E choisi. Elle est de plus bien définie sur les modèles* $(\mathcal{E})$ *et est aussi indépendante du modèle* $\mathcal{E}$ *choisi.*

DÉMONSTRATION. L'invariance plus générale par le changement de variables $(**)$ est un calcul direct. Cela avait déjà été remarqué dans **[Kau99]** proposition 2.2 page 43. □

On va donc utiliser la section η pour calculer la hauteur de Faltings exprimée comme en (3.2). On a besoin pour cela du résultat de la proposition suivante. L'invariance par les deux classes de changements de variables nous permettra de tirer parti du meilleur modèle aux places finies (suivant **[Kau99]**) et d'un modèle pratique au regard des fonctions thêta aux places archimédiennes (suivant **[Loc94]**).

On commence par définir :

DÉFINITION 6.4. *Soient k un corps de nombres et C/k une courbe algébrique lisse à réduction semi-stable définie sur k et de genre g. On note* $p : \mathcal{C} \to S$ *un* *modèle entier sur* $S = \mathrm{Spec}(\mathcal{O}_k)$ *semi-stable de la courbe* C. *On appelle hauteur de Faltings de* C/k *la quantité :*

$$h_{\mathrm{F}^+}(C) = \frac{1}{[k:\mathbb{Q}]}\widehat{\deg}(\det p_*\omega_{C/S}),$$

où le choix de métriques hermitiennes est $\|\alpha\|_v^2 = \frac{i^{g^2}}{(2\pi)^{2g}}\int \alpha\wedge\overline{\alpha}$.

On remarquera qu'il s'agit bien de la hauteur stable car on travaille sur un modèle semi-stable de C.

PROPOSITION 6.5. *Soient* k *un corps de nombres et* C/k *une courbe algébrique lisse à réduction semi-stable. On a alors :*

$$h_{\mathrm{F}^+}(J_C) = h_{\mathrm{F}^+}(C).$$

DÉMONSTRATION. Notons $S = \mathrm{Spec}\,\mathcal{O}_k$ et soit $p : C \longrightarrow S$ un modèle entier semi-stable de la courbe C, de section neutre ε. On considère $A = \mathrm{Pic}^0_{C/S}$. On a alors :

$$\mathrm{Lie}(A) \simeq R^1p_*O_C.$$

De plus par dualité de Grothendieck (on pourra consulter le paragraphe 6.4.3 page 243 de [**Liu02**]) :

$$(R^1p_*O_C)^\vee \simeq p_*\omega_{C/S}.$$

On calcule alors :

$$\varepsilon^*\Omega^1_{A/S} \simeq \mathrm{Lie}(A)^\vee \simeq p_*\omega_{C/S},$$

d'où :

$$\varepsilon^*\Omega^g_{A/S} \simeq \det p_*\omega_{C/S},$$

et cet isomorphisme est une isométrie d'après le 4.15 de l'exposé II de [**SPA**]. Il suffit alors de prendre le degré d'Arakelov de chaque côté pour obtenir la proposition. □

Partie non archimédienne. On garde les notations des paragraphes précédents. La section η correspond au choix de section Λ fait dans [**Kau99**]. Dans cet article, Kausz analyse les contributions en chaque place finie, ce qui permet d'obtenir la proposition suivante :

PROPOSITION 6.6. *La section* η *s'étend en une section globale entière sur le modèle de la courbe* C. *Il existe des entiers naturels* e_v *tels que la somme des contributions aux places finies s'exprime par :*

$$\log\Big(\mathrm{Card}(\det p_*\omega_{C/S}/\eta\,\mathcal{O}_k)\Big) =$$

$$g\log \mathrm{N}_{k/\mathbb{Q}}(\Delta_{\mathrm{Liu}}(C)) - \sum_{v|\Delta_{\mathrm{Liu}}(C)} d_v(8g+4)e_v\log \mathrm{N}_{k/\mathbb{Q}}(v).$$

DÉMONSTRATION. La section s'étend sur le modèle entier d'après le théorème 3.1 page 44 de [**Kau99**], on a ainsi $\mathrm{ord}_v(\eta)\geq 0$. En reprenant les calculs à l'équation (1) de la preuve de la proposition 5.5 page 57 de *loc.cit.*, on peut déduire que pour toute place finie v il existe un entier $e_v\in\mathbb{N}$ dépendant du modèle hyperelliptique et tel que :

$$\mathrm{ord}_v(\eta) = g\,\mathrm{ord}_v(\Delta_{\mathcal{E}}) - (8g+4)e_v.$$

Notons qu'en combinant avec l'article de Maugeais [**Mau03**], théorème 1.1 (voir aussi [**Mau06**] pour une version corrigée) on peut supprimer l'hypothèse de bonne réduction en 2 faite dans [**Kau99**]. □

Partie archimédienne. On se base ici sur les travaux de Lockhart [**Loc94**], dont les calculs s'appuient en bonne partie sur [**Mum84**], lequel nécessite un modèle hyperelliptique de degré impair. Rappelons ici les notations de l'introduction. Tout d'abord pour $m \in \frac{1}{2}\mathbb{Z}^{2g}$ on pose :

$$\varphi_m(\tau) = \theta_m(0,\tau)^8,$$

où $\theta_m(z,\tau)$ est la fonction thêta de caractéristique m associée au réseau de dimension g dont la définition est rappelée en (3.1). Si S est un sous-ensemble de $\{1,2,...,2g+1\}$ on définit alors $m_S = \sum_{i\in S} m_i \in \frac{1}{2}\mathbb{Z}^{2g}$ avec :

$$m_{2i-1} = \begin{bmatrix} {}^t(0 & \dots & 0 & \frac{1}{2} & 0 & \dots & 0) \\ {}^t(\frac{1}{2} & \dots & \frac{1}{2} & 0 & 0 & \dots & 0) \end{bmatrix}, \quad 1 \leq i \leq g+1,$$

$$m_{2i} = \begin{bmatrix} {}^t(0 & \dots & 0 & \frac{1}{2} & 0 & \dots & 0) \\ {}^t(\frac{1}{2} & \dots & \frac{1}{2} & \frac{1}{2} & 0 & \dots & 0) \end{bmatrix}, \quad 1 \leq i \leq g,$$

où le coefficient non nul de la première ligne est en i-ème position. Soit alors $\mathcal{T}$ la collection des sous-ensembles de $\{1,...,2g+1\}$ de cardinal $g+1$. Soit $U = \{1,3,...,2g+1\}$ et notons $\circ$ l'opérateur de différence symétrique. On définit alors :

$$\varphi(\tau) = \prod_{T\in\mathcal{T}} \varphi_{m_{T\circ U}}(\tau). \tag{6.2}$$

On pose enfin $l = \binom{2g+1}{g+1}$ et $n = \binom{2g}{g+1} = \frac{g}{2g+1}l$. On a alors la proposition suivante dont la preuve figure dans [**Loc94**] :

Proposition 6.7. *(Lockhart) Soient $C/\mathbb{C}$ une courbe hyperelliptique et $P \in C(\mathbb{C})$. Soit E un modèle de Weierstrass de (C,P). On uniformise $J_C(\mathbb{C}) \simeq \mathbb{C}^g/\Lambda_E$ avec le réseau $\Lambda_E = \Omega_1\mathbb{Z}^g + \Omega_2\mathbb{Z}^g$ et $\tau_E = \Omega_1^{-1}\Omega_2$. Soit $V(\Lambda_E)$ le covolume du réseau Λ_E dans $\mathbb{C}^g$. Alors la quantité $|\Delta_E|V(\Lambda_E)^{4+\frac{2}{g}}$ ne dépend pas du choix de E et on a :*

$$|\Delta_E|V(\Lambda_E)^{4+\frac{2}{g}} = 2^{4g}\pi^{8g+4}\Big(|\varphi(\tau_E)|\det(\operatorname{Im}\tau_E)^{2l}\Big)^{\frac{1}{n}}.$$

Nous allons ainsi pouvoir calculer la contribution archimédienne à la hauteur de Faltings positive (voir [**deJo07**] pour un calcul similaire) :

Proposition 6.8. *Soient C/k une courbe de genre g donnée dans un modèle hyperelliptique E de discriminant Δ_E et $\omega_i = \frac{x^{i-1}dx}{2y+q(x)}$, $i \in \{1,...,g\}$, une base des formes différentielles sur C. Soit $v \in M_k^\infty$. Soit $\eta = \Delta_E^g(\omega_1\wedge ... \wedge\omega_g)^{\otimes 8g+4}$. Alors on a :*

$$\log||\eta||_v = (8g+4)\log\Big(2^{-\frac{g}{4g+2}}|\varphi(\tau_E)|^{\frac{1}{4l}}\det(\operatorname{Im}\tau_E)^{\frac{1}{2}}\Big).$$

DÉMONSTRATION. Il suffit de calculer :

$$
\begin{aligned}
||\eta||_v^2 &= |\Delta_E|^{2g}(||\omega_1 \wedge \ldots \wedge \omega_g||_v^2)^{8g+4} \\
&= |\Delta_E|^{2g} \frac{1}{(2\pi)^{2g(8g+4)}} \left(\int_{A_v(\mathbb{C})} i^{g^2} \omega_1 \wedge \ldots \wedge \omega_g \wedge \overline{\omega_1} \wedge \ldots \wedge \overline{\omega_g} \right)^{8g+4} \\
&= |\Delta_E|^{2g} \frac{1}{(2\pi)^{2g(8g+4)}} (2^g V(\Lambda_E))^{8g+4} \\
&= \frac{2^{g(8g+4)}}{(2\pi)^{2g(8g+4)}} |\Delta_E|^{2g} V(\Lambda_E)^{8g+4} \\
&= \frac{2^{g(8g+4)}}{(2\pi)^{2g(8g+4)}} 2^{8g^2} \pi^{2g(8g+4)} (|\varphi(\tau_E)| \det(\operatorname{Im} \tau_E)^{2l})^{\frac{2g}{n}} \\
&= 2^{-4g} (|\varphi(\tau_E)| \det(\operatorname{Im} \tau_E)^{2l})^{\frac{2g}{n}},
\end{aligned}
$$

donc

$$||\eta||_v = 2^{-2g} |\varphi(\tau_E)|^{\frac{g}{n}} \det(\operatorname{Im} \tau_E)^{\frac{2lg}{n}}.$$

La formule est établie en observant que $\frac{g}{n} = \frac{8g+4}{4l}$ et $\frac{2lg}{n} = \frac{8g+4}{2}$. On obtient lorsque $g = 1$, comme $\varphi(\tau) = 2^8 \Delta(\tau)$ (voir **[Loc94]** page 740), l'expression $||\eta||_v = 2^6 \Delta(\tau)(\operatorname{Im} \tau)^6$. □

Preuve du théorème 2.3.

DÉMONSTRATION. On déduit la preuve du théorème en mettant bout à bout les calculs des deux paragraphes précédents, rassemblés dans les propositions 6.6 et 6.8. On peut choisir le discriminant minimal dans la formule finale en minimisant le modèle place par place. On définit donc le discriminant $\Delta_{\min}$ comme le produit des discriminants minimaux locaux. On a donc par définition que $\operatorname{ord}_v(\Delta_{\min/k}) = \operatorname{ord}_v(\Delta_{\mathrm{Liu}}(C/k_v))$ pour chaque place v finie de k. Le calcul donne naturellement l'expression de $(8g+4)\, h_{\mathrm{F}^+}(\mathrm{Jac}(C))$. □

7. Calculs explicites en dimension 1

On montre dans les paragraphes qui suivent que la formule de hauteur d'une courbe elliptique est bien la même dans toutes ces différentes approches.

7.1. Formule hyperelliptique. La spécialisation au cas du genre $g = 1$ du théorème 2.3 fournit $l = \binom{2g+1}{g+1} = 3$. Soit E une courbe elliptique définie sur un corps de nombres k. Pour toute place archimédienne v, on note τ_v un nombre complexe tel que $E_v(\mathbb{C}) \simeq \mathbb{C}/(\mathbb{Z} + \tau_v \mathbb{Z})$ et $\Delta_{\min}$ le discriminant minimal de E/k. Alors $e_v = 0$ et $12 f_v = \operatorname{ord}_v(\Delta_{\min})$, donc

$$
\begin{aligned}
d \cdot h_{\mathrm{F}^+}(E) = & \sum_{v | \Delta_{\min}} d_v \frac{1}{12} \operatorname{ord}_v(\Delta_{\min}) \log \mathrm{N}_{k/\mathbb{Q}}(v) \\
& - \sum_{v \in M_k^\infty} d_v \log \left(2^{-\frac{2}{12}} |\varphi(\tau_v)|^{\frac{1}{4 \cdot 3}} \det(\operatorname{Im} \tau_v)^{\frac{1}{2}} \right),
\end{aligned}
$$

de plus il y a trois sous-ensembles de cardinal 2 dans $\{1,2,3\}$, qui fournissent après un rapide calcul $\varphi(\tau)=\theta_2^8\theta_3^8\theta_4^8=2^8\Delta(\tau)$. On obtient bien

$$12d\cdot h_{\mathrm{F}^+}(E)=\sum_{v|\Delta_{\min}} d_v \mathrm{ord}_v(\Delta_{\min})\log \mathrm{N}_{k/\mathbb{Q}}(v)-\sum_{v\in M_k^\infty} d_v\log\left(|2^6\Delta(\tau_v)|\det(\mathrm{Im}\,\tau_v)^6\right).$$

7.2. Formule d'Autissier. La spécialisation au cas des courbes elliptiques de la formule d'Autissier donnée au théorème 2.2 fournit :

$$h_{\mathrm{F}^+}(E)=2\hat{h}_L(\Theta)+\frac{2}{d}\sum_{v\in M_k^\infty} d_v I(E_v,\lambda_v).$$

On a $\hat{h}_L(\Theta)=0$ (voir la partie de définition page 1453 de **[Aut13]**) car le diviseur est de torsion dans le cas elliptique. De plus par la proposition 2.1 page 1452 de **[Aut13]** il vient

$$I(E,\lambda)=-\frac{1}{24}\log\left(|\Delta(\tau)|(2\,\mathrm{Im}\,\tau)^6\right).$$

On obtient donc bien la même formule que celle du théorème 2.1 pour le cas de bonne réduction partout. Dans le cas général, pour les contributions locales aux places finies, le résultat est donné dans le théorème 7 de Faltings **[Fa84]**.

Références

[Aut06] P. Autissier, *Hauteur de Faltings et hauteur de Néron-Tate du diviseur thêta* (French, with English and French summaries), Compos. Math. **142** (2006), no. 6, 1451–1458, DOI 10.1112/S0010437X0600234X. MR2278754

[Aut13] P. Autissier, *Un lemme matriciel effectif* (French, with English and French summaries), Math. Z. **273** (2013), no. 1-2, 355–361, DOI 10.1007/s00209-012-1008-x. MR3010164

[BiLa04] C. Birkenhake and H. Lange, *Complex abelian varieties*, 2nd ed., Grundlehren der Mathematischen Wissenschaften [Fundamental Principles of Mathematical Sciences], vol. 302, Springer-Verlag, Berlin, 2004. MR2062673

[Bo96a] J.-B. Bost, *Arakelov geometry of abelian varieties*, Proceedings of a conference on arithmetic geometry, technical report, Max-Planck Institut für Mat., Bonn, Vol. **96–51** (1996).

[Bo96b] J.-B. Bost, *Intrinsic heights of stable varieties and abelian varieties*, Duke Math. J. **82** (1996), no. 1, 21–70, DOI 10.1215/S0012-7094-96-08202-2. MR1387221

[BoDa99] J.-B. Bost et D. David, Notes on the comparison of heights of abelian varieties, *Manuscrit*, (1999).

[BoMeMB90] J.-B. Bost, J.-F. Mestre, and L. Moret-Bailly, *Sur le calcul explicite des "classes de Chern" des surfaces arithmétiques de genre* 2 (French), Astérisque **183** (1990), 69–105. Séminaire sur les Pinceaux de Courbes Elliptiques (Paris, 1988). MR1065156

[Br83] L. Breen, *Fonctions thêta et théorème du cube* (French), Lecture Notes in Mathematics, vol. 980, Springer-Verlag, Berlin, 1983. MR823233

[Bru13] P. Bruin, *Bornes optimales pour la différence entre la hauteur de Weil et la hauteur de Néron-Tate sur les courbes elliptiques sur* $\overline{\mathbb{Q}}$ (French), Acta Arith. **160** (2013), no. 4, 385–397, DOI 10.4064/aa160-4-5. MR3119786

[Col98] P. Colmez, *Sur la hauteur de Faltings des variétés abéliennes à multiplication complexe* (French, with English summary), Compositio Math. **111** (1998), no. 3, 359–368, DOI 10.1023/A :1000390105495. MR1617134

[CoSi86] G. Cornell and J. H. Silverman (eds.), *Arithmetic geometry*, Springer-Verlag, New York, 1986. Papers from the conference held at the University of Connecticut, Storrs, Connecticut, July 30–August 10, 1984. MR861969

[CrPrSi06] J. E. Cremona, M. Prickett, and S. Siksek, *Height difference bounds for elliptic curves over number fields*, J. Number Theory **116** (2006), no. 1, 42–68, DOI 10.1016/j.jnt.2005.03.001. MR2197860

[DaPh02] S. David and P. Philippon, *Minorations des hauteurs normalisées des sous-variétés de variétés abeliennes. II* (French, with English and French summaries), Comment. Math. Helv. **77** (2002), no. 4, 639–700, DOI 10.1007/PL00012437. MR1949109

[deJo05] R. de Jong, *On the Arakelov theory of elliptic curves*, Enseign. Math. (2) **51** (2005), no. 3-4, 179–201. MR2214884

[deJo07] R. de Jong, *Explicit Mumford isomorphism for hyperelliptic curves*, J. Pure Appl. Algebra **208** (2007), no. 1, 1–14, DOI 10.1016/j.jpaa.2005.11.002. MR2269824

[Del85] P. Deligne, *Preuve des conjectures de Tate et Shafarevitch* (French), Séminaire Bourbaki Vol. 1983/84, Exposés 615–632, Astérisque No. 121-122 (1985), 1–368. MR768950

[Fa84] G. Faltings, *Calculus on arithmetic surfaces*, Ann. of Math. (2) **119** (1984), no. 2, 387–424, DOI 10.2307/2007043. MR740897

[FlySm97] E. V. Flynn and N. P. Smart, *Canonical heights on the Jacobians of curves of genus 2 and the infinite descent*, Acta Arith. **79** (1997), no. 4, 333–352, DOI 10.4064/aa-79-4-333-352. MR1450916

[Fre83] E. Freitag, *Siegelsche Modulfunktionen* (German), Grundlehren der Mathematischen Wissenschaften [Fundamental Principles of Mathematical Sciences], vol. 254, Springer-Verlag, Berlin, 1983. MR871067

[GaRé14] É. Gaudron and G. Rémond, *Théorème des périodes et degrés minimaux d'isogénies* (French, with English and French summaries), Comment. Math. Helv. **89** (2014), no. 2, 343–403, DOI 10.4171/CMH/322. MR3225452

[HiSi00] M. Hindry and J. H. Silverman, *Diophantine geometry*, Graduate Texts in Mathematics, vol. 201, Springer-Verlag, New York, 2000. An introduction. MR1745599

[Kau99] I. Kausz, *A discriminant and an upper bound for ω^2 for hyperelliptic arithmetic surfaces*, Compositio Math. **115** (1999), no. 1, 37–69, DOI 10.1023/A :1000580901251. MR1671741

[Liu96] Q. Liu, *Modèles entiers des courbes hyperelliptiques sur un corps de valuation discrète* (French, with English summary), Trans. Amer. Math. Soc. **348** (1996), no. 11, 4577–4610, DOI 10.1090/S0002-9947-96-01684-4. MR1363944

[Liu02] Q. Liu, *Algebraic geometry and arithmetic curves*, Oxford Graduate Texts in Mathematics, vol. 6, Oxford University Press, Oxford, 2002. Translated from the French by Reinie Erné ; Oxford Science Publications. MR1917232

[Loc94] P. Lockhart, *On the discriminant of a hyperelliptic curve*, Trans. Amer. Math. Soc. **342** (1994), no. 2, 729–752, DOI 10.2307/2154650. MR1195511

[Mau03] S. Maugeais, *Relèvement des revêtements p-cycliques des courbes rationnelles semi-stables* (French, with French summary), Math. Ann. **327** (2003), no. 2, 365–393, DOI 10.1007/s00208-003-0458-1. MR2015076

[Mau06] S. Maugeais, *Erratum à l'article "Relèvement des revêtements p-cycliques des courbes rationnelles semi-stables"*, Math. Ann. **336** (2006), 239–246.

[MB85] L. Moret-Bailly, *Pinceaux de variétés abéliennes* (French, with English summary), Astérisque **129** (1985), 266. MR797982

[MB89] L. Moret-Bailly, *La formule de Noether pour les surfaces arithmétiques* (French), Invent. Math. **98** (1989), no. 3, 491–498, DOI 10.1007/BF01393833. MR1022303

[MB90] L. Moret-Bailly, *Sur l'équation fonctionnelle de la fonction thêta de Riemann* (French), Compositio Math. **75** (1990), no. 2, 203–217. MR1065206

[Mum66] D. Mumford, *On the equations defining abelian varieties. I*, Invent. Math. **1** (1966), 287–354, DOI 10.1007/BF01389737. MR0204427

[Mum83] D. Mumford, *Tata lectures on theta. I*, Progress in Mathematics, vol. 28, Birkhäuser Boston, Inc., Boston, MA, 1983. With the assistance of C. Musili, M. Nori, E. Previato and M. Stillman. MR688651

[Mum84] D. Mumford, *Tata lectures on theta. II*, Progress in Mathematics, vol. 43, Birkhäuser Boston, Inc., Boston, MA, 1984. Jacobian theta functions and differential equations; With the collaboration of C. Musili, M. Nori, E. Previato, M. Stillman and H. Umemura. MR742776

[Nér65] A. Néron, *Quasi-fonctions et hauteurs sur les variétés abéliennes* (French), Ann. of Math. (2) **82** (1965), 249–331, DOI 10.2307/1970644. MR0179173

[Paz12] F. Pazuki, *Theta height and Faltings height* (English, with English and French summaries), Bull. Soc. Math. France **140** (2012), no. 1, 19–49. MR2903770

[Ra85] M. Raynaud, *Hauteurs et isogénies* (French), Astérisque **127** (1985), 199–234. Seminar on arithmetic bundles : the Mordell conjecture (Paris, 1983/84). MR801923

[Sil86] J. H. Silverman, *The arithmetic of elliptic curves*, Graduate Texts in Mathematics, vol. 106, Springer-Verlag, New York, 1986. MR817210

[SPA] L. Szpiro (ed.), *Séminaire sur les pinceaux arithmétiques : la conjecture de Mordell*, Société Mathématique de France, Paris, 1985. Papers from the seminar held at the École Normale Supérieure, Paris, 1983–84; Astérisque No. 127 (1985) (1985). MR801916

[Sto02] M. Stoll, *On the height constant for curves of genus two. II*, Acta Arith. **104** (2002), no. 2, 165–182, DOI 10.4064/aa104-2-6. MR1914251

[Uen88] K. Ueno, *Discriminants of curves of genus* 2 *and arithmetic surfaces*, Algebraic geometry and commutative algebra, Vol. II, Kinokuniya, Tokyo, 1988, pp. 749–770. MR977781

[ZaMa72] Ju. G. Zarhin and Ju. I. Manin, *Height on families of abelian varieties* (Russian), Mat. Sb. (N.S.) **89(131)** (1972), 171–181, 349. MR0332801

Théorie des nombres, IMB, Université de Bordeaux, 351, cours de la Libération, 33 405 Talence Cedex, France—et—Department of Mathematical Sciences, University of Copenhagen, Universitetsparken 5, DK-2100 Copenhagen Ø, Denmark

Email address: `fabien.pazuki@math.u-bordeaux.fr`

Contemporary Mathematics
Volume **722**, 2019
https://doi.org/10.1090/conm/722/14528

Using zeta functions to factor polynomials over finite fields

Bjorn Poonen

Abstract. In 2005, Kayal suggested that Schoof's algorithm for counting points on elliptic curves over finite fields might yield an approach to factor polynomials over finite fields in deterministic polynomial time. We present an exposition of his idea and then explain details of a generalization involving Pila's algorithm for abelian varieties.

1. Introduction

Factoring univariate polynomials over finite fields is a solved problem in practice. Known algorithms are fast, and they are proved to run in polynomial time if granted access to a source of randomness. But the theoretical question of whether there exists a *deterministic* polynomial-time algorithm remains open. See the survey articles [**Len82**], [**Len90**], and [**vzGP01**]; the last of these contains a very extensive bibliography.

In 1985, Schoof gave a deterministic polynomial-time algorithm to compute the number of points on a given elliptic curve over a finite field [**Sch85**, Section 3]. At the Mathematisches Forschungsinstitut Oberwolfach in July 2005, Neeraj Kayal suggested a way to use Schoof's algorithm to attempt to factor polynomials over finite fields in deterministic polynomial time. The author, who was present, responded that one could use higher genus curves or higher-dimensional abelian varieties in place of elliptic curves, and that these heuristically had a greater chance of success.

It seems that the only written record of the ideas of Kayal and the author before now is the 2006 master's thesis of Amalaswintha Wolfsdorf [**Wol06**]. She describes Kayal's idea for elliptic curves in detail, and writes a few sentences on the higher genus case based on a November 25, 2005 email from the present author. Our purpose is to present a brief exposition of Kayal's idea and to explain details of the generalization, which is Theorem 5.1 in this article.

2. Schoof's algorithm

Understanding Kayal's idea requires some knowledge of Schoof's algorithm, which we now recall, in the special case of a prime field $\mathbb{F}_p$.

2010 *Mathematics Subject Classification.* Primary 11Y05; Secondary 14G10, 14G15, 14K15.
Key words and phrases. Factoring polynomials, zeta function, abelian variety, finite field.
This research was supported in part by National Science Foundation grant DMS-1601946 and Simons Foundation grants #402472 (to Bjorn Poonen) and #550033.

THEOREM 2.1 ([**Sch85**, Section 3]). *There exists a deterministic polynomial-time algorithm that takes as input a prime p and a Weierstrass equation of an elliptic curve E over $\mathbb{F}_p$, and outputs $\#E(\mathbb{F}_p)$.*

Polynomial-time means polynomial in the size of the input, which is of order $\log p$.

SKETCH OF PROOF. Hasse proved that $\#E(\mathbb{F}_p) = p - a + 1$ for some $a \in \mathbb{Z}$ satisfying $|a| \leq 2\sqrt{p}$. If one can compute $a \bmod \ell$ for all primes $\ell \neq p$ up to some bound L, then an effective Chinese remainder theorem lets one compute $a \bmod \prod_{\ell<L} \ell$. If L is chosen as a sufficiently large constant multiple of $\log p$, then $\prod_{\ell<L} \ell > 4\sqrt{p}$, so $a \bmod \prod_{\ell<L} \ell$ determines a.

The Frobenius endomorphism F of E satisfies $F^2 - aF + p = 0$ in $\operatorname{End} E$. In particular, $F^2 - aF + p$ acts as 0 on the ℓ-torsion subscheme $E[\ell]$, and this condition uniquely determines $a \bmod \ell$. It remains to explain how to compute with these objects. First, $E[\ell]$ is $\operatorname{Spec} R$ for some $\mathbb{F}_p$-algebra R defined by $O(1)$ explicit equations of degree polynomial in ℓ, and these equations can be computed from the group law on E; from this, one can compute an explicit multiplication table for R with respect to an $\mathbb{F}_p$-basis. (With a little more work, following Schoof, one can work even more explicitly by using division polynomials, but this does not generalize as easily.) The action of F on $E[\ell]$ is given by the pth power map on R, whose action on $\mathbb{F}_p$-algebra generators can be computed explicitly by writing the exponent p in binary and using repeated squaring and multiplication. Similarly, the action of p (or any smaller integer) on $E[\ell]$ can be computed by writing p in binary and using repeated doubling and addition on E. Combining these lets one compute the action of $F^2 - aF + p$ on $E[\ell]$ in time bounded by $P(\ell, \log p)$ for some polynomial P. For each ℓ, try $a = 0, 1, \ldots, \ell - 1$ until the value mod ℓ is found that makes $F^2 - aF + p$ kill $E[\ell]$. The total running time is at most $\sum_{\ell<L} \sum_{a=0}^{\ell-1} P(\ell, \log p)$, which is polynomial in $\log p$. $\square$

REMARK 2.2. In [**Sch85**, Section 4], Schoof gave an application of his algorithm to computing, for any fixed integer a, the square root of a in $\mathbb{F}_p$ in deterministic polynomial time (involving a constant depending on a) for primes p such that $\left(\frac{a}{p}\right) = 1$. This itself is a case of polynomial factoring over a finite field!

3. Kayal's factoring idea

For simplicity, suppose that p is a large prime and suppose that we are given the product $f(t) = (t - r_1)(t - r_2) \in \mathbb{F}_p[t]$ for some unknown distinct $r_1, r_2 \in \mathbb{F}_p$. Let $B = \mathbb{F}_p[t]/(f(t))$, so B is secretly isomorphic to $\mathbb{F}_p \times \mathbb{F}_p$. Elements of B are represented by polynomials of degree ≤ 1 in $\mathbb{F}_p[t]$. A Weierstrass equation over B with discriminant in $B^\times$ defines an elliptic scheme E over B, which secretly specializes to two elliptic curves over $\mathbb{F}_p$, say E_1 and E_2.

What happens if we blithely run Schoof's algorithm on E, as if B were $\mathbb{F}_p$? If $\#E_1(\mathbb{F}_p) = \#E_2(\mathbb{F}_p)$, then there exists $a \in \mathbb{Z}$ such that $F^2 - aF + p = 0$ in $\operatorname{End} E$, and the algorithm runs as usual, and outputs the common value $\#E_1(\mathbb{F}_p) = E_2(\mathbb{F}_p)$, but we learn nothing about the factorization of $f(t)$. Now suppose instead that $\#E_1(\mathbb{F}_p) \neq \#E_2(\mathbb{F}_p)$. Write $\#E_i(\mathbb{F}_p) = p - a_i + 1$ for $i = 1, 2$, so $a_1 \neq a_2$. Then for some ℓ, we have $a_1 \not\equiv a_2 \pmod{\ell}$. Thus, when we check integers a to see if $F^2 - aF + p$ kills $E[\ell]$, which amounts to certain elements of B vanishing, we instead find an integer a_1 for which these elements of B vanish mod $t - r_1$ but do not all vanish mod $t - r_2$. Hence we discover a nontrivial factor of $f(t)$.

Heuristically it is likely that $\#E_1(\mathbb{F}_p) \neq \#E_2(\mathbb{F}_p)$, since there are about $4\sqrt{p}$ possible values for the order of an elliptic curve over $\mathbb{F}_p$. If we are unlucky enough to have chosen E so that $\#E_1(\mathbb{F}_p) = E_2(\mathbb{F}_p)$, we can try again with a different E, or use the same linear polynomials as Weierstrass coefficients while replacing $f(t)$ by $f(t+1)$. We do not have a proof, however, that a deterministic sequence of such trials will succeed after polynomially many attempts.

REMARK 3.1. The same approach can be tried to factor a polynomial $f(t) := (t - r_1) \cdots (t - r_d)$ for distinct $r_1, \ldots, r_d \in \mathbb{F}_p$, by induction on d. If, using obvious notation, $\#E_1(\mathbb{F}_p), \ldots, \#E_d(\mathbb{F}_p)$ are not all equal, then Schoof's algorithm will find a nontrivial factor g of f, and then we can apply the inductive hypothesis to factor g and f/g.

REMARK 3.2. Berlekamp [**Ber70**] showed that there is a deterministic polynomial-time reduction of the problem of factoring polynomials in $\mathbb{F}_q[t]$ for arbitrary prime powers q to the problem of factoring polynomials in $\mathbb{F}_p[t]$ with distinct roots all in $\mathbb{F}_p$; see also [**Len82**, Sections 3 and 4] for another exposition of this.

4. Pila's algorithm

To generalize Kayal's approach to abelian varieties, we need Pila's generalization of Schoof's algorithm.

Let A be a g-dimensional abelian variety over $\mathbb{F}_p$. Let F be the Frobenius endomorphism of A. For each prime $\ell \neq p$, we may form the ℓ-adic Tate module $T_\ell A := \varprojlim_n A[\ell^n]$. Let $P(t)$ be the characteristic polynomial of F acting on $T_\ell A$, so $\deg P = 2g$. A priori the coefficients of P are in $\mathbb{Z}_\ell$, but in fact they lie in $\mathbb{Z}$ and are independent of the choice of ℓ. Knowledge of P is equivalent to knowledge of the zeta function Z_A.

An abelian variety A over $\mathbb{F}_p$ can be described explicitly by giving a positive integer N and a finite list of homogeneous polynomials in $\mathbb{F}_p[x_0, \ldots, x_N]$ whose common zero locus in $\mathbb{P}^N$ is A, together with the addition morphism $A \times A \to A$ (and also the inversion morphism if desired) in terms of explicit polynomial mappings on affine patches. Pila's algorithm accepts such data as input, and outputs $P(t) \in \mathbb{Z}[t]$. Its running time is bounded by a polynomial in $\log p$ whose degree and coefficients depend only on N and the number and degrees of the polynomials defining A and the addition law [**Pil90**, Theorem A].

The general outline of Pila's algorithm is similar to that of Schoof's algorithm: it computes $P(t) \bmod \ell$ for many small primes ℓ by studying the action of F on $A[\ell]$, and then reconstructs $P(t)$ by using an effective Chinese remainder theorem. For each ℓ, it tries each monic degree $2g$ polynomial in $(\mathbb{Z}/\ell\mathbb{Z})[t]$ and tests whether it equals $P(t) \bmod \ell$. Each test involves a deterministic sequence of arithmetic operations on elements of $\mathbb{F}_p$ controlled by queries: each query asks whether some previously computed element is 0, and the result dictates which arithmetic operation is to be carried out next. This is all that we will need to know about Pila's algorithm.

REMARK 4.1. Just as Schoof used his algorithm to give a deterministic polynomial-time algorithm for computing square roots modulo p of a fixed integer, Pila used his algorithm to find ℓth roots of unity in $\mathbb{F}_p$ in deterministic polynomial time for fixed ℓ, when they exist [**Pil90**, Theorem D]; this is another example of polynomial factoring over a finite field.

5. Generalization of Kayal's factoring idea

We will prove that we can replace elliptic curves by abelian varieties in Kayal's approach. Also, instead of using only the order of the group of points, we can use the whole zeta function. The advantage of using higher-dimensional abelian varieties is that there are many more possible zeta functions, so success becomes very likely, at least heuristically: see Section 6.

Given a variety V over a finite field, let Z_V be its zeta function, viewed as a rational function in $\mathbb{Q}(T)$.

Let U be a dense open subscheme of $\mathbb{A}^1_{\mathbb{Z}} := \operatorname{Spec} \mathbb{Z}[t]$. Let $\mathcal{A} \to U$ be an abelian scheme. For each prime p and $u \in U(\mathbb{F}_p)$, let $\mathcal{A}_u$ be the fiber above u. Concretely, $\mathcal{A}$ can be thought of as a family of abelian varieties defined by equations with coefficients in $\mathbb{Z}[t]$; specializing t to a suitably general value $u \in \mathbb{F}_p$ produces an abelian variety $\mathcal{A}_u$ over $\mathbb{F}_p$; here "suitably general" means outside a certain bad locus, which may be taken to be of the form $\Delta(t) = 0$ for some "discriminant" $\Delta(t) \in \mathbb{Z}[t]$ that is not identically zero but vanishes at any u for which the specialization $\mathcal{A}_u$ is degenerate.

Theorem 5.1 below will involve the following:

Hypothesis Z. There exist a dense open subscheme U of $\mathbb{A}^1_{\mathbb{Z}}$ and an abelian scheme $\mathcal{A} \to U$ such that for every sufficiently large prime p, the $Z_{\mathcal{A}_u}$ for the different $u \in U(\mathbb{F}_p)$ are distinct.

Theorem 5.1. *There is a deterministic algorithm that takes as input a finite field $\mathbb{F}_q$ and a nonzero polynomial $f \in \mathbb{F}_q[t]$, and outputs the factors of f in $\mathbb{F}_q[t]$, such that if Hypothesis Z holds, then the running time is polynomial in $\log q$ and $\deg f$.*

Remark 5.2. Our proof will show that an algorithm as in Theorem 5.1 not only exists, but also can be written down explicitly, even if we do not know in advance the abelian scheme $\mathcal{A} \to U$ in Hypothesis Z.

Proof. By Remark 3.2, we may assume that q is a prime p and that f has distinct roots all in $\mathbb{F}_p$. Of course we also assume that $\deg f \geq 2$.

First, we give an algorithm depending on explicit knowledge of an abelian scheme $\mathcal{A} \to U$ as in Hypothesis Z. More precisely, we may shrink U to assume that $U = \operatorname{Spec} T$, where $T = \mathbb{Z}[t][1/\Delta]$ for some nonzero polynomial $\Delta \in \mathbb{Z}[t]$, and we may assume that we are given explicit polynomials describing $\mathcal{A}$ over T in the same way that we described abelian varieties over $\mathbb{F}_p$ in Section 4.

There are at most $(\deg f)(\deg \Delta)$ values $c \in \mathbb{F}_p$ such that $f(t+c)$ and $\Delta(t)$ have a nontrivial gcd, so by trying $c = 0, 1, \dots$ in turn, we quickly find such a c (of course, we may assume that $p > (\deg f)(\deg \Delta)$). Replace $f(t)$ by $f(t+c)$ to assume that $\gcd(f, \Delta) = 1$. Let $B = \mathbb{F}_p[t]/(f(t))$. Then $\operatorname{Spec} B$ is a closed subscheme of U, and the base change $\mathcal{A}_B$ is an abelian scheme over B. It consists of a disjoint union of abelian varieties $\mathcal{A}_u$ over $\mathbb{F}_p$, one for each zero u of f.

Apply Pila's algorithm to $\mathcal{A}_B$, but each time it queries an element of B to test whether it is 0, instead compute a gcd with f to test whether it is 0, a unit, or a nonzero zerodivisor. By Hypothesis Z, the zeta functions $Z_{\mathcal{A}_u}$ for two different zeros

u of f are distinct in $\mathbb{Q}(T)$, so there exists a prime ℓ such that the characteristic polynomials mod ℓ of $\mathcal{A}_u$ for these two u are distinct. Therefore the computations in Pila's algorithm must eventually diverge for these two values of u, which can happen only if a nonzero zerodivisor in B is encountered. At that point, we have found a nontrivial factor f_0 of f. Apply induction to the factors f_0 and f/f_0. This completes the description of the algorithm when we are given $\mathcal{A} \to U$ explicitly. In particular, there exists an algorithm to factor polynomials, even though we might not know which algorithm it is that does it.

We now describe a new program Ω that does not rely on knowledge of $\mathcal{A} \to U$. Program Ω enumerates all computer programs and runs them in parallel, devoting a fraction 2^{-n} of its computing power to the nth program; at each step of each program, Ω tests whether what that program has printed so far is a list of linear polynomials over $\mathbb{F}_p$ whose product is f, and if so, Ω terminates the whole computation with this answer. If Hypothesis Z is true, and n is the number of the program described in earlier paragraphs using an abelian scheme $\mathcal{A} \to U$ as in Hypothesis Z, then Ω finds the factorization in time bounded by 2^n times a polynomial, but 2^n is a constant, so this is still polynomial in the size of the input. If Hypothesis Z is false, then Ω still terminates with the correct factorization because there exists N such that program N factors polynomials by trial division, but the running time of Ω is not guaranteed to be bounded by a polynomial in this case. □

6. A heuristic for Hypothesis Z

For a g-dimensional abelian variety A over $\mathbb{F}_p$, the complex zeros of the characteristic polynomial $P(t)$ have absolute value $p^{1/2}$, so the coefficient of t^{2g-m} in $P(t)$ is $O_g(p^{m/2})$, with the implied constant depending on g but not p. Also, the functional equation of Z_A shows that the coefficient of t^m in $P(t)$ is determined by the coefficient of t^{2g-m}. Thus $P(t)$ is determined by coefficients of t^{2g-m} for $m = 1, 2, \ldots, g$, so there are at most $\prod_{m=1}^{g} O_g(p^{m/2}) = O_g(p^{g(g+1)/4})$ possibilities for $P(t)$. Equivalently, if $\mathcal{Z}_{g,p}$ is the set of zeta functions of all g-dimensional abelian varieties over $\mathbb{F}_p$, then $\#\mathcal{Z}_{g,p} = O_g(p^{g(g+1)/4})$ as $p \to \infty$. In fact, DiPippo and Howe **[DH98]** prove that for fixed g, we have $\#\mathcal{Z}_{g,p} \sim p^{g(g+1)/4}$ as $p \to \infty$, where in this section we use the notation $f(p) \sim h(p)$ to mean that $f(p)/h(p)$ tends to a positive constant depending only on g as $p \to \infty$.

If we sample about p zeta functions from $\mathcal{Z}_{g,p}$ at random, then the expected number of equal pairs is $\sim \binom{p}{2} \frac{1}{p^{g(g+1)/4}} \sim p^{2-g(g+1)/4}$. If we do this for all primes p greater than or equal to some large integer p_0, then the expected total number of equal pairs for all p is $\sim \sum_{p \geq p_0} p^{2-g(g+1)/4}$, which tends to 0 as $p_0 \to \infty$, provided that $2 - g(g+1)/4 < -1$, which holds for $g \geq 4$.

Now let U be a dense open subscheme of $\mathbb{A}^1_{\mathbb{Z}}$, and let $\mathcal{A} \to U$ be an abelian scheme of relative dimension $g \geq 4$. The previous paragraph suggests that if we model the zeta functions of the fibers of $\mathcal{A} \to U$ above $\mathbb{F}_p$-points of U as being independent random elements of $\mathcal{Z}_{g,p}$, then for sufficiently large p_0 it is true for every $p \geq p_0$ that these zeta functions will be distinct; in other words, $\mathcal{A} \to U$ should satisfy the condition in Hypothesis Z, unless there is some extra structure to the family that the model fails to reflect.

It even seems reasonable to guess that for a typical 1-parameter family of genus 4 curves, the family of Jacobians will satisfy the condition in Hypothesis Z. See **[SV19]** for some specific candidate families.

7. Weakening Hypothesis Z

Something slightly weaker than Hypothesis Z would suffice to obtain a polynomial running time in Theorem 5.1:

HYPOTHESIS Z′. There exist a dense open subscheme U of $\mathbb{A}^1_{\mathbb{Z}}$ and an abelian scheme $\mathcal{A} \to U$ such that for every sufficiently large prime p, there are at least $p - (\log p)^{O(1)}$ distinct zeta functions $Z_{\mathcal{A}_u}$ as u varies over $U(\mathbb{F}_p)$.

Under Hypothesis Z′, given $f \in \mathbb{F}_p[t]$ that factors completely, one can attempt to factor the polynomials $f(t+c)$ for $(\log p)^{O(1)}$ different values $c \in \mathbb{F}_p$ by running Pila's algorithm as in the proof of Theorem 5.1. As long as the $O(1)$ here is larger than the $O(1)$ in Hypothesis Z′, and as long as p is sufficiently large, there will be at least one such c such that all the zeros of $f(t+c)$ mod p lie in $U(\mathbb{F}_p)$ and the zeta functions of the fibers above these zeros are pairwise distinct. Thus the algorithm will succeed in factoring $f(t+c)$ for at least one c, and evaluating the factors at $t-c$ recovers the factorization of $f(t)$. The running time of the algorithm is still polynomial in $\log p$, albeit possibly with a larger exponent.

8. Using varieties other than abelian varieties

Suppose that instead of an abelian scheme as in Hypothesis Z, one had an arbitrary finite-type morphism $\mathcal{X} \to U$ for a dense open subscheme U of $\mathbb{A}^1_{\mathbb{Z}}$ such that for any sufficiently large prime p and distinct $u_1, u_2 \in U(\mathbb{F}_p)$, there exists a prime ℓ bounded by a polynomial in $\log p$ and a nonnegative integer i such that the characteristic polynomials of Frobenius acting on $\mathrm{H}^i_{\mathrm{et}}(\mathcal{X}_{u_1} \times \overline{\mathbb{F}}_p, \mathbb{Z}/\ell\mathbb{Z})$ and $\mathrm{H}^i_{\mathrm{et}}(\mathcal{X}_{u_2} \times \overline{\mathbb{F}}_p, \mathbb{Z}/\ell\mathbb{Z})$ are different. Then again one could factor polynomials over finite fields in deterministic polynomial time, provided that one had an analogue of Pila's algorithm that could compute these characteristic polynomials using a deterministic sequence of arithmetic operations and queries whose number is bounded by a polynomial in ℓ whose degree and coefficients depend only on $\mathcal{X} \to U$.

Madore and Orgogozo [**MO15**, Théorème 0.1] gave an algorithm for computing such characteristic polynomials, but their bound on the running time is only primitive recursive, not polynomial in ℓ.

Acknowledgements

I thank Andrew Sutherland and José Felipe Voloch for encouraging me to write this article, I thank Amalaswintha Wolfsdorf for sharing her master's thesis with me, and I thank the referee for helpful suggestions.

References

[Ber70] E. R. Berlekamp, *Factoring polynomials over large finite fields*, Math. Comp. **24** (1970), 713–735, DOI 10.2307/2004849. MR0276200 ↑143

[DH98] Stephen A. DiPippo and Everett W. Howe, *Real polynomials with all roots on the unit circle and abelian varieties over finite fields*, J. Number Theory **73** (1998), no. 2, 426–450, DOI 10.1006/jnth.1998.2302. MR1657992 ↑145

[Len82] A. K. Lenstra, *Factorization of polynomials*, Computational methods in number theory, Part I, Math. Centre Tracts, vol. 154, Math. Centrum, Amsterdam, 1982, pp. 169–198. MR700263 ↑141, 143

[Len90] H. W. Lenstra Jr., *Algorithms for finite fields*, Number theory and cryptography (Sydney, 1989), London Math. Soc. Lecture Note Ser., vol. 154, Cambridge Univ. Press, Cambridge, 1990, pp. 76–85. MR1055400 ↑141

[MO15] David A. Madore and Fabrice Orgogozo, *Calculabilité de la cohomologie étale modulo ℓ* (French, with English and French summaries), Algebra Number Theory **9** (2015), no. 7, 1647–1739, DOI 10.2140/ant.2015.9.1647. MR3404650 ↑146

[Pil90] J. Pila, *Frobenius maps of abelian varieties and finding roots of unity in finite fields*, Math. Comp. **55** (1990), no. 192, 745–763, DOI 10.2307/2008445. MR1035941 ↑143, 144

[Sch85] René Schoof, *Elliptic curves over finite fields and the computation of square roots mod p*, Math. Comp. **44** (1985), no. 170, 483–494, DOI 10.2307/2007968. MR777280 ↑141, 142

[SV19] Andrew V. Sutherland and José Felipe Voloch, *Maps between curves and arithmetic obstructions*, Arithmetic Geometry: Computation and Applications, Contemp. Math., vol. 722, Amer. Math. Soc., Providence, RI, 2019. ↑146

[vzGP01] Joachim von zur Gathen and Daniel Panario, *Factoring polynomials over finite fields: a survey*, J. Symbolic Comput. **31** (2001), no. 1-2, 3–17, DOI 10.1006/jsco.1999.1002. Computational algebra and number theory (Milwaukee, WI, 1996). MR1806203 ↑141

[Wol06] Amalaswintha Wolfsdorf, *Factorising polynomials over finite fields*, March 2006. Master's thesis, University of Oxford, `arXiv:1709.05513v1`. ↑141

Department of Mathematics, Massachusetts Institute of Technology, Cambridge, MA 02139-4307

Email address: `poonen@math.mit.edu`

URL: `http://math.mit.edu/~poonen/`

Contemporary Mathematics
Volume **722**, 2019
https://doi.org/10.1090/conm/722/14530

Canonical models of arithmetic $(1;\infty)$-curves

Jeroen Sijsling

ABSTRACT. In 1983 Takeuchi showed that up to conjugation there are exactly 4 arithmetic subgroups of $\mathrm{PSL}_2(\mathbb{R})$ with signature $(1;\infty)$. Shinichi Mochizuki gave a purely geometric characterization of the corresponding arithmetic $(1;\infty)$-curves, which also arise naturally in the context of his recent work on inter-universal Teichmüller theory.

Using Belyĭ maps, we explicitly determine the canonical models of these curves. We also study their arithmetic properties and modular interpretations.

CONTENTS

Let (E, O) be a pointed curve of genus 1 over $\mathbb{C}$, and let $e \in \mathbb{Z}_{\geq 2} \cup \{\infty\}$. We can then construct the universal ramified cover $U \to E$ that is unramified away from O and ramifies over that point with index e. As a complex analytic space U is isomorphic to the upper half plane $\mathcal{H}$, and the covering map $\mathcal{H} \to E$ can be described as a quotient by a Fuchsian subgroup $\mathrm{P}\Gamma$ of $\mathrm{PSL}_2(\mathbb{R})$. As a subgroup of $\mathrm{PSL}_2(\mathbb{R})$, the group $\mathrm{P}\Gamma$ is well-defined up to conjugacy. We can ask when the pair (E, O) and the ramification index e determine an arithmetic group $\mathrm{P}\Gamma$. Alternatively, we can ask which arithmetic subgroups $\mathrm{P}\Gamma$ of $\mathrm{PSL}_2(\mathbb{R})$ have signature $(1;e)$, which is to say that it admits a presentation of the form

$$\mathrm{P}\Gamma = \langle \alpha, \beta \mid [\alpha, \beta]^e = 1 \rangle. \tag{0.1.1}$$

This question was answered by Takeuchi [**21**], who provided the finite list of arithmetic groups of signature $(1;e)$. When e is finite, then the group $\mathrm{P}\Gamma$ has no cusps and comes from a non-trivial quaternion algebra. The resulting quotient curves

$$X(\mathrm{P}\Gamma) = \mathrm{P}\Gamma \backslash \mathcal{H} \tag{0.1.2}$$

2010 *Mathematics Subject Classification.* Primary 14H10.
Key words and phrases. Canonical models, Shimura curves, modular curves.

were studied in [**17**], where their canonical models in the sense of Shimura [**14**] were determined.

The present article considers the curves in Takeuchi's list that were not treated in [**17**], namely those for which $e = \infty$. This is the case for 4 conjugacy classes of groups $\mathrm{P}\Gamma$. These groups have a single conjugacy class of cusps, and are commensurable with the modular group $\mathrm{P}\Gamma(1) = \mathrm{PSL}_2(\mathbb{Z})$. This means that classical methods are available to compute canonical models of the corresponding quotient curves $X(\mathrm{P}\Gamma)$.

It should be mentioned that determining these 4 curves did not require any fundamentally new theory or concepts to be developed. Still, our computations were rather involved in practice, and there seems to be no way to determine equations for these curves by a short theoretical detour. We have used the computer algebra system Magma [**2**] to facilitate these calculations. All the code needed to obtain the result in this article can be found online at [**18**]. Along the way, we find an explicit Shimura curve one of whose Atkin-Lehner quotients has a smaller canonical field of moduli than the original curve (case III below, further discussed in section 3.1), and a modular description of the cover corresponding to the commutator subgroup of $\mathrm{SL}_2(\mathbb{Z})$ (in case IV below, discussed in section 3.2 and 4.6).

However, the main motivation for this computation is not simply that these 4 curves "are there", but that they play a role in the work of Shinichi Mochizuki. In the articles [**11**, **12**] he showed that the property of these curves being arithmetic can be rephrased purely geometrically, in terms of a certain category $\overline{\mathrm{Loc}}(X)$ introduced in [**12**, Definition 2.1] not having a terminal object (or core). In Mochizuki's later inter-universal Teichmüller theory these curves are exceptional, in the sense that the theory developed does not apply to these curves and the corresponding categories, requiring them to be excluded from consideration. For instances of this happening in the literature on this theory, we refer to Fesenko's expository work [**6**, Footnote 27] as well as the original work of Mochizuki [**13**, Definition 3.1(d)].

Because of this, the question of determining these curves was raised by Fesenko at the workshop on inter-universal Teichmüller theory at the University of Oxford in December 2015. This question then went unanswered until the models in this article were found.

We will use classical methods, as well as the more recent methods by Shimura [**14**], to determine the canonical models of the 4 arithmetic $(1;\infty)$-curves in Takeuchi's list. Moreover, we describe the correspondence of these curves with the classical modular curve $X(1)$, and we give the arithmetic and modular properties of these canonical models.

A main tool that we will use is that of Belyĭ maps. For a survey that shows the ubiquity of these maps, we refer to [**19**]; they also play a significant role in Mochizuki's work [**13**]. Expressed briefly, given an arithmetic group $\mathrm{P}\Gamma$ commensurable with a triangle group $\mathrm{P}\Delta$, the inclusion of groups $\mathrm{P}\Gamma \cap \mathrm{P}\Delta \subset \mathrm{P}\Delta$ gives rise to a Belyĭ map $X(\mathrm{P}\Gamma \cap \mathrm{P}\Delta) \to X(\mathrm{P}\Delta)$, which is a cover of a projective line ramified above at most three points. The combinatorics of this inclusion yield equations for $X(\mathrm{P}\Gamma)$ via algorithms such as those in [**8**], and the correspondence with the curve $X(\mathrm{P}\Delta)$ can be used to obtain significant arithmetic information regarding the curve $X(\mathrm{P}\Gamma)$, for instance properties of its canonical model. These considerations apply in particular to the classical case $\mathrm{P}\Delta = \mathrm{P}\Gamma(1) = \mathrm{PSL}_2(\mathbb{Z})$ in which we find ourselves.

Section 1 introduces the relevant groups and their associated orders. Section 2 gives the resulting Belyĭ maps, which are used in section 3 to determine the canonical models and their arithmetic properties. .Finally, the modular interpretation of these models is described in section 4.

Acknowledgments. I cordially thank Ivan Fesenko for pointing out the relevance of the curves in this article; without his interest this paper would not have existed. Further heartfelt thanks go out to John Voight for sharing his wisdom on dessins, canonical models and conductors.

1. Uniformizations and orders

In this section our exposition will be rather condensed, as a more comprehensive exposition of these topics is in [**15**, **17**, **21**].

In [**21**, Proposition 3.1, Theorem 3.4], Takeuchi describes his $(1;\infty)$-groups as follows.

LEMMA 1.1.1. *Let $P\Gamma \subset \mathrm{PSL}_2(\mathbb{R})$ be an arithmetic $(1;\infty)$-group. Then the inverse image Γ of $P\Gamma$ in $\mathrm{SL}_2(\mathbb{R})$ is generated by two elements α and β whose commutator is parabolic and for which the traces of α, β and $\alpha\beta$ are given in Table 1. Conversely, these traces determine Γ up to $\mathrm{SL}_2(\mathbb{R})$-conjugacy.*

Case	$\mathrm{Tr}(\alpha)$	$\mathrm{Tr}(\beta)$	$\mathrm{Tr}(\alpha\beta)$
I	$\sqrt{5}$	$2\sqrt{5}$	5
II	$\sqrt{6}$	$2\sqrt{3}$	$3\sqrt{2}$
III	$2\sqrt{2}$	$2\sqrt{2}$	4
IV	3	3	3

TABLE 1. Trace triples

As in [**15**, §1.2], we construct α and β by taking

$$\alpha = \begin{pmatrix} \lambda & 0 \\ 0 & \lambda^{-1} \end{pmatrix}, \beta = \begin{pmatrix} a & b \\ b & d \end{pmatrix}. \tag{1.1.2}$$

Here λ is the largest real solution of $\lambda^{-1} + \lambda = \mathrm{Tr}(\alpha)$. Having determined λ, we recover a and d by from $\mathrm{Tr}(\beta)$ and $\mathrm{Tr}(\alpha\beta)$ by solving a system of linear equations. Finally, we choose b to be the positive real solution of $ad - b^2 = 1$.

The matrices α, β thus obtained are far from having entries in $\mathbb{Z}$. However, in the proof of [**21**, Theorem 3.4] (and see also [**20**]) Takeuchi shows the following more precise statement.

LEMMA 1.1.3. *Let S be the set $\{1, \alpha^2, \beta^2, \alpha^2\beta^2\}$, and let $\Gamma^{(2)} = \langle \gamma^2 : \gamma \in \Gamma \rangle$.*

- (i) *The group $\Gamma^{(2)}$ has signature $(1;\infty^4)$, and the image $P\Gamma^{(2)}$ of $\Gamma^{(2)}$ in $\mathrm{PSL}_2(\mathbb{R})$ is normal in the image $P\Gamma \subset \mathrm{PSL}_2(\mathbb{R})$. The quotient group $P\Gamma \,/\, P\Gamma^{(2)}$ is a Klein Vierergruppe generated by α and β.*
- (ii) *The $\mathbb{Q}$-vector space generated by S is a quaternion algebra over the rationals.*
- (iii) *The $\mathbb{Z}$-module generated by $\Gamma^{(2)}$ is an order $\mathcal{O}$ of B, and the group $\mathcal{O}^1$ of elements of $\mathcal{O}$ with reduced norm 1 contains $\Gamma^{(2)}$.*

REMARK 1.1.4. In [**21**, Proposition 3.1] Takeuchi also gives a list of generators of $\Gamma^{(2)}$, so that we can explicitly construct the order $\mathcal{O}$ in part (ii) of the Lemma.

In our cases Γ has a cusp, hence so does $\Gamma^{(2)}$. Therefore $\mathcal{O}$ is in fact conjugate to an order in the trivial quaternion algebra $M_2(\mathbb{Q})$ over $\mathbb{Q}$. This simplifies our considerations, since the original matrices α and β will have entries in an extension of $\mathbb{Q}$; in case II this extension is even a biquadratic number field. To find an order in $M_2(\mathbb{Q})$ to which $\mathcal{O}$ is conjugate, we can let Γ act on S by left multiplication and identify the resulting associative algebra with $M_2(\mathbb{Q})$ by using the algorithms in [**22**].

We have slightly tweaked this approach; instead of taking $\mathcal{O} = \mathbb{Z}[\Gamma^{(2)}]$, we take $\mathcal{O}' = \mathbb{Z}[\Gamma']$, where Γ' is the largest subgroup between $\Gamma^{(2)}$ and Γ that generates some order in the algebra B.

PROPOSITION 1.1.5. *In the cases I-IV, the largest groups Γ' between $\Gamma^{(2)}$ and Γ that generate an order $\mathcal{O}' = \mathbb{Z}[\Gamma']$ in the matrix algebra B in Lemma* 1.1.3 *are given in Table* 2.

Case	Largest Γ' generating an order
I	$\langle\Gamma^{(2)}, \alpha\beta\rangle$
II	$\Gamma^{(2)}$
III	$\langle\Gamma^{(2)}, \alpha\beta\rangle$
IV	Γ

TABLE 2. Groups used to generate orders

PROOF. The quotient $P\Gamma \,/\, P\Gamma^{(2)}$ is a Klein Vierergruppe generated by α and β, so it suffices to see which of $\alpha, \beta, \alpha\beta$ can be written as a $\mathbb{Q}$-linear combination of the basis S of B, and if so, which possible additions of these elements give rise to an order again. This is a straightforward computation as S is also an $\mathbb{R}$-basis of $M_2(\mathbb{R})$. Note that a priori there is no reason for there to be a largest Γ' at all; this just happens to be the case in all our examples. □

The groups Γ' are indicated in Table 2. After trivializing the quaternion algebras B involved, we have conjugated the corresponding orders $\mathcal{O}'$ of $M_2(\mathbb{Q})$ into $M_2(\mathbb{Z})$ by an *ad hoc* calculation. In general, one could determine a maximal order of $M_2(\mathbb{Q})$ containing $\mathcal{O}'$. Such a maximal order is conjugate to $M_2(\mathbb{Z})$, so that an explicit conjugating element can be used to map $\mathcal{O}'$ into $M_2(\mathbb{Z})$.

Identifying the groups Γ' and Γ with their images under the preceding conjugations, we now have reduced to a simpler situation, where we have an inclusion $\Gamma^{(2)} \subset \Gamma'$ as well as inclusions

$$\Gamma \supset \Gamma' \subset \mathbb{Z}[\Gamma']^1 = \mathcal{O}'^1 \subset \mathrm{SL}_2(\mathbb{Z}). \tag{1.1.6}$$

Proposition 2.1.3 will show that in fact $\Gamma' = \Gamma \cap \mathrm{SL}_2(\mathbb{Z})$. Note that it is *not* clear that Γ' equals $\mathbb{Z}[\Gamma']^1$; in fact in case IV the order $\mathcal{O}'$ is a non-Eichler order of index 4 whose group of units is of index 2 in $\mathrm{SL}_2(\mathbb{Z})$ and hence properly contains Γ. In all the other cases, Proposition 2.1.2 will show that indeed $\mathcal{O}'^1 = \Gamma'$. In fact, in case IV we still have that $\mathbb{Z}[\Gamma^{(2)}]^1 = \Gamma^{(2)}$, so that we could have worked with the original order $\mathcal{O} = \mathbb{Z}[\Gamma^{(2)}]$ defined by $\Gamma^{(2)}$ instead. Regardless, we will see below that case IV is less complicated than the others.

2. Belyĭ maps

2.1. Monodromy triples. The reason for our using the unit group $\mathcal{O}'^1$ is that it is less complicated to determine whether a given element $\gamma \in \mathrm{SL}_2(\mathbb{Z})$ belongs to this group; it suffices to see whether $\gamma \in \mathcal{O}'$, which gives a straightforward additive criterion. In this way, we can quickly determine the monodromy of the cover (and in fact Belyĭ map) corresponding to the inclusion $\mathcal{O}'^1 \subset \mathrm{SL}_2(\mathbb{Z})$ by using the standard generators S, T and $(ST)^{-1}$ of $\mathrm{SL}_2(\mathbb{Z})$ of order $2, 3, \infty$ and using the methods of [**8**, §3].

REMARK 2.1.1. It is also possible to work directly with one of the groups $\Gamma^{(2)}$, Γ' or Γ by using Dirichlet domains. For details, we again refer to [**8**, §3], especially Algorithms 3.8 and 3.14 therein.

In case IV the group Γ is itself a subgroup of $\mathrm{SL}_2(\mathbb{Z})$. Considering the possible ramification then shows that it is of index 6, and that the cover $X(\Gamma) \to X(1)$ has ramification type $(3^2), (2^3), (6^1)$ above $j = 0, 1728, \infty$ respectively. Geometrically, this is nothing but the quotient of an elliptic curve with j-invariant 0 by its automorphism group; in the next section, we determine the correct model of this cover over its canonical field of definition $\mathbb{Q}$.

In the other cases, we have to put in more effort. The permutation triples $(\sigma_0, \sigma_1, \sigma_\infty)$ that describe the monodromy of the covers defined by the inclusion $\mathcal{O}'^1 \subset \mathrm{SL}_2(\mathbb{Z})$ above $j = 0, 1728, \infty$ are given in Table 3. We have followed the convention $\sigma_\infty \sigma_1 \sigma_0 = 1$. Note that as mentioned in the previous section we get a subgroup of index 2, not 6, in case IV. For all other cases, we have the following proposition.

Case			Monodromy
I	σ_0	$=$	$(1,3,8)(2,7,4)(5,11,9)(6,12,10)$
	σ_1	$=$	$(1,2)(3,5)(4,6)(7,10)(8,9)(11,12)$
	σ_∞	$=$	$(1,4,10,2,8,11,6,7,12,5)(3,9)$
II	σ_0	$=$	$(1,3,8)(2,7,4)(5,14,9)(6,15,10)$ $(11,16,12)(13,19,17)(18,23,20)(21,22,24)$
	σ_1	$=$	$(1,2)(3,5)(4,6)(7,11)(8,12)(9,13)$ $(10,14)(15,17)(16,18)(19,21)(20,22)(23,24)$
	σ_∞	$=$	$(1,4,10,5)(2,8,16,20,21,13,14,15,19,24,18,11)$ $(3,9,17,6,7,12)(22,23)$
III	σ_0	$=$	$(1,3,8)(2,7,4)(5,11,9)(6,12,10)$
	σ_1	$=$	$(1,2)(3,5)(4,6)(7,11)(8,10)(9,12)$
	σ_∞	$=$	$(1,4,10,3,9,6,7,5)(2,8,12,11)$
IV	σ_0	$=$	(1)
	σ_1	$=$	$(1,2)$
	σ_∞	$=$	$(1,2)$

TABLE 3. Monodromy generators for $\mathcal{O}'^1 \subset \mathrm{SL}_2(\mathbb{Z})$

PROPOSITION 2.1.2. *In the cases I-III, we have* $\mathbb{Z}[\Gamma']^1 = \Gamma'$ *in Table* 2.

PROOF. This follows from Table 3 since clearly $\Gamma' \subset \mathbb{Z}[\Gamma']^1$ and the index of $\mathbb{Z}[\Gamma']^1$, as read off from that table, coincides with the index of Γ', as determined by the covolume of this group. □

COROLLARY 2.1.3. *In all cases, we have* $\Gamma' = \Gamma \cap \mathrm{SL}_2(\mathbb{Z})$.

PROOF. The inclusion $\Gamma' \subset \Gamma \cap \mathrm{SL}_2(\mathbb{Z})$ clearly holds, and if the intersection were larger, then the corresponding group would generate an order in $M_2(\mathbb{Z})$ larger than that generated by Γ', a contradiction with the definition of the latter group. □

We have to determine the Belyĭ maps corresponding to the monodromy triples in Table 3. These maps are of sufficiently low degree to be accessible directly by general methods such as those developed in [**8**], and the functionality under development in [**7**] quickly returns explicit formulas for them. Still, we indicate a useful technique in the calculation of covers that facilitates these computations.

2.2. Decomposing covers. Let $\overline{f} : \overline{Z} \to \overline{X}$ be a ramified cover of degree d of a known base space $\overline{X}$ that is the projective completion of an unramified cover $f : Z \to X$. We wish to determine whether f can be written as a non-trivial composition $Z \to Y \to X$ (which will imply a similar statement for $\overline{f}$), and if so to describe the maps $Z \to Y$ and $Y \to X$ combinatorially. When studying our Belyĭ maps, we are of course considering $\overline{X} = \mathbb{P}^1$, $X = \mathbb{P}^1 - \{0, 1728, \infty\}$.

The cover f itself corresponds to a homomorphism $\varphi : \pi_1(X, x) \to S_d$, where x is a chosen base point of X. Let G be the image of φ, and let H be the stabilizer of $1 \in \{1, \ldots, d\}$. Then G is canonically isomorphic to a quotient of $\pi_1(X, x)$, and in this way the set of cosets G/H becomes a $\pi_1(X, x)$-set. After having chosen a presentation of $\pi_1(X, x)$, its generators are mapped by φ to elements of S_d, which group we can identify with $\mathrm{Sym}(G/H)$. In the case where such a generator represents a simple loop around a branch point in the completion $\overline{X}$ of X, its image under φ describes the local monodromy.

Via the dictionary of covering theory [**8**, **16**], a factorization of f as $Z \to Y \to X$ corresponds to a subgroup K between H and G. In our cases $X = \mathbb{P}^1 - \{0, 1, \infty\}$, and we have determined the representation $\pi_1(X, x) \to S_d$ up to conjugacy in Table 3. Since the groups involved are very small, we can calculate the lattice of subgroups of G that contain H. In practice we do not merely find a factorization in one step, but a succession of inclusions $H \subset K \subset K' \subset \cdots \subset G$. We wish to describe the resulting covers, for which we may without loss of generality consider the case $H \subset K \subset G$ of a single intermediary subgroup.

For a start, the inclusion $K \subset G$ corresponds to the $\pi_1(X, x)$-set G/K. This means that we can reuse the generators of $\pi_1(X, x)$ on this smaller set of cosets to describe the monodromy, which furnishes a combinatorial description of the cover $Y \to X$.

The cover corresponding to the inclusion $H \subset K$ can be more difficult to describe, since in this case we have to describe a cover of Y; in particular, the fundamental group used changes from $\pi_1(X, x)$ to $\pi_1(Y, y)$, where y is some element of Y that is in the fiber of Y over x. Using the methods of [**8**], one can calculate a presentation of $\pi_1(Y, y)$ in terms of the elements of $\pi_1(X, x)$, and with it the representation of this group on the set of cosets K/H. In the coming lemma, we determine some weaker invariants, namely the local monodromy around the points of Y. In all of the cases under consideration in this article, this suffices for the calculation of the covers involved.

LEMMA 2.2.1. *Let $f : Z \to X$ be a cover described by the monodromy morphism $\varphi : \pi_1(X, x) \to G \subset S_d$, and let K be a group between $H = \mathrm{Stab}(1)$ and G. Let $Z \to Y \to X$ be the resulting factorization of f. Let γ be an element of*

$\pi_1(X,x)$ that represents a simple loop around an element $\overline{x}$ of $\overline{X}-X$, and let $\sigma=\varphi(\gamma)$ be the corresponding local monodromy. Then the monodromy of the cover $Z\to Y$ around the points $\overline{y}$ of $\overline{Y}$ over $\overline{x}$ can be described as follows.

(i) *The points $\overline{y}$ correspond to the equivalence classes in G/K under left multiplication by σ.*

(ii) *Let $c\in G/K$ be representative of an equivalence class in (i). Define $\tau\in K$ by*

$$\tau=(c^{-1}\sigma c)^e=c^{-1}\sigma^e c, \tag{2.2.2}$$

where e is the cardinality of the equivalence class. Then the monodromy around the point $\overline{y}$ is described by τ.

PROOF. Consider the simple loop around $\overline{x}$ in $\pi_1(X,x)$. It lifts to various paths on Y, which need not again be loops. However, the smallest powers described τ described in the Lemma correspond to the simple loops on Y with various centers in the fiber of $\overline{Y}$ over $\overline{x}$ obtained by concatenating the paths thus obtained. Since all loops on Y around a point in such a fiber come from exactly one such orbit under powering, our statement is proved. □

We can now determine the covers that we need by the decomposition procedure described above.

LEMMA 2.2.3. *In case II, the Belyĭ map $X(\Gamma)\to X(1)$ is described by the map*

$$(x,y)\mapsto 6912\frac{(2x^3-6x^2-1)^3}{(x-2)^6(x+1)^3x^2(x-3)} \tag{2.2.4}$$

from the curve $y^2=x(x-1)(3x-2)(3x+1)$.

PROOF. In this case the monodromy group has cardinality 288 and trivial center. We use Lemma 2.2.1. There are multiple ways in which we can use the subgroup lattice between H and G. We have chosen a particularly simple one; between H and G there is a group that contains H of index 2 and that gives rise to a cover of $\mathbb{P}^1-\{0,1,\infty\}$ with permutation triple

$$\begin{aligned}\sigma_1&=(1,2)(3,4)(5,10)(6,9)(7,8)(11,12)\\ \sigma_0&=(1,3,6)(2,5,4)(7,9,10)(8,11,12)\\ \sigma_\infty&=(1,4)(2,6,7,12,8,10)(3,5,9).\end{aligned} \tag{2.2.5}$$

Riemann-Hurwitz shows that the resulting cover has genus 0. Moreover, since all ramification above the elliptic points of index 2 and 3 has been absorbed, we see that the corresponding curve has signature $(0;\infty^4)$, where the cusps correspond to the 4 elements above ∞. We can then recover $X(\Gamma^{(2)})$ by taking a degree 2 cover that ramifies above these 4 points. So we instead consider the cover $Z\to X$ associated to (2.2.5) and the associated groups G and H.

Using the subgroup lattice shows that once more there is an intermediate subgroup K between G and H, generated by the elements

$$\begin{gathered}(1,4)(2,6,7,12,8,10)(3,5,9),\\ (1,11,4)(2,3,12)(5,6,8)(7,9,10),\\ (2,8,7)(3,5,9)(6,10,12).\end{gathered} \tag{2.2.6}$$

It contains H of index 3. We get a factorization $Z \to Y \to X$. Considering cosets of K shows that $Y \to X$ is described by the triple

$$\begin{aligned} \sigma_1 &= (1,2)(3,4) \\ \sigma_0 &= (1,2,3) \\ \sigma_\infty &= (2,3,4). \end{aligned} \tag{2.2.7}$$

This gives the cover

$$x \mapsto 6912\frac{x^3(x+2)}{4x-1} \tag{2.2.8}$$

of the j-line.

The cover $Z \to Y$ ramifies over the points $x = -2$ in the fiber over $j = 0$ and the points $x = 1/4, \infty$ in the fiber over $j = \infty$. Applying Lemma 2.2.1 shows that the ramification over these points is given by $(3^1, 2^1 1^1, 2^1 1^1)$. The Belyĭ map $b : x \mapsto (-27/4)x^2(x-1)$ ramifies in this way over 0, 1 and ∞, so by moving these branch points we get our cover $Z \to Y$, which we can take to be $(-8b-1)/(4b-4)$. We obtain $Z \to X$ by substituting this latter map in the former map (2.2.8) and then polishing by substituting $x/3$ for x. □

Similarly, we obtain:

LEMMA 2.2.9. *In case I, the Belyĭ map $X(\Gamma) \to X(1)$ is described by the map*

$$(x,y) \mapsto \frac{-(x^2-10x+5)^3}{x} \tag{2.2.10}$$

from the curve $y^2 = x(x^2 - 22x + 125)$.

PROOF. In this case we can use Lemma 2.2.1, but we dispense with it; the cover in question has an automorphism, represented on the level of permutations by the element $(1,11)(2,12)(3,9)(4,6)(5,8)(7,10)$ of order 2. After quotienting out, we get a degree 6 cover with ramification type $(3^2, 2^2 1^2, 5^1 1^1)$. This is a cover of genus 0 and degree 6 which we can find in [**1**]; it is nothing but the rational function in (2.2.10) considered as a map from $\mathbb{P}^1$ to $\mathbb{P}^1$.

To obtain our genus 1 curve, we only have to take a further degree 2 cover that corresponds to the automorphism that we just quotiented out. This cover branches at the two points of ramification index 1 in the fiber over $j = 1728$ and the two elements in the fiber over $j = 0$. Since the former points satisfy $x^2 - 22x + 125 = 0$ and the latter are $x = 0, \infty$, we see that we obtain the elliptic curve in the statement of the lemma. □

LEMMA 2.2.11. *In case III, the Belyĭ map $X(\Gamma) \to X(1)$ is described by the map*

$$(x,y) \mapsto 256\frac{(x^2+1)^3}{x^4} \tag{2.2.12}$$

from the curve $y^2 = x(4x^2 + 1)$.

PROOF. In this case the monodromy group has cardinality 96 and trivial center. We again use Lemma 2.2.1. One more we obtain a degree 2 cover of a Belyĭ map of genus 0, which this time has ramification triple

$$
\begin{aligned}
\sigma_1 &= (1,2)(5,6)\\
\sigma_0 &= (1,3,6)(2,5,4)\\
\sigma_\infty &= (1,4,5,3)(2,6).
\end{aligned}
\tag{2.2.13}
$$

We can find this cover directly in [**1**]. Alternatively, the lemma shows that it factors as $Z \to Y \to X$, where $Y \to X$ has ramification triple $(3^1, 2^1 1^1, 2^1 1^1)$ and where $Z \to Y$ has ramification $(2^1, 2^1)$ above the points in the fiber above $j = \infty$ (there is a point above $j = 1728$ that could a priori also gives rise to ramification but does not do so in the end).

The cover $Y \to X$ is given by

$$x \mapsto \frac{-256}{x^2(x-1)}. \tag{2.2.14}$$

By our description of the map $Z \to Y$, we see that we have to precompose with a cover of the form $x \mapsto x^2/(x^2+c)$. To ensure correct ramification we take $c = 1$, and we end up with the rational function in the Lemma. □

LEMMA 2.2.15. *In case IV, the Belyĭ map $X(\Gamma) \to X(1)$ is described by the map*

$$(x, y) \mapsto x^3 \tag{2.2.16}$$

from the curve $y^2 = x^3 - 1728$.

PROOF. As mentioned below, this group correponds to the commutator subgroup of $\mathrm{SL}_2(\mathbb{Z})$. This gives rise to a Galois morphism with group $\mathbb{Z}/6\mathbb{Z}$, which perforce comes by taking the elliptic curve with j-invariant equal to 0 and quotienting out its automorphism group. We have taken the particular twist above because it gives the canonical model considered in the next section. □

3. Canonical models

We now determine canonical models for our $(1;\infty)$-curves, using two different methods.

3.1. First approach: q-expansions. The first method is elementary and arguably still the most insightful; we choose the defining equations that we encounter in such a way that the resulting q-expansions of the coordinates are rational.

To illustrate this method, we consider case II, where we first try the model furnished in Lemma 2.2.3, namely $y^2 = x(x+1)(x-2)(x-3)$. In this case the fiber of the Belyĭ map above ∞ has 4 elements with distinct ramification indices. All of these will therefore give rise to a branch that gives a rational q-expansion of x. However, we also need to consider the square root that we have to draw when determining y, and this only gives a rational q-expansion for that coordinate if we switch to the quadratic twist by -1. This fully determines the canonical model over $\mathbb{Q}$ of the curve $X(\Gamma^{(2)})$. It is independent of the particular point over ∞ that is chosen.

Since $\Gamma^{(2)} \neq \Gamma$, it still remains to determine canonical model of $X(\Gamma)$ in this case. This is obtained by dividing out the 2-torsion of the Jacobian, which in fact leads to an isomorphic curve. Case IV corresponds to a subgroup of $\mathrm{SL}_2(\mathbb{Z})$, so we do not have to take any further isogeny, while in case I we can recover a model of $X(\Gamma)$ from the given one for $X(\Gamma')$ be taking an isogeny with kernel $(0,0)$.

The case III leads to a subtlety, due to an automorphism of the cover obtained from the group Γ' (which is the map $x \mapsto -x$ in Lemma 2.2.11). In this case there is no common quadratic twist that makes both branches over ∞ have rational q-expansions. Either choice of branch leads to a model of $X(\Gamma')$ over $\mathbb{Q}$. Neither of them can be called canonical over $\mathbb{Q}$, however, but only over $\mathbb{Q}(i)$, where either model over $\mathbb{Q}$ leads to all branches being rational. On the other hand, the model of the codomain $X(\Gamma)$ of the resulting 2-isogeny (which again has kernel $(0,0)$) does not depend on the choice of branch, and we get rational q-expansions at the single cusp of this curve. Therefore also in this case we obtain a canonical model for $X(\Gamma)$ over $\mathbb{Q}$, even though $X(\Gamma')$ admits no such model.

Table 4 summarizes the canonical models found; the parabolic point corresponds to the point at infinity in all cases. Besides a reference to the LMFDB [**9**], where more detailed information on these curves can be found, we have also included their j-invariant and Faltings height h_{Falt}. Table 5 gives some of the resulting q-expansions of the coordinates x and y.

Case	Curve	LMFDB label	j-invariant	h_{Falt}
I	$y^2 = x^3 - 44x^2 - 16x$	20.a1	$2^{14}31^3/5^3$	3.814
II	$y^2 = x^3 - 4x^2 - 384x - 2304$	24.a3	$2^2 73^3/3^4$	3.152
III	$y^2 = x(x^2 - 256)$	32.a4	1728	-1.311
IV	$y^2 = x^3 - 1728$	36.a3	0	-1.321

TABLE 4. Canonical models

Case	Expansions of x and y
I	$x = q^{-1/5} + 16 + 134q^{1/5} + 760q^{2/5} + 3345q^{3/5} + 12256q^{4/5} + \dots$
	$y = q^{-3/10} + 2q^{-1/10} - 129q^{1/10} - 1778q^{3/10} - 13725q^{1/2} - \dots$
II	$x = q^{-1/6} + 2 + 79q^{1/6} + 352q^{1/3} + 1431q^{1/2} + 4160q^{2/3} + \dots$
	$y = q^{-1/4} + q^{-1/12} - 76q^{1/12} - 778q^{1/4} - 5224q^{5/12} + \dots$
III	$x = q^{-1/4} + 52q^{1/4} + 834q^{3/4} + 4760q^{5/4} + 24703q^{7/4} + \dots$
	$y = q^{-3/8} - 50q^{1/8} - 2599q^{5/8} - 29154q^{9/8} - 238728q^{13/8} + \dots$
IV	$x = q^{-1/3} + 824q^{2/3} - 613348q^{5/3} + 831470016q^{8/3} + \dots$
	$y = q^{-1/2} + 372q^{1/2} + 29250q^{3/2} - 134120q^{5/2} + \dots$

TABLE 5. q-expansions

The curves that we found have minimal conductor among all their possible quadratic twists over $\mathbb{Q}$. Note that while case III gives a curve that is isomorphic to $X_0(32)$ over $\mathbb{Q}$, the latter curve is in fact a $(1;\infty^8)$-curve. The corresponding groups are commensurable, with neither being a subgroup of the other up to conjugacy.

3.2. Second approach: Shimura's viewpoint. A second approach to obtain canonical models is to use work by Shimura [**4, 14**] that gives a more general description of canonical models and their properties. In [**17**], this served as an essential tool to determine canonical models in the presence of a non-split quaternion algebra.

In this approach we further enlarge the orders $\mathcal{O}'$ from section 1 to orders $\mathcal{O}'' \supset \mathcal{O}'$ with $\mathcal{O}''^1 = \mathcal{O}'^1$ but with the property that the norm surjects to as large a subset of $\mathbb{Z}$ as possible. Shimura then describes the canonical field of definition of $X(\mathcal{O}''^1)$ and indicates how the traces of Frobenius of the canonical model over this field can be determined. We briefly sketch the results here.

In cases I and II the overorder $\mathcal{O}''$ can be chosen in such a way that the norm surjects to $\mathbb{Z}$. In this case the canonical field of definition equals $\mathbb{Q}$. The corresponding curve has good reduction outside the primes that divide the index of $\mathcal{O}''$ in $M_2(\mathbb{Z})$, and its traces of Frobenius can be determined using Shimura congruence, as in [**15**, Chapter 4]. In combination with the known j-invariant, these data suffice to determine the canonical model completely. We checked that the results obtained agree with the first approach.

In case III we can at most obtain an order $\mathcal{O}''$ whose norm map attains all primes that are congruent to 1 modulo 4. Shimura's results then show that the canonical field of definition of the curve $X(\mathcal{O}'^1) = X(\langle\Gamma^{(2)}, \alpha\beta\rangle)$ is $\mathbb{Q}(i)$, which is in line with the subtlety encountered when using the first method above. The methods from [**15**] can again be used to compute the traces of Frobenius of Deligne's non-connected model at all primes, and once more the results agree with the more elementary method above.

Case IV is quite surprising. In this case $\mathrm{P}\Gamma$ is the image of the commutator subgroup of $\mathrm{SL}_2(\mathbb{Z})$ in $\mathrm{PSL}_2(\mathbb{R})$. Indeed, like the former group, the latter is abelian of index 6 in $\mathrm{PSL}_2(\mathbb{R})$. By the universal property of the commutator subgroup, these groups therefore have to coincide.

The commutator subgroup of $\mathrm{SL}_2(\mathbb{Z})$ is described in [**3**, Remark 3.9]. Let

$$\gamma_1 = \begin{pmatrix} 1 & 1 \\ 1 & 2 \end{pmatrix}, \gamma_2 = \begin{pmatrix} 1 & -1 \\ -1 & 2 \end{pmatrix} \tag{3.2.1}$$

be generators of the commutator subgroup of $\mathrm{SL}_2(\mathbb{Z})$. Recall that the order $\mathcal{O}' = \mathbb{Z}[\Gamma]$ had the property $\Gamma \subsetneq \mathcal{O}'^1$ and is therefore too large to study Γ. We can instead consider the order $\mathcal{O} = \mathbb{Z}[\Gamma^{(2)}]$. Then

$$\Gamma = \langle \mathcal{O}^1, \gamma_1, \gamma_2 \rangle. \tag{3.2.2}$$

It turns out that $\mathcal{O}$ can be enlarged to an order of index 36 in $M_2(\mathbb{Z})$ with unit group $\Gamma^{(2)}$. None of the elements $\gamma_1, \gamma_2, \gamma_1\gamma_2$ belongs to this group. Yet the groups $\langle\Gamma^{(2)}, \alpha\rangle$, $\langle\Gamma^{(2)}, \beta\rangle$, $\langle\Gamma^{(2)}, \alpha\beta\rangle$ all generate additional orders, also of index 36 in $M_2(\mathbb{Z})$, to which exactly one of these commutators belongs. All these 4 orders are distinct, but have the same index in $M_2(\mathbb{Z})$ and give rise to curves with canonical field of definition $\mathbb{Q}$ with isogenous Jacobians.

3.3. Orders obtained. Bases of the orders $\mathcal{O}''$ can be found by running the code at [**18**]. None of these orders is Eichler, and in all cases the index in $M_2(\mathbb{Z})$ equals the corresponding conductor in Table 4. In the next section we will construct these orders $\mathcal{O}''$ as intersections.

4. Modular interpretations

The curves $X(\Gamma)$ also parametrize the points of certain moduli functors, which we describe in this section. A more advanced general exposition is given in [**5**].

4.1. Over $\mathbb{C}$. Consider pairs

$$(E,(\lambda_1,\lambda_2)), \tag{4.1.1}$$

where E is an elliptic curve over $\mathbb{C}$ and where (λ_1,λ_2) is an ordered basis of the homology group $H_1(E,\mathbb{Z})$, which we assume to be positively oriented in the sense that $\mathrm{im}(\lambda_1/\lambda_2) > 0$ in the one-dimensional $\mathbb{C}$-vector space $V = H_1(E,\mathbb{Z}) \otimes \mathbb{R} = H^0(E,\omega_E)^\vee$ that is the universal cover of E.

The group $\mathrm{SL}_2(\mathbb{Z})$ acts on the left on the set of objects (4.1.1) as follows: if $\gamma' = \left(\begin{smallmatrix} a & b \\ c & d \end{smallmatrix}\right) \in \mathrm{SL}_2(\mathbb{Z})$, then

$$\gamma'(E,(\lambda_1,\lambda_2)) = (E,(a\lambda_1 + b\lambda_2, c\lambda_1 + d\lambda_2)). \tag{4.1.2}$$

Let Γ' be a subgroup of $\mathrm{SL}_2(\mathbb{Z})$ (not necessarily congruence). We say that two pairs $(E,(\lambda_1,\lambda_2))$, $(E',(\lambda_1',\lambda_2'))$ are *Γ'-equivalent* if and only if there exists an isomorphism $\varphi : E \to E'$ such that

$$(E',(\varphi_*(\lambda_1),\varphi_*(\lambda_2))) = \gamma'(E',(\lambda_1',\lambda_2')) \tag{4.1.3}$$

for some γ' in Γ'.

PROPOSITION 4.1.4. *The points of the quotient space $X(\Gamma')$ bijectively correspond to the Γ'-equivalence classes of pairs* (4.1.1).

PROOF. Given a pair as in (4.1.1), there exists a unique $\tau \in \mathcal{H}$ such that

$$(E,(\lambda_1,\lambda_2)) \cong (E_\tau,(\tau,1)). \tag{4.1.5}$$

Here $E_\tau = V/\Lambda_\tau$. The action of Γ' then factors through the usual action of $\mathrm{PSL}_2(\mathbb{R})$ on the upper half plane $\mathcal{H}$. Moreover, two pairs $(E_\tau,(\tau,1))$, $(E_{\tau'},(\tau',1))$ are equivalent under if and only if τ and τ' are related by this action. □

In the cases I-III, the group Γ is not itself a subgroup of $\mathrm{SL}_2(\mathbb{Z})$, and we first restrict to $\Gamma' = \Gamma \cap \mathrm{SL}_2(\mathbb{Z})$. In all of these cases, the group Γ can be obtained from Γ' by adjoining elements $w \in M_2(\mathbb{Z})$ that are not in $\mathrm{SL}_2(\mathbb{Z})$ but whose square is a still scalar. These elements w normalize the order $\mathcal{O}'$ defined by Γ'.

Such elements w induce involutions by conjugation, which can be interpreted as follows in terms of the pairs (4.1.1). Let s be the square of w considered as a scalar, let $w' = w^{-1} = w/s$, and let $\Lambda^w = w'(\Lambda)$ be the overlattice of Λ defined by w'. Then Λ^w contains Λ as a sublattice with index s. We define

$$E^w = V/\Lambda^w \tag{4.1.6}$$

and define

$$(\lambda_1^w,\lambda_2^w) = w'(\lambda_1,\lambda_2) \tag{4.1.7}$$

as in (4.1.2). Then the association

$$(E,(\lambda_1,\lambda_2)) \mapsto (E^w,(\lambda_1^w,\lambda_2^w)) \tag{4.1.8}$$

gives rise to a well-defined map on Γ'-orbits because $w^{-1}\Gamma' w = \Gamma'$.

Restricting to pairs of the form $(E_\tau,(\tau,1))$ as above, we see the following.

PROPOSITION 4.1.9. *Suppose that $\Gamma = \langle \Gamma', S\rangle$, where S is a set of involutions w represented by matrices whose square is a scalar and that satisfy $w\Gamma' w^{-1} = \Gamma'$.*

Then the points of the quotient space $X(\Gamma)$ bijectively correspond to the equivalence classes of pairs (4.1.1), *up to the additional identifications generated by*

$$(E,(\lambda_1,\lambda_2)) \sim (E^w,(\lambda_1^w,\lambda_2^w)) \tag{4.1.10}$$

for $w \in S$.

4.2. Over general base fields. Now suppose that $\Gamma' \subset \mathrm{SL}_2(\mathbb{Z})$ is a congruence subgroup, a situation that applies to our examples. Then another description is possible, which lends itself to generalization to arbitrary fields. We take a relatively elementary approach; a more high-brow exposition can be found in [**10**] and the references mentioned therein.

Suppose that the full-level subgroup $\Gamma(N)$ is a subgroup of Γ', and let $\overline{\Gamma}'$ be the image of Γ' in $\mathrm{SL}_2(\mathbb{Z}/N\mathbb{Z})$. Consider pairs

$$(E,(\overline{\lambda}_1,\overline{\lambda}_2)), \tag{4.2.1}$$

where E is an elliptic curve and where $(\overline{\lambda}_1,\overline{\lambda}_2)$ is an ordered basis of the N-torsion group $E[N]$.

The group $\mathrm{SL}_2(\mathbb{Z}/N\mathbb{Z})$ acts on the left on the set of objects (4.2.1) as follows: if $\gamma' = \left(\begin{smallmatrix} a & b \\ c & d \end{smallmatrix}\right) \in \mathrm{SL}_2(\mathbb{Z}/N\mathbb{Z})$, then

$$\gamma'(E,(\overline{\lambda}_1,\overline{\lambda}_2)) = (E,(a\overline{\lambda}_1 + b\overline{\lambda}_2, c\overline{\lambda}_1 + d\overline{\lambda}_2)). \tag{4.2.2}$$

Via this action, we also get an action of Γ', which projects to $\mathrm{SL}_2(\mathbb{Z}/N\mathbb{Z})$. We say that two pairs $(E,(\overline{\lambda}_1,\overline{\lambda}_2))$, $(E',(\overline{\lambda}_1',\overline{\lambda}_2'))$ are Γ'-equivalent if and only if there exists an isomorphism $\varphi : E \to E'$ such that

$$(E',(\varphi(\overline{\lambda}_1),\varphi(\overline{\lambda}_2))) = \gamma'(E',(\overline{\lambda}_1',\overline{\lambda}_2')) \tag{4.2.3}$$

for some γ' in Γ'. Then we again have the following.

PROPOSITION 4.2.4. *The points of the algebraic curve $X(\Gamma')$ over an algebraically closed field k bijectively correspond to the Γ'-equivalence classes of pairs* (4.2.1) *over k.*

Note that for a non-algebraically closed field k, such an equivalence class can define a k-rational point of $X(\Gamma')$ without its constituents being k-rational; if k is perfect, then it suffices that the class is closed under conjugation by the absolute Galois group of k.

We can also give a description of the involution defined by an element $w \in \Gamma$ with integral scalar square s as above. We define $w'' = Nw^{-1} \in M_2(\mathbb{Z}/N\mathbb{Z})$. Say $w'' = \left(\begin{smallmatrix} a & b \\ c & d \end{smallmatrix}\right)$. Then defining

$$w''(\overline{\lambda}_1,\overline{\lambda}_2) = (a\overline{\lambda}_1 + b\overline{\lambda}_2, c\overline{\lambda}_1 + d\overline{\lambda}_2), \tag{4.2.5}$$

we let

$$E^w = E/\langle w''(\overline{\lambda}_1,\overline{\lambda}_2)\rangle. \tag{4.2.6}$$

We choose a basis $(\widetilde{\overline{\lambda}}_1,\widetilde{\overline{\lambda}}_2)$ of $E[Ns]$ in such a way that $s\widetilde{\overline{\lambda}}_i = \overline{\lambda}_i$. Let $w''' = sw^{-1} = w$ and define

$$(\overline{\lambda}_1^w,\overline{\lambda}_2^w) = w'''(\widetilde{\overline{\lambda}}_1,\widetilde{\overline{\lambda}}_2). \tag{4.2.7}$$

This is a basis of $E^w[N]$, and the association

$$(E,(\overline{\lambda}_1,\overline{\lambda}_2)) \mapsto (E^w,(\overline{\lambda}_1^w,\overline{\lambda}_2^w)) \tag{4.2.8}$$

gives rise to a well-defined map on Γ'-orbits, and gives the requested modular description. The reader will readily formulate the analogue of Proposition 4.1.9. Again, over perfect fields k the rational points of $X(\Gamma)$ parametrize equivalence classes that are closed under Galois conjugation.

In our concrete examples, we can give slightly more elegant prime-by-prime descriptions of the moduli problems, essentially because when $\Gamma' = \Gamma_0(N)$ we obtain the classical description in terms of torsion subgroups (instead of bases). We now proceed to give these descriptions, up to conjugacy.

4.3. Case I. In this case the order $\mathcal{O}''$ from the end of the previous section is the intersection of the index 4 order

$$\mathcal{O}_4 = \langle \left(\begin{smallmatrix} 1 & 0 \\ 0 & 1 \end{smallmatrix}\right), \left(\begin{smallmatrix} 1 & 0 \\ 0 & -1 \end{smallmatrix}\right), \left(\begin{smallmatrix} 0 & 1 \\ 1 & 1 \end{smallmatrix}\right), \left(\begin{smallmatrix} 0 & 1 \\ -1 & 1 \end{smallmatrix}\right) \rangle \tag{4.3.1}$$

and the index 5 order

$$\mathcal{O}_5 = \langle \left(\begin{smallmatrix} 1 & 0 \\ 0 & 0 \end{smallmatrix}\right), \left(\begin{smallmatrix} 0 & 1 \\ 0 & 0 \end{smallmatrix}\right), \left(\begin{smallmatrix} 0 & 0 \\ 5 & 0 \end{smallmatrix}\right), \left(\begin{smallmatrix} 0 & 0 \\ 0 & 1 \end{smallmatrix}\right) \rangle. \tag{4.3.2}$$

We obtain a congruence subgroup of level 10.

The corresponding modular description admits the following simplification: the curve $X(\Gamma')$ parametrizes equivalence classes of triples

$$(E, H_5, (\overline{\lambda}_1, \overline{\lambda}_2)), \tag{4.3.3}$$

where E is an elliptic curve, H_5 is a subgroup of $E[5]$ of order 5, and $(\overline{\lambda}_1, \overline{\lambda}_2)$ is a basis of $E[2]$. Two such triples $(E, H_5, (\overline{\lambda}_1, \overline{\lambda}_2))$ and $(E', H_5', (\overline{\lambda}_1', \overline{\lambda}_2'))$ are equivalent if and only there exists an isomorphism $\varphi : E \to E'$ that maps H_5 into H_5' and such that $\varphi(\overline{\lambda}_1, \overline{\lambda}_2) = \alpha(\overline{\lambda}_1', \overline{\lambda}_2')$, where α is a power of $\left(\begin{smallmatrix} 0 & 1 \\ 1 & 1 \end{smallmatrix}\right) \in \mathrm{SL}_2(\mathbb{F}_2)$.

There is a single involution defined by the element

$$w_5 = \left(\begin{smallmatrix} 0 & -1 \\ 5 & 0 \end{smallmatrix}\right) \tag{4.3.4}$$

of determinant 5. In terms of the triples (4.3.3), its action can be described by

$$w_5(E, H_5, (\overline{\lambda}_1, \overline{\lambda}_2)) = (E', H_5', (\overline{\lambda}_1', \overline{\lambda}_2')), \tag{4.3.5}$$

where E' is the quotient of E by H_5, H_5' is the image of $E[5]$ on E', and $(\overline{\lambda}_1', \overline{\lambda}_2')$ is the image of $(\overline{\lambda}_2, \overline{\lambda}_1)$ on E'.

4.4. Case II. In this case the order $\mathcal{O}''$ from the end of the previous section is the intersection of the index 8 order

$$\mathcal{O}_8 = \langle \left(\begin{smallmatrix} 1 & 0 \\ 0 & 1 \end{smallmatrix}\right), \left(\begin{smallmatrix} 1 & 0 \\ 0 & -1 \end{smallmatrix}\right), \left(\begin{smallmatrix} 0 & 0 \\ 2 & 0 \end{smallmatrix}\right), \left(\begin{smallmatrix} 0 & 2 \\ 1 & 0 \end{smallmatrix}\right) \rangle \tag{4.4.1}$$

and the index 3 order

$$\mathcal{O}_3 = \langle \left(\begin{smallmatrix} 1 & 0 \\ 0 & 0 \end{smallmatrix}\right), \left(\begin{smallmatrix} 0 & 1 \\ 0 & 0 \end{smallmatrix}\right), \left(\begin{smallmatrix} 0 & 0 \\ 3 & 0 \end{smallmatrix}\right), \left(\begin{smallmatrix} 0 & 0 \\ 0 & 1 \end{smallmatrix}\right) \rangle. \tag{4.4.2}$$

We obtain a congruence subgroup of level 12. Its elements parametrize triples

$$(E, H_3, (\overline{\lambda}_1, \overline{\lambda}_2)), \tag{4.4.3}$$

where this time H_3 is a subgroup of order 3 of H_3, and where $(\overline{\lambda}_1, \overline{\lambda}_2)$ is a basis of $E[4]$ up to the equivalence defined by the elements of the order in $\mathrm{SL}_2(\mathbb{Z}/4\mathbb{Z})$ determined by the elements in the basis of $\mathcal{O}_8$.

There are three involutions, defined by the elements

$$\begin{aligned} w_2 &= \left(\begin{smallmatrix} -2 & -2 \\ 3 & 2 \end{smallmatrix}\right), \\ w_3 &= \left(\begin{smallmatrix} -3 & -4 \\ 3 & 3 \end{smallmatrix}\right), \\ w_6 &= \left(\begin{smallmatrix} 0 & 2 \\ -3 & 0 \end{smallmatrix}\right), \end{aligned} \tag{4.4.4}$$

of determinant 2, 3, 6 respectively. Note that these elements only commute modulo the group $\mathcal{O}''^1$.

The action of the involution w_2 is given by

$$w_2(E, H_3, (\overline{\lambda}_1, \overline{\lambda}_2)) = (E', H_3', (\overline{\lambda}_1', \overline{\lambda}_2')), \tag{4.4.5}$$

where E' is the quotient of E by $2\overline{\lambda}_1$, where H_3' is the image of H_3 on E', and where $(\overline{\lambda}_1', \overline{\lambda}_2')$ is some point on E' such that $(\overline{\lambda}_1', 2\overline{\lambda}_2')$ equals the image of $(\overline{\lambda}_2, -\overline{\lambda}_1)$ on E'.

The action of the involution w_2 is given by

$$w_3(E, H_3, (\overline{\lambda}_1, \overline{\lambda}_2)) = (E', H_3', (\overline{\lambda}_1', \overline{\lambda}_2')), \tag{4.4.6}$$

where E' is the quotient of E by H_3, where H_3' is the image of $E[3]$ on E', and where $(\overline{\lambda}_1', \overline{\lambda}_2')$ is the image of $(\overline{\lambda}_1, \overline{\lambda}_1 - \overline{\lambda}_2)$ on E'.

The action of the involution w_6 is the composition of that of w_2 and w_3. It is given by

$$w_6(E, H_3, (\overline{\lambda}_1, \overline{\lambda}_2)) = (E', H_3', (\overline{\lambda}_1', \overline{\lambda}_2')), \tag{4.4.7}$$

where E' is the quotient of E by the group generated by H_3 and $2\overline{\lambda}_1$, where H_3' is the image of $E[3]$ on E', and where $(\overline{\lambda}_1', \overline{\lambda}_2')$ is some point on E' such that $(\overline{\lambda}_1', 2\overline{\lambda}_2')$ equals the image of $(\overline{\lambda}_2, \overline{\lambda}_1)$ on E'.

4.5. Case III. In this case the order $\mathcal{O}''$ is of index 32, given by

$$\mathcal{O}_{32} = \langle \left(\begin{smallmatrix} 1 & 0 \\ 0 & 1 \end{smallmatrix}\right), \left(\begin{smallmatrix} 0 & 2 \\ -2 & 0 \end{smallmatrix}\right), \left(\begin{smallmatrix} 2 & -1 \\ 1 & -2 \end{smallmatrix}\right), \left(\begin{smallmatrix} 0 & 1 \\ 3 & 0 \end{smallmatrix}\right) \rangle. \tag{4.5.1}$$

We obtain a congruence subgroup of level 8. There is a single involution defined by the element

$$w_2 = \left(\begin{smallmatrix} -1 & -3 \\ 1 & 1 \end{smallmatrix}\right) \tag{4.5.2}$$

of determinant 2. In this case we get a modular description in terms of the 8-torsion for which a simpler description than the generic one in section 4.2 does not seem to be readily available.

4.6. Case IV. We consider the order $\mathcal{O}''$ obtained by enlarging $\Gamma^{(2)}$. This is the intersection of the index 4 order

$$\mathcal{O}_4 = \langle \left(\begin{smallmatrix} 1 & 0 \\ 0 & 1 \end{smallmatrix}\right), \left(\begin{smallmatrix} 1 & 0 \\ 0 & -1 \end{smallmatrix}\right), \left(\begin{smallmatrix} 0 & 1 \\ 1 & 1 \end{smallmatrix}\right), \left(\begin{smallmatrix} 0 & 1 \\ -1 & 1 \end{smallmatrix}\right) \rangle \tag{4.6.1}$$

and the index 9 order

$$\mathcal{O}_9 = \langle \left(\begin{smallmatrix} 1 & 0 \\ 0 & 1 \end{smallmatrix}\right), \left(\begin{smallmatrix} 0 & 1 \\ 0 & 1 \end{smallmatrix}\right), \left(\begin{smallmatrix} 1 & 1 \\ 0 & -1 \end{smallmatrix}\right), \left(\begin{smallmatrix} 0 & 0 \\ 3 & 0 \end{smallmatrix}\right) \rangle. \tag{4.6.2}$$

We obtain a congruence subgroup of level 6, from which we can obtain Γ by adding the commutators $\gamma_1 = \left(\begin{smallmatrix} 1 & 1 \\ 2 & 1 \end{smallmatrix}\right)$ and $\gamma_2 = \left(\begin{smallmatrix} 1 & -1 \\ -2 & 1 \end{smallmatrix}\right)$. This group corresponds to the unique normal subgroup of $\mathrm{SL}_2(\mathbb{Z}/6\mathbb{Z})$ that gives rise to a quotient that is cyclic of order 6. Using the corresponding character makes it easier to check the equivalence (4.2.3).

References

[1] Bryan Birch, *Noncongruence subgroups, covers and drawings*, The Grothendieck theory of dessins d'enfants (Luminy, 1993), London Math. Soc. Lecture Note Ser., vol. 200, Cambridge Univ. Press, Cambridge, 1994, pp. 25–46. MR1305392

[2] Wieb Bosma, John Cannon, and Catherine Playoust, *The Magma algebra system. I. The user language*, J. Symbolic Comput. **24** (1997), no. 3-4, 235–265, DOI 10.1006/jsco.1996.0125. Computational algebra and number theory (London, 1993). MR1484478

[3] Keith Conrad, $SL_2(\mathbf{Z})$, Expository note available at `http://www.math.uconn.edu/~kconrad/blurbs`.

[4] Pierre Deligne, *Variétés de Shimura: interprétation modulaire, et techniques de construction de modèles canoniques* (French), Automorphic forms, representations and L-functions (Proc. Sympos. Pure Math., Oregon State Univ., Corvallis, Ore., 1977), Proc. Sympos. Pure Math., XXXIII, Amer. Math. Soc., Providence, R.I., 1979, pp. 247–289. MR546620

[5] P. Deligne and M. Rapoport, *Les schémas de modules de courbes elliptiques* (French), Modular functions of one variable, II (Proc. Internat. Summer School, Univ. Antwerp, Antwerp, 1972), Springer, Berlin, 1973, pp. 143–316. Lecture Notes in Math., Vol. 349. MR0337993

[6] Ivan Fesenko, *Arithmetic deformation theory via arithmetic fundamental groups and nonarchimedean theta-functions, notes on the work of Shinichi Mochizuki*, Eur. J. Math. **1** (2015), no. 3, 405–440, DOI 10.1007/s40879-015-0066-0. MR3401899

[7] Michael Klug, Michael Musty, Sam Schiavone, Jeroen Sijsling, and John Voight, *Computing a database of Belyĭ maps*, Unpublished preprint, 2017.

[8] Michael Klug, Michael Musty, Sam Schiavone, and John Voight, *Numerical calculation of three-point branched covers of the projective line*, LMS J. Comput. Math. **17** (2014), no. 1, 379–430, DOI 10.1112/S1461157014000084. MR3356040

[9] The LMFDB Collaboration, *The L-functions and Modular Forms Database*, `http://www.lmfdb.org`, 2017, [Online; accessed 30 June 2017].

[10] James Milne, *Canonical models of shimura curves*, Notes at `http://www.jmilne.org/math/articles/2003a.pdf`, 2003.

[11] Shinichi Mochizuki, *Correspondences on hyperbolic curves*, J. Pure Appl. Algebra **131** (1998), no. 3, 227–244, DOI 10.1016/S0022-4049(97)00078-9. MR1637015

[12] Shinichi Mochizuki, *The absolute anabelian geometry of canonical curves*, Doc. Math. **Extra Vol.** (2003), 609–640. Kazuya Kato's fiftieth birthday. MR2046610

[13] Shinichi Mochizuki, *Inter-universal Teichmüller theory I: Construction of Hodge theaters*, Preprint available at `http://www.kurims.kyoto-u.ac.jp/~motizuki/papers-english.html`, 2017.

[14] Goro Shimura, *On canonical models of arithmetic quotients of bounded symmetric domains*, Ann. of Math. (2) **91** (1970), 144–222, DOI 10.2307/1970604. MR0257031

[15] Jeroen Sijsling, *Equations for arithmetic pointed tori*, Ph.D. thesis, Universiteit Utrecht, 2010, Available at `https://sites.google.com/site/sijsling/research/thesis-sijsling.pdf`.

[16] Jeroen Sijsling, *Arithmetic* $(1;e)$*-curves and Belyĭ maps*, Math. Comp. **81** (2012), no. 279, 1823–1855, DOI 10.1090/S0025-5718-2012-02560-9. MR2904604

[17] Jeroen Sijsling, *Canonical models of arithmetic* $(1;e)$*-curves*, Math. Z. **273** (2013), no. 1-2, 173–210, DOI 10.1007/s00209-012-1000-5. MR3010156

[18] Jeroen Sijsling, *Canonical models of arithmetic (1; inf)-curves*, Code available at `https://github.com/JRSijsling/canmod-1inf`, 2017.

[19] J. Sijsling and J. Voight, *On computing Belyi maps* (English, with English and French summaries), Numéro consacré au trimestre "Méthodes arithmétiques et applications", automne 2013, Publ. Math. Besançon Algèbre Théorie Nr., vol. 2014/1, Presses Univ. Franche-Comté, Besançon, 2014, pp. 73–131. MR3362631

[20] Kisao Takeuchi, *A characterization of arithmetic Fuchsian groups*, J. Math. Soc. Japan **27** (1975), no. 4, 600–612, DOI 10.2969/jmsj/02740600. MR0398991

[21] Kisao Takeuchi, *Arithmetic Fuchsian groups with signature* $(1;e)$, J. Math. Soc. Japan **35** (1983), no. 3, 381–407, DOI 10.2969/jmsj/03530381. MR702765

[22] John Voight, *Identifying the matrix ring: algorithms for quaternion algebras and quadratic forms*, Quadratic and higher degree forms, Dev. Math., vol. 31, Springer, New York, 2013, pp. 255–298, DOI 10.1007/978-1-4614-7488-3_10. MR3156561

Institut für Reine Mathematik, Universität Ulm, Helmholtzstrasse 18, 89081 Ulm, Germany

Email address: jeroen.sijsling@uni-ulm.de

Contemporary Mathematics
Volume **722**, 2019
https://doi.org/10.1090/conm/722/14532

Maps between curves and arithmetic obstructions

Andrew V. Sutherland and José Felipe Voloch

ABSTRACT. Let X and Y be curves over a finite field. In this article we explore methods to determine whether there is a rational map from Y to X by considering L-functions of certain covers of X and Y and propose a specific family of covers to address the special case of determining when X and Y are isomorphic. We also discuss an application to factoring polynomials over finite fields.

1. Introduction

Given two algebraic curves X, Y of genus at least two over a finite field we would like to decide if there is a rational map from Y to X. Hess and Möhlmann [**7, 12**] have given an algorithm to decide if such a map exists by performing an optimised search for the map up to a known bound; if the search is unsuccessful then no such map exists. This procedure works well when the map exists, but it may be very time consuming when it does not. The purpose of this paper is to investigate methods for deciding when there is no such map without performing an exhaustive search. We concentrate on the case of isomorphisms but briefly touch on the general case.

Our methods can also provide a short certificate of the non-existence of a rational map, a feature not available with existing algorithms.

A result of Poonen [**13**] extending an idea of Kayal shows that, given a one-parameter family of curves over a finite field with distinct L-polynomials for distinct values of the parameter and a suitable bound on the genus, one can construct a deterministic polynomial time algorithm for factoring polynomials over that field. Our investigations suggest some candidate families of such curves, but unfortunately we cannot prove that they work.

2. The fundamental group

Let X/K be a smooth geometrically connected variety over a field K. Let G_K be the absolute Galois group of K and $\bar{X}$ the base-change of X to an algebraic closure of K. We denote by $\pi_1(.)$ the algebraic fundamental group functor on (geometrically pointed) schemes and we omit base-points from the notation. We have the fundamental exact sequence

$$1 \to \pi_1(\bar{X}) \to \pi_1(X) \to G_K \to 1. \tag{2.1}$$

2010 *Mathematics Subject Classification.* Primary 11T06, 14H25.

The map $p_X : \pi_1(X) \to G_K$ from the above sequence is obtained by functoriality from the structural morphism $X \to \operatorname{Spec} K$. Grothendieck's anabelian program is to specify a class of varieties, termed anabelian, for which the varieties and morphisms between them can be recovered from the corresponding fundamental groups together with the corresponding maps p_X when the ground field is finitely generated over $\mathbb{Q}$. There has been some work done over finite fields as well, although the anabelian program will not work in the same way (the analogue of the section conjecture is false, for example).

For the rest of the paper we restrict to the case where K is a finite field. As usual, $\mathbb{F}_q$ is the field of q elements and we denote by p its characteristic. Here is a positive result.

THEOREM 2.1. *(Mochizuki–Tamagawa) Let X, Y be smooth projective curves of genus at least two over a finite field $\mathbb{F}_q$. If there is an isomorphism from $\pi_1(X)$ to $\pi_1(Y)$ inducing the identity on $G_{\mathbb{F}_q}$ via p_X, p_Y, then X is isomorphic to Y.*

The fundamental group is a mysterious object. What kind of information can we extract from it? First of all, if J_X denotes the Jacobian of a curve X we have the fundamental exact sequence (2.1) for J_X also, and $\pi_1(\bar{J_X})$ is the abelianisation of $\pi_1(\bar{X})$; thus the prime-to-p part of $\pi_1(\bar{J_X})$ is the product of the Tate modules of J_X. The fundamental exact sequence describes the Galois action on the Tate module, so its description is equivalent to the L-function of X, which we can compute by counting points on X over suitable extensions of $\mathbb{F}_q$. But by the very nature of the fundamental group, we can count points on covers as well. Since knowing J_X alone up to isogeny is not enough to recover X, we need to pass to covers. According to J. Stix (personal communication) the proof of Theorem 2.1 requires only solvable covers. The most natural covers come from the Hilbert class field tower. Let $\mathrm{Fr} : J_X \to J_X$ denote the $\mathbb{F}_q$-Frobenius map. Define $H(X) := (I - \mathrm{Fr})^*(X) \subset J_X$; it is an unramified abelian cover of X with Galois group $J_X(\mathbb{F}_q)$, well defined up to a twist that corresponds to a choice of divisor of degree one embedding X into J_X. Define $H_0(X) := X$, $H_1(X) := H(X)$ and successively define $H_{n+1}(X) := H_n(H(X))$ for integers $n \geq 1$. These covers can be derived from $\pi_1(X)$ but are perhaps more computationally accessible.

CONJECTURE 2.2. *Let X, Y be smooth projective curves of genus at least two over a finite field $\mathbb{F}_q$. If, for each n, there are choices of twists such that the L-function of $H_n(X)$ is equal to the L-function of $H_n(Y)$ for all $n \geq 0$, then X is isomorphic to an $\mathbb{F}_q/\mathbb{F}_p$ conjugate of Y.*

Up to isomorphism there are exactly 8 smooth projective curves of positive genus over finite fields for which $H(X) = X$, equivalently, curves whose function fields have class number one [**10**, **16**], of which 5 have genus at least two. We have verified that Conjecture 2.2 holds when X and Y are among these 5 curves, and it therefore holds if either $H(X) = X$ or $H(Y) = Y$. We may thus assume henceforth that X and Y have non-trivial Hilbert class field towers.

The basis for our heuristic is the following consequence of the usual calculation leading to the birthday paradox. If $\mathcal{M}$ and $\mathcal{I}$ are finite sets with cardinalities M and I respectively, then for $I \preceq M^2$ the probability that a random map $\mathcal{M} \to \mathcal{I}$ is non-injective is bounded above zero, but for I asymptotically larger than M^2 this probability decays rapidly to zero. Explicitly, this probability is

$$1 - \textstyle\prod_{j=0}^{M-1}(1 - j/I) \sim 1 - e^{-M(M-1)/2I}.$$

We first apply this to the set $\mathcal{M}$ of isomorphism classes of curves of some fixed genus $g > 1$ over a finite field $\mathbb{F}_q$ and the set $\mathcal{I}$ of isogeny classes of abelian varieties of dimension g over a finite field $\mathbb{F}_q$. For fixed g and large q we have $M \sim q^{3g-3}$ and $I \sim q^{g(g+1)/4}$, hence it is reasonable to expect that there will be distinct curves with isogenous Jacobians when $6g - 6 \leq g(g+1)/4$, that is, $g \leq 22$ (see [**3**] and [**15**] where this kind of question is discussed in more detail). For larger genus one expects this to be very rare, but we note a result of Mestre [**11**] that allows one to construct, for every $g > 1$, pairs of non-isomorphic genus g curves with isogenous Jacobians (over sufficiently large finite fields).

Provided $H(X) \neq X$ (as we now assume), passing from X to $H(X)$ increases the genus and thus makes it more likely that we can use isogeny invariants to distinguish non-isomorphic curves. For large q and g, the genus of $H(X)$ is much larger than the genus g of X; indeed, it is on the order of gq^g. However, the Jacobian of $H(X)$ is not an arbitrary abelian variety of this dimension; it decomposes up to isogeny as a product of abelian varieties of smaller dimension. The precise shape of the decomposition depends on the group structure of $J(\mathbb{F}_q)$ but it falls into a small set of possibilities. Thus, given its huge dimension, barring any additional constraints, the isogeny class of the Jacobian of $H(X)$ still varies in a huge space. So, on probabilistic grounds, it is still likely that the map $X \mapsto L(H(X), T)$ is injective when g and q are large. This makes it plausible that, up to a finite set of exceptions, Conjecture 2.2 holds even when we restrict to $n \leq 1$. The set of exceptions is non-empty as the examples in the next section show.

In addition to $H(X)$, we can also consider the cover of X obtained by pulling back via multiplication by 2 on the Jacobian (assuming the characteristic is not 2). This gives a cover of X of degree 2^{2g}. Note that, over the ground field this cover is not abelian, but it is abelian after base change to the algebraic closure. In general, this cover is not a subcover of the $H_n(X)$ considered above. Its Jacobian $J_{X^{(2)}}$ is not a random abelian variety of its dimension, since it decomposes (after base change to the algebraic closure of the ground field) up to isogeny into a product of the Jacobian J_X of X and $2^{2g} - 1$ abelian varieties of dimension $g - 1$, the Prym varieties of X. Typically, these Prym varieties are not defined over the ground field, only those coming from the rational 2-torsion of the Jacobian are. Even in the case of full rational 2-torsion, if we assume that the isogeny classes of these factors are random, we are picking them out of a set of size $\sim q^{(2^{2g}-1)g(g-1)/4}$ which is much smaller than if we regarded $J_{X^{(2)}}$ as random, but still very large. If we iterate this construction, as we did with $H(X)$, we quickly reach a point where the corresponding 2-torsion is not rational, hence this construction gives information which is different from what comes from $H(X)$ and its iterates. On the other hand we should note that the construction of Mestre [**11**] mentioned above produces curves that not only have isogenous Jacobians, but a few of their Prym varieties (which are defined over the ground field) will also be isogenous.

Everett Howe provided us with the following family of examples:

$$X(t)\colon y^2 = (x^3 - 1/t^3)(x^3 - t^3).$$

This is a one parameter family of genus 2 curves with geometrically split Jacobians having full rational 2-torsion over $\mathbb{F}_q$ for $q \equiv 1 \bmod 3$. Among elements of this family, for many values of q one can find distinct values of t for which the curves $X(t)$ are geometrically non-isomorphic and the L-functions of $X^{(2)}(t)$ coincide (as do those of $X(t)$). For example, one can take $t = 3, 13$ in $\mathbb{F}_{97}$, or $t = 3, 11$ in $\mathbb{F}_{127}$.

Similar examples can be obtained using

$$X(t)\colon y^2 = x(x-1)(x-t)(x-1/(1-t))(x-(t-1)/t)$$

over any finite field $\mathbb{F}_q$. Here one can take $t = 4, 5$ in $\mathbb{F}_{31}$, or $t = 4, 13$ or $t = 6, 12$ in $\mathbb{F}_{41}$, for example. The fact that these curves have split Jacobian increases the chances of instances where the L-functions of $X^{(2)}(t)$ coincide. In our computations of these examples, we noticed that these coincidences were more common than expected. We do not have an explanation for this. But we have not found any examples of genus larger than two.

QUESTION 2.3. *Are there non-isomorphic curves X, Y over $\mathbb{F}_q$ of genus at least three and $p \neq 2$ with $J_{X^{(2)}}, J_{Y^{(2)}}$ isogenous?*

For maps between curves of different genera, it is less clear what to expect. In particular, we do not have a result generalizing Theorem 2.1. But one can consider the following:

QUESTION 2.4. *Let X, Y be smooth projective curves of genus at least two over a finite field $\mathbb{F}_q$, with $H(X) \neq X$ and $H(Y) \neq Y$. Suppose the L-function of $H_n(X)$ divides the L-function of $H_n(Y)$ for all $n \geq 0$. Does this imply the existence of a dominant map $Y \to X$?*

As shown by an example of Brendan Creutz, the answer to Question 2.4 is no if we allow $H(X) = X$. A generalization of his idea (which is the case $n = 0$) is as follows. Start with X such that $X(\mathbb{F}_q) = \emptyset$. Consider the Jacobian J of $H_n(X)$ and by slicing with suitable hypersurfaces, construct a smooth curve $D \subset J$ with $0 \in D$, hence in particular $D(\mathbb{F}_q) \neq \emptyset$. This D cannot map to X (as $X(\mathbb{F}_q) = \emptyset$) but $L(H_n(X), T) | L(D, T) | L(H_n(D), T)$ by construction. So the n in Question 2.4 cannot be uniformly bounded.

3. Certifying non-isomorphism

If two curves can be distinguished by the L-polynomials of low degree covers then a succinct certificate can be given in the form of a prime ℓ for which the corresponding two L-polynomials are distinct modulo ℓ, together with the calculation of these polynomials; note that we can assume $\ell = O(g \log q)$, since otherwise the L-polynomials must coincide. For fixed g the Schoof-Pila algorithm can be used to determine ℓ and compute the L-polynomials modulo ℓ with a running time that is polynomial in $\log q$, but exponential in g. When g is large relative to $\log p$, where p is the characteristic of $\mathbb{F}_q$, one is better off using algorithms based on p-adic cohomology to compute the L-polynomials over $\mathbb{Z}$ and then reduce modulo a suitable prime ℓ. The complexity of the p-adic approach is polynomial in g but exponential in $\log p$. The most general algorithm of this type is due to Tuitman [**17**], and is applicable to all curves that admit a suitable lift to characteristic zero; its complexity is quasi-linear in p and polynomial in g. When $q = p$ is prime one can instead apply Harvey's result for arithmetic schemes [**4**], which improves the dependence on p to $O(p^{1/2+o(1)})$. At present there is no algorithm known with a running time that is polynomial in both g and $\log q$, thus in general, it may be costly to verify this certificate. But typically the degrees of the covers and the value of ℓ will be quite small (much smaller than $g \log q$), in which case computing the L-polynomials modulo ℓ (or even just enough terms to distinguish them) may be feasible.

4. Examples

The simplest case to consider in Conjecture 2.2 is when $g = 2$ and $q = 2$; in this case there are 20 isomorphism classes of curves, all of which have distinct L-functions, so one could take $n = 0$ in Conjecture 2.2. The next simplest case is $g = 2$ and $q = 3$; now there are 69 isomorphism classes of curves, but only 50 isogeny classes of Jacobians. Of the 50 isogeny classes of Jacobians, 31 contain a unique Jacobian, while 19 contain a pair of Jacobians of non-isomorphic curves. Among these 19 all but 4 pairs are distinguished by considering the L-functions of $H_1(X)$. These 4 pairs are considered in the first 3 examples below, each of which demonstrates that Conjecture 2.2 does not hold if we restrict to $n \leq 1$.

EXAMPLE 4.1. The genus two curves:

$$C_1\colon y^2 = 2x^6 + 2x^4 + 2x^3 + 2, \qquad C_2\colon y^2 = 2x^6 + 2x^5 + x^4 + x^2 + 2x + 2$$

over $\mathbb{F}_3$ are non-isomorphic, but they have isogenous Jacobians J_1, J_2 with L-polynomial:

$$9T^4 - 6T^3 + 3T^2 - 2T + 1.$$

The corresponding Hilbert class fields have degree $\#J_1(\mathbb{F}_3) = \#J_2(\mathbb{F}_3) = 5$, and the Riemann-Hurwitz theorem implies that the curves $H_1(C_1)$, $H_1(C_2)$ both have genus 6. The function fields of $H_1(C_1)$ and $H_1(C_2)$ both have exactly the same number of degree $1, 2, 3, 4, 5, 6$ places (the counts are $5, 0, 10, 15, 60, 140$, respectively, as computed using the function `NumberOfPlacesOfDegree`, which is implemented in Magma [**1**], along with the function `HilbertClassField` which we used to compute $H_1(C_1)$ and $H_1(C_2)$), which implies that their L-polynomials coincide. The computation of $H_2(C_i)$ seems out of reach so we cannot verify whether these distinguish the two curves. Instead we look at 2-power covers in the setting of Question 2.3.

The polynomial $f_1(x)$ in the equation $y^2 = f_1(x)$ for C_1 is irreducible over $\mathbb{F}_3$, while the polynomial $f_2(x)$ in the equation $y^2 = f_2(x)$ for C_2 splits into irreducible cubic factors; this implies that the Jacobian J_2 has full 2-torsion over $\mathbb{F}_{27}$, while J_1 does not. This is already enough to show that the two Jacobians J_1 and J_2 (and therefore the curves C_1 and C_2) are non-isomorphic, but this does not immediately fit our approach of computing L-polynomials.

However, by taking double covers over $\mathbb{F}_{27}$ and looking at the corresponding elliptic curves, (see Example 4.5 below for a similar calculation), we get elliptic curves with 2-torsion for the second genus 2 curve but not for the first, so the isomorphism classes of C_1 and C_2 are distinguished by the L-functions of these double covers.

EXAMPLE 4.2. The genus two curves:

$$C_1\colon y^2 = x^5 + x^4 + 2x + 1, \qquad C_2\colon y^2 = x^5 + x^3 + x^2 + 2x + 2$$

over $\mathbb{F}_3$ both have L-polynomial $L(T) = 9T^4 - 3T^3 + T^2 - T + 1$, and the curves $H_1(C_1)$, $H_1(C_2)$ of genus 8 also have equal L-polynomials.

This example is particularly interesting, in that the corresponding Jacobians J_1 and J_2 are isomorphic; indeed, after a linear change of variable this is precisely the example given by Howe in [**8**].

As in the previous example, verifying Conjecture 2.2 seems to be computationally out of reach, but we can distinguish them by taking double covers. We need to work over $\mathbb{F}_{3^5}$ and each curve has 15 étale double covers lying in 3 orbits of 5 curves under Frobenius. The Jacobians of the double covers have an additional elliptic curve factor and we get, as trace of Frobenius for these factors, the values (up to sign) of: $28, 28, 8$ for the first curve and $28, 20, 8$ for the second. The appearance of the 20 shows these elliptic curve factors are not isogenous, so the curves are not isomorphic.

A similar example is the pair of curves $C_1\colon y^2 = 2x^6+x^4+x^3+1$ and $C_2\colon y^2 = x^6+x^4+x^3+2$ over $\mathbb{F}_3$, with L-polynomial $L(T) = 9T^4-3T^3+3T^2-T+1$. Traces of Frobenius for the elliptic curve factors are $4, 16, 28$ for first curve and $4, 4, 16$ for the second.

EXAMPLE 4.3. The fourth and final example for $g = 2$ and $q = 3$ is the pair of curves

$$C_1\colon y^2 = x^5 - 1, \qquad C_2\colon y^2 = x^5 + 1,$$

which are non-isomorphic quadratic twists. Their Jacobians are both supersingular with L-polynomial $9T^4+1$, and the genus 11 curves $H_1(C_1)$, $H_1(C_2)$ have the same L-polynomial.

The curves C_1 and C_2 both have 4 points over $\mathbb{F}_3$ and admit a unique (up to twist) unramified double cover. We pin down the double cover by insisting that it have 6 points over $\mathbb{F}_3$ (the other twist has 2 points). Then we look at an unramified triple cover of the double cover, of which there are three, all twists of each other. Finally, we see how the 6 points split on these covers and use this information to distinguish the curves.

We have double covers $X_1\colon w^2 = x^4+x^3+x^2+x+1$ and $X_2\colon w^2 = x^4-x^3+x^2-x+1$ of C_1 and C_2 respectively. Here and below, we just present the equation defining the cover, so the full equation of the curve includes the previous equations as well. The polynomials in x defining the covers are factors of the corresponding polynomials in the definition of the curves, as required to get unramified covers. To get the triple covers below we follow the classical method of Hasse and Witt.

Triple covers of X_1 are given by $Y_{1,a} : z^3 - z = (x+1)w + a$, for $a = 0, 1, 2$.

Triple covers of X_2 are given by $Y_{2,a} : z^3 - z = (x-1)w + a$, for $a = 0, 1, 2$.

The distinguishing feature is that while all the $Y_{1,a}$ have $\mathbb{F}_3$-points $(12, 3, 3$, respectively), $Y_{2,0}$ is pointless (the curves $Y_{2,1}$ and $Y_{2,2}$ both have 9 $\mathbb{F}_3$-points). This implies that the L-polynomials of $H_2(C_1)$, $H_2(C_2)$ differ and confirms Conjecture 2.2 in this case.

EXAMPLE 4.4. We did an exhaustive search over $\mathbb{F}_2$ and found that there is exactly one pair of non-isomorphic smooth plane quartics C_1, C_2 over $\mathbb{F}_2$ with the same L-polynomial for which $H_1(C_1), H_1(C_2)$ also have the same L-polynomial:

$$C_1\colon x^3z + xyz^2 + y^4 + y^2z^2 + yz^3, \qquad C_2\colon x^3z + xy^2z + y^4 + y^2z^2 + yz^3.$$

Both curves have L-polynomial $8T^6-4T^5+2T^3-T+1$, with 6 rational points on their Jacobians, and the Hilbert class curves $H_1(C_1)$, $H_1(C_2)$ have genus 13.

The curves C_1, C_2 both have a unique (up to twist) quadratic unramified cover, say D_1, D_2 of genus 5. By the Deuring-Shafarevich formula, D_1, D_2 themselves have a unique (up to twist) quadratic unramified cover, and they have distinct L-polynomials, even up to quartic twists, which is enough to show C_1, C_2 are non-isomorphic and distinguished by the L-polynomials of $H_2(C_1)$, $H_2(C_2)$, confirming Conjecture 2.2 for this example.

EXAMPLE 4.5. Another example is the pair of genus two curves

$$C_1 \colon y^2 = x^6 + 3x^2 + 4, \qquad C_2 \colon y^2 = x^6 + 5x^4 + 5x^2 + 1$$

over $\mathbb{F}_7$, which have the same L-function. To show that they are not isomorphic one can look at the respective double covers and show, by counting points, that there cannot be a matching between the double covers of the two curves. Specifically, both curves have three unramified double covers defined over $\mathbb{F}_7$. The Jacobians of these covers split as the product of the Jacobian of the original curve with an additional elliptic curve. For the first curve, all three of these elliptic curves have trace of Frobenius -4. For the second curve, the elliptic curve obtained from the cover $z^2 = x^2 + 1$ has trace of Frobenius 0.

5. Factoring polynomials over finite fields

As mentioned in the introduction, the existence of a one-parameter family X_t of curves of genus g over $\mathbb{F}_p(t)$ with g bounded (or growing very slowly with p) such that the L-polynomials $L(X_t, T)$ are all distinct (or the number of collisions is bounded independent of p) for varying $t \in \mathbb{F}_p$ (excluding the t of bad reduction, those for which X_t is singular) leads to a deterministic polynomial-time algorithm for factoring polynomials in $\mathbb{F}_p[t]$. There are well-known randomized algorithms to solve this problem whose expected running times are polynomially-bounded that are quite fast in practice, so this question is primarily of theoretical interest. But even for polynomials of degree two, no deterministic polynomial-time algorithm is known, unless one assumes the Generalized Riemann Hypothesis (GRH), and for general polynomials the question remains open even under GRH.

Using the same heuristic as in section 2.1, there are p choices of t and $p^{g(g+1)/4}$ possible values for the L-polynomial so one would expect this to hold for "most" families as soon as $g > 2$, since $p^2 < p^{g(g+1)/4}$. Buium [**2**] has shown that most families (in a differential algebraic sense) in characteristic zero have finitely many isogeny correspondences, however, even if this result extends to characteristic p, it does not rule out sporadic isogenies. Conjecture 2.2 does not give the result either, as the genus of the resulting covers grows too quickly.

We first considered the family $y^2 = x^7 + (t-1)x^3 + tx^2 + (t+1)x + 1$ of curves of genus 3. One expects the number of isogeny classes of 3-dimensional abelian varieties over $\mathbb{F}_p$ to be about p^3. So under our probabilistic heuristic, a one parameter family of curves (with about p elements) has a probability of about $1/(2p)$ of containing isogenous Jacobians. Using the algorithms in [**5, 6, 9**] we have verified that for all primes $p \leq 10000$ the L-polynomials in this family are distinct for all t of good reduction (for each p, at most 9 values of $t \in \mathbb{F}_p$ yield singular curves). Now $\sum 1/(2p) = O(\log\log p)$ diverges (albeit slowly), so one might expect a collision of L-polynomials to occur in this family for some $p > 10^5$ (but one would expect the number of collisions for each p to be bounded by a constant).

To obtain a more compelling example, we instead consider the genus 4 hyperelliptic family:

$$X_t \colon y^2 = x^9 + (t-1)x^3 + tx^2 + (t+1)x + 1.$$

Now the number of isogeny classes is on the order of p^5, and our heuristic model predicts a probability of roughly $1/(2p^3)$ that two L-polynomials $L(X_t, T)$ in our family coincide for some pair of $t \in \mathbb{F}_p$. The sum $\sum 1/(2p^3)$ now converges.

We have verified that for primes $p \leq 2^{17}$ the L-polynomials arising in this family are distinct for all t of good reduction (now at most 11 values of $t \in \mathbb{F}_p$ yield singular X_t), and it seems quite likely that the L-polynomials $L(X_t, T)$ arising in the family are distinct for all primes p. Indeed, if $\pi(t) = t/\log(t) + \varepsilon(t)$ denotes the prime counting function, we can bound the tail of our sum $\sum 1/(2p^3)$ using

$$\sum_{p>2^{17}} \frac{1}{2p^3} = \int_{2^{17}}^{\infty} \frac{d\pi(t)}{2t^3} = \int_{2^{17}}^{\infty} \frac{dt}{2t^3 \log t} + \frac{\varepsilon(t)}{2t^3}\Big|_{2^{17}}^{\infty} + \int_{2^{17}}^{\infty} \frac{3\varepsilon(t)dt}{2t^4},$$

and applying the bound $\varepsilon(t) \leq (3t)/(2\log(t)^2)$ (valid for $t \geq 59$) from [**14**] yields

$$\sum_{p>2^{17}} \frac{1}{2p^3} < 1.187 \times 10^{-12} + 4.36 \times 10^{-13} + 3.15 \times 10^{-13} < 2 \times 10^{-12}.$$

Thus under our heuristic model, the probability that the L-polynomials $L(X_t, T)$ at good values of t are not all distinct for every prime p is less than 2×10^{-12}.

These two families were chosen essentially at random by writing a plausible family with no specializations having the same L-polynomial for small primes. We note that the similar looking families $y^2 = x^7 + (t-1)x^3 + (t+1)x + 1, y^2 = x^9 + (t-1)x^3 + (t+1)x + 1$ have specializations with the same L-polynomial for some small primes.

Acknowledgements: Both authors would like to thank Kiran Kedlaya for mentioning the results of [**13**] at AGCT, and Bjorn Poonen for writing up that account at our instigation, as well as Brendan Creutz, Noam Elkies, and Jakob Stix for helpful discussions, and the organizers of AGCT. We also thank Everett Howe for his feedback on an early draft of this article. The first author thanks the National Science Foundation for financial support under grant DMS-152256 and the Simons Foundation for financial support under grant #550033, and the second author thanks the Simons Foundation for financial support under grant #234591 .

References

[1] W. Bosma, J.J. Cannon, C. Fieker, and A. Steel (Eds.), *Handbook of Magma functions*, v2.23, 2017. 171

[2] A. Buium, *A finiteness theorem for isogeny correspondences*, Astérisque **218** (1993), 35–60. Journées de Géométrie Algébrique d'Orsay (Orsay, 1992). MR1265308 ↑173

[3] S. A. DiPippo and E. W. Howe, *Real polynomials with all roots on the unit circle and abelian varieties over finite fields*, J. Number Theory **73** (1998), no. 2, 426–450, DOI 10.1006/jnth.1998.2302. MR1657992 ↑169

[4] D. Harvey, *Computing zeta functions of arithmetic schemes*, Proc. Lond. Math. Soc. (3) **111** (2015), no. 6, 1379–1401, DOI 10.1112/plms/pdv056. MR3447797 ↑170

[5] D. Harvey and A. V. Sutherland, *Computing Hasse-Witt matrices of hyperelliptic curves in average polynomial time*, LMS J. Comput. Math. **17** (2014), no. suppl. A, 257–273, DOI 10.1112/S1461157014000187. MR3240808 ↑173

[6] D. Harvey and A. V. Sutherland, *Computing Hasse-Witt matrices of hyperelliptic curves in average polynomial time, II*, Frobenius distributions: Lang-Trotter and Sato-Tate conjectures, Contemp. Math., vol. 663, Amer. Math. Soc., Providence, RI, 2016, pp. 127–147, DOI 10.1090/conm/663/13352. MR3502941 ↑173

[7] F. Hess, *An algorithm for computing isomorphisms of algebraic function fields*, Algorithmic number theory, Lecture Notes in Comput. Sci., vol. 3076, Springer, Berlin, 2004, pp. 263–271, DOI 10.1007/978-3-540-24847-7_19. MR2137359 ↑167

[8] E. W. Howe, *Constructing distinct curves with isomorphic Jacobians*, J. Number Theory **56** (1996), no. 2, 381–390, DOI 10.1006/jnth.1996.0026. MR1373560 ↑171

[9] K. S. Kedlaya and A. V. Sutherland, *Computing L-series of hyperelliptic curves*, Algorithmic number theory, Lecture Notes in Comput. Sci., vol. 5011, Springer, Berlin, 2008, pp. 312–326, DOI 10.1007/978-3-540-79456-1_21. MR2467855 ↑173

[10] P. Mercuri and C. Stirpe, *Classification of algebraic function fields with class number one*, J. Number Theory **154** (2015), 365–374, DOI 10.1016/j.jnt.2015.02.008. MR3339577 ↑168

[11] J.-F. Mestre, *Couples de jacobiennes isogènes de courbes hyperelliptiques de genre arbitraire* arxiv 0902.3470 169

[12] G. Möhlmann, *Einbettungen globaler Funktionenkörper* Diplomarbeit TU Berlin 2008 167

[13] B. Poonen, *Using zeta functions to factor polynomials over finite fields*, Arithmetic Geometry: Computation and Applications, Contemp. Math., vol. 722, Amer. Math. Soc., Providence, RI, 2019. 167, 174

[14] J. B. Rosser and L. Schoenfeld, *Approximate formulas for some functions of prime numbers*, Illinois J. Math. **6** (1962), 64–94. MR0137689 ↑174

[15] A. S. Shankar and J. Tsimerman, *Unlikely intersections in finite characteristic* arxiv 1610.03552 169

[16] Q. Shen and S. Shi, *Function fields of class number one*, J. Number Theory **154** (2015), 375–379, DOI 10.1016/j.jnt.2015.02.005. MR3339578 ↑168

[17] J. Tuitman, *Counting points on curves using a map to* $\mathbf{P}^1$*, II*, Finite Fields Appl. **45** (2017), 301–322, DOI 10.1016/j.ffa.2016.12.008. MR3631366 ↑170

Department of Mathematics, Massachusetts Institute of Technology, 77 Massachusetts Avenue, Cambridge, Massachusetts 02139

Email address: drew@math.mit.edu

URL: http://math.mit.edu/~drew

School of Mathematics and Statistics, University of Canterbury, Private Bag 4800, Christchurch 8140, New Zealand

Email address: felipe.voloch@canterbury.ac.nz

URL: http://www.math.canterbury.ac.nz/~f.voloch

Selected Published Titles in This Series

722 **Yves Aubry, Everett W. Howe, and Christophe Ritzenthaler, Editors,** Arithmetic Geometry: Computation and Applications, 2019

718 **David Ayala, Daniel S. Freed, and Ryan E. Grady, Editors,** Topology and Quantum Theory in Interaction, 2018

717 **Federico Bonetto, David Borthwick, Evans Harrell, and Michael Loss, Editors,** Mathematical Problems in Quantum Physics, 2018

716 **Alex Martsinkovsky, Kiyoshi Igusa, and Gordana Todorov, Editors,** Surveys in Representation Theory of Algebras, 2018

715 **Sergio R. López-Permouth, Jae Keol Park, S. Tariq Rizvi, and Cosmin S. Roman, Editors,** Advances in Rings and Modules, 2018

714 **Jens Gerlach Christensen, Susanna Dann, and Matthew Dawson, Editors,** Representation Theory and Harmonic Analysis on Symmetric Spaces, 2018

713 **Naihuan Jing and Kailash C. Misra, Editors,** Representations of Lie Algebras, Quantum Groups and Related Topics, 2018

712 **Nero Budur, Tommaso de Fernex, Roi Docampo, and Kevin Tucker, Editors,** Local and Global Methods in Algebraic Geometry, 2018

711 **Thomas Creutzig and Andrew R. Linshaw, Editors,** Vertex Algebras and Geometry, 2018

710 **Raphaël Danchin, Reinhard Farwig, Jiří Neustupa, and Patrick Penel, Editors,** Mathematical Analysis in Fluid Mechanics, 2018

709 **Fernando Galaz-García, Juan Carlos Pardo Millán, and Pedro Solórzano, Editors,** Contributions of Mexican Mathematicians Abroad in Pure and Applied Mathematics, 2018

708 **Christian Ausoni, Kathryn Hess, Brenda Johnson, Ieke Moerdijk, and Jérôme Scherer, Editors,** An Alpine Bouquet of Algebraic Topology, 2018

707 **Nitya Kitchloo, Mona Merling, Jack Morava, Emily Riehl, and W. Stephen Wilson, Editors,** New Directions in Homotopy Theory, 2018

706 **Yeonhyang Kim, Sivaram K. Narayan, Gabriel Picioroaga, and Eric S. Weber, Editors,** Frames and Harmonic Analysis, 2018

705 **Graham J. Leuschke, Frauke Bleher, Ralf Schiffler, and Dan Zacharia, Editors,** Representations of Algebras, 2018

704 **Alain Escassut, Cristina Perez-Garcia, and Khodr Shamseddine, Editors,** Advances in Ultrametric Analysis, 2018

703 **Andreas Malmendier and Tony Shaska, Editors,** Higher Genus Curves in Mathematical Physics and Arithmetic Geometry, 2018

702 **Mark Grant, Gregory Lupton, and Lucile Vandembroucq, Editors,** Topological Complexity and Related Topics, 2018

701 **Joan-Carles Lario and V. Kumar Murty, Editors,** Number Theory Related to Modular Curves, 2018

700 **Alexandre Girouard, Dmitry Jakobson, Michael Levitin, Nilima Nigam, Iosif Polterovich, and Frédéric Rochon, Editors,** Geometric and Computational Spectral Theory, 2017

699 **Mark L. Agranovsky, Matania Ben-Artzi, Catherine Bénéteau, Lavi Karp, Dmitry Khavinson, Simeon Reich, David Shoikhet, Gilbert Weinstein, and Lawrence Zalcman, Editors,** Complex Analysis and Dynamical Systems VII, 2017

698 **Alexander M. Blokh, Leonid A. Bunimovich, Paul H. Jung, Lex G. Oversteegen, and Yakov G. Sinai, Editors,** Dynamical Systems, Ergodic Theory, and Probability: in Memory of Kolya Chernov, 2017

For a complete list of titles in this series, visit the AMS Bookstore at **www.ams.org/bookstore/conmseries/**.